Quotes from reviews

'This book will undoubtedly be welcomed by the extensive engineering community concerned with the impact of ocean waves on ships, off-shore structures, coastal protection, dikes, harbours, beaches and tidal basins . . . The book contains a trove of practical information on all aspects of waves in the open ocean and coastal regions . . . providing an invaluable source of information.'
K. Hasselmann, Director (retired) of the Max-Planck-Institut für Meteorologie, Hamburg, and Emeritus Professor of Theoretical Geophysics, University of Hamburg, Germany

'The author, well-known for his work in wave modeling and the development of the SWAN model, provides a valuable introduction to ocean wave statistics, generation by wind, and modeling in deep and shallow water. . . . The book will be very helpful to students, as well as professionals, interested in wind-wave wave modeling. All SWAN users will want a copy.'
R.A. Dalrymple, Williard & Lillian Hackerman Professor of Civil Engineering, Johns Hopkins University, USA

'. . . the best introduction to practical engineers to grasp the directional spectral wave approach. . . . The book is excellent not only as a textbook for students but also as a reference book for professionals.'
Y. Goda, Executive Advisor to ECOH CORPORATION, Emeritus Professor of Civil Engineering, Yokohama National University, Director-General (retired) of the Port and Airport Research Institute, Japan

'. . . ideally suited as a reference work for advanced undergraduate and graduate students and researches. . . . The book is a "must have" for engineers and scientists interested in the ocean. . . . The book explains quite complex processes with remarkable clarity and the use of informative examples. Drawing on the author's international reputation as a researcher in the field, the book brings together classical theory and state of the art techniques in a consistent framework. It is an invaluable reference for students, researchers and practitioners.'
I. Young, Vice-Chancellor and President of Swinburne University of Technology, Australia

'This is a great book. The author is one of the leading experts in the field of waves who has taught the subject for over 20 years – and it shows. The book has a broad scope, which would be of interest to students just learning the subject, as well as professionals who wish to broaden their range of knowledge or who want to refresh their memory . . . recommended for introductory as well as advanced students and professionals.'
J. W. Kamphuis, Emeritus Professor of Civil Engineering, Queen's University, Canada

'This book presents an original and refreshing view on nearly all topics which are required nowadays to deal with wind generated waves at the sea surface. . . . The logical structure . . . and the fact that it avoids complex numbers and vector notation will . . . facilitate its comprehension.'
A. Sánchez-Arcilla, Professor of Coastal Engineering, Universitat Politècnica de Catalunya, Spain

'. . . highlights key concepts, unites seemingly unconnected theories, and unlocks the complexity of the sea. [This book] will become an important reference for students, coastal and ocean engineers, and oceanographers.'
J. Smith, Editor, International Conference on Coastal Engineering, US Army Engineer Research and Development Center, USA

'This book is exceptionally well organized for teachers who want a thorough introduction to ocean waves in nature. It fills a key gap in text books, between overly simplistic treatments of ocean waves and detailed theoretical/mathematical treatises beyond the needs of most students. I found the text very clear and readable. Explanations and derivations within this book are both innovative and instructive and the focus on key elements required to build a strong foundation in ocean waves remains strong throughout the book.'
D. T. Resio, Chief Research and Development Advisor, US Army Engineer Research & Development Center, USA

. . . an excellent source of information about wind-generated, ocean-surface gravity waves. . . nicely illustrated, well written, contains many references, and will be of interest to scientists and engineers.
S. Elgar, Oceanography (the magazine of The Oceanography Society)

WAVES IN OCEANIC AND COASTAL WATERS

Waves in Oceanic and Coastal Waters describes the observation, analysis and prediction of wind-generated waves in the open ocean, in shelf seas, and in coastal regions. The book brings graduate students, researchers and engineers up-to-date with the science and technology involved, assuming only a basic understanding of physics, mathematics and statistics.

Most of this richly illustrated book is devoted to the physical aspects of waves. After introducing observation techniques for waves, both at sea and from space, the book defines the parameters that characterize waves. Using basic statistical and physical concepts, the author discusses the prediction of waves in oceanic and coastal waters, first in terms of generalized observations, and then in terms of the more theoretical framework of the spectral energy balance: their origin (generation by wind), their transformation to swell (dispersion), their propagation into coastal waters (shoaling, refraction, diffraction and reflection), the interaction amongst themselves (wave-wave interactions) and their decay (white-capping, bottom friction, and surf-breaking). He gives the results of established theories and also the direction in which research is developing. The book ends with a description of SWAN (Simulating Waves Nearshore), the preferred computer model of the engineering community for predicting waves in coastal waters.

Early in his career, the author was involved in the development of techniques to measure the directional characteristics of wind-generated waves in the open sea. He contributed to various projects, in particular the Joint North Sea Wave Project (JONSWAP), which laid the scientific foundation for modern wave prediction. Later, he concentrated on advanced research and development for operational wave prediction and was thus involved in the initial development of the computer models currently used for global wave prediction at many oceanographic and meteorological institutes in the world. More recently, he initiated, supervised and co-authored SWAN, the computer model referred to above, for predicting waves in coastal waters. For ten years he co-chaired the Waves in Shallow Environments (WISE) group, a world wide forum for research and development underlying operational wave prediction. He has published widely on the subject and teaches at the Delft University of Technology and UNESCO-IHE in the Netherlands.

WAVES IN OCEANIC AND COASTAL WATERS

LEO H. HOLTHUIJSEN

Delft University of Technology and
UNESCO-IHE

CAMBRIDGE
UNIVERSITY PRESS

CAMBRIDGE UNIVERSITY PRESS
Cambridge, New York, Melbourne, Madrid, Cape Town, Singapore,
São Paulo, Delhi, Dubai, Tokyo

Cambridge University Press
The Edinburgh Building, Cambridge CB2 8RU, UK

Published in the United States of America by Cambridge University Press, New York

www.cambridge.org
Information on this title: www.cambridge.org/9780521129954

First published 2007
Reprinted with corrections 2008
This digitally printed version 2009

A catalogue record for this publication is available from the British Library

ISBN 978-0-521-86028-4 Hardback
ISBN 978-0-521-12995-4 Paperback

Additional resources for this publication at www.cambridge.org/9780521129954

Contents

Preface

In my position as associate professor at Delft University of Technology and as a guest lecturer at UNESCO-IHE (Delft, the Netherlands), I have for more than 20 years, with great pleasure, supported students and professionals in their study of ocean waves. At Delft University I have had, in addition, the opportunity to work with colleagues, notably Nico Booij, on developing numerical wave models, one of which (SWAN) has widely been accepted as an operational model for predicting waves in coastal waters.

Over the years, I have made notes to assist these professionals, students and myself, during courses, workshops and training sessions. With the growing interest and willingness of others to formalise these (mostly handwritten) notes, I found that I should make the effort myself. The result is this book *Waves in Oceanic and Coastal Waters*, which provides an introduction to the observation, analysis and prediction of wind-generated waves in the open ocean, in shelf seas and in coastal regions. The title of the book is a little prosaic because I want to focus directly on the subject matter of the book. A more poetic title would be *Waves of The Blue Yonder*, which would convey better the awe and mystery that I feel when watching waves at sea, wondering where they come from and what they have seen on their journey across the oceans. The cover photo illustrates this feeling beautifully.

Understanding the text of the book requires some basic knowledge of physics, mathematics and statistics. The text on *observing* waves (Chapter 22) is descriptive; no mathematics or statistics is used. Understanding the text on *describing* ocean waves (Chapters 33 and 44) does require some knowledge of mathematics and statistics, since concepts of analytical integration and probabilities are used. The text on the linear *theory* of surface gravity waves (Chapters 55 and 77) and the text on *modelling* wind-generated waves (Chapters 66 and 88) rely heavily on the concepts of conservation of mass, momentum and energy. Therefore, some background in physics is needed. These concepts are expressed with partial differential equations, so some background in mathematics is also needed. Finally, the book ends in Chapter 99 with a description of the fundamentals of SWAN (both its physical principles and numerical techniques).

I first treat waves in oceanic waters and later in coastal waters. The reason for this separation is both didactic and practical: the physical processes increase in number and complexity as waves move from the ocean into coastal waters. Describing waves in the oceans therefore gives a good introduction to the more challenging subject of waves in coastal waters. In addition, many readers will be interested only in the ocean environment and need not be bothered with the coastal environment.

I am well aware that many formulations in this book can be written in vector or complex notation. Such notation would make for compact reading for those who are familiar with it. However, students who are not familiar with it would not readily absorb the material presented, so I have chosen not to use it. With a few exceptions, I have written in terms of components rather than vectors and real quantities rather than complex quantities. Concerning the references in the book: I have used a fair number of these, to (a) refer to specific information, (b) indicate where issues are being discussed and (c) refer to books and articles for further reading. I have not tried to be complete in this. That would be nearly impossible, if only because of the continual appearance of new publications. Moreover, any subject is accessible on the Internet, which is completely up to date, including electronic versions of scientific and engineering journals.

If this book helps professionals to enjoy their work more, students to pursue their interest in waves and others to look at waves with an informed eye, it has more than served its purpose.

L. H. Holthuijsen, Delft

Acknowledgements

I was supported in writing this book by three close friends and colleagues: Luigi Cavaleri of the Istituto di Scienze Marine in Venice (Italy), whom I visited so often (memories of Venice waking up in the early morning sunlight, when it is still a cool and quiet place); Masataka Yamaguchi of the Ehime University in Matsuyama (Japan), who introduced me to the many charms of Japan (memories of the mountains and quiet villages along the rugged Pacific coast of his home island Shikoku); and Nico Booij, with whom I shared, almost daily, my professional enthusiasm, ideas and dreams in such diverse places as Delft, Reykjavík and Beijing. They read the book from cover to cover (and back, more than once) and they gave their comments and suggestions freely. This was not a trivial effort. They saved me from embarrassing errors and helped achieve a balance between scope, reliability and accessibility on the one hand and detail, accuracy and formalism on the other. I am very grateful to them and I am proud that they are my friends, and have been for 25 years now. I also want to thank Linwood Vincent of the US Office of Naval Research, whose inspiring words encouraged me to write this book.

In addition, I have had the privilege to be assisted by several colleagues with specific information, in particular on the subject of wave statistics: Akira Kimura of the University of Tottori, Japan; Evert Bouws and Sofia Caires of the Royal Netherlands Meteorological Institute; Ulla Machado of Oceanor, Norway; Sverre Haver of Statoil, Norway; Agnieszka Herman of the Lower Saxonian Central State Board for Ecology in Norderney, Germany; Mercè Casas Prat of the Universitat Politècnica de Catalunya, Spain; and Ad Reniers, Pieter van Gelder, André van der Westhuysen and Marcel Zijlema of the Delft University of Technology. Mrs. Paula Delhez and her colleagues of the Delft University Library helped me find the references in this book. I am very grateful to all of them as it greatly improved the quality of the book. Still, any errors that are left (and fate dictates that some will be) are wholly mine.

In the book I have used data provided by the Royal Netherlands Meteorological Institute (the Netherlands), Fugro Oceanor AS (Norway), the National Oceanic and Atmospheric Administration (USA), Statoil Norge AS (Norway) and XIOM (Xarxa d'Instruments Oceanogràfics i Meteorologics, of the Generalitat de Catalunya, Spain). I am grateful for their permission to use these data (further acknowledgements are given in the text). I am also grateful to the copyright holders for permission to use the figures listed below.

Datawell, the Netherlands: Fig. 2.3.
Institute of Marine Sciences, Italy: Fig. 2.4.
Det Norske Veritas, Norway: Fig. 3.3.
American Society of Civil Engineers, USA: Fig. 4.1.
Springer Science and Business Media, Germany: Fig. 5.12.
World Scientific, Singapore, www.worldscibooks.com/engineering/4064.html: Fig. 5.12.
Elsevier, the Netherlands: Figs. 6.18 and 8.9.

I am deeply indebted to Philip Plisson for his gracious permission to use his poetic photo for the cover of the book.

1

Introduction

1.1 Key concepts

- This book offers an *introduction* to observing, analysing and predicting ocean waves for university students and professional engineers and, of course, others who are interested. Understanding the text of the book requires some basic knowledge of physics (mechanics), mathematics (analytical integrals and partial differential equations) and statistics (probabilities).
- The book is structured from *observing* to *describing* to *modelling* ocean waves. It closes with a description of the physics and numerics of the freely available, open-source computer model SWAN for predicting waves in coastal waters.
- Ocean waves (or rather: wind-generated surface gravity waves) can be described at several *spatial* scales, ranging from hundreds of metres or less to thousands of kilometres or more and at several *time* scales, ranging from seconds (i.e., one wave period) to thousands of years (wave climate).
  (a) On *small* space and time scales (less than a dozen wave lengths or periods, e.g., the surf zone at the beach or a flume in a hydraulic laboratory), it is possible to describe the actual sea-surface motion in detail. This is called the phase-*resolving* approach.
  (b) On *intermediate* space and time scales (from dozens to hundreds of wave lengths or periods, e.g., a few kilometres or half an hour at sea), the wave conditions are described with average characteristics, the most important of which is the wave spectrum. This requires the wave conditions to be constant in a statistical sense (stationary and homogeneous).
  (c) On *large* space and time scales (from hundreds to hundreds of thousands of wave lengths or periods, e.g., oceans or shelf seas), space and time should be divided into segments, with the waves in each described with one spectrum. The sequence of segments allows the spectrum to be treated as varying in space and time.
  (d) On a *climatological* time scale (dozens of years or more), usually only the statistical properties of a characteristic wave height (the significant wave height) are considered.

1.2 This book and its reader

Waves at the surface of the ocean are among the most impressive sights that Nature can offer, ranging from the chaotic motions in a violent hurricane to the tranquillity of a gentle swell on a tropical beach. Everyone will appreciate this poetic aspect but scientists and engineers have an additional, professional interest. The scientist is interested in the dynamics and kinematics of the waves: how they are generated by the wind, why they break and how they interact with currents and the sea bottom. The engineer (variously denoted as ocean engineer, naval architect, civil engineer, hydraulic engineer, etc.) often has to design, operate or manage structures or natural systems in the marine environment such as offshore platforms, ships, dykes, beaches

1

and tidal basins. To a greater or lesser extent, the behaviour of such structures and systems is affected by the waves and some basic knowledge of these waves is therefore required. This book offers an *introduction* to this fascinating subject for engineers and university students, particularly those who need to operate numerical wave models. Others may be interested too, if only out of pure curiosity.

The book starts where anyone interested in ocean waves should start: with observing waves as they appear in Nature, either in the open sea or along the shore.[1] Take the opportunity to go out to sea or wander along the shores of the ocean to experience the beauty and the cruelty of waves, and to question the 'where and why' of these waves. The book therefore starts with *observation* techniques, before continuing with the question of how to describe these seemingly random motions of the sea, which we call waves. Only then does the book present a truly theoretical concept. It is the variance density spectrum of the waves that is used to *describe* the waves. This, in its turn, is followed by the linear *theory* of surface gravity waves (as they are formally called). This theory gives the interrelation amongst such physical characteristics as the surface motion, the wave-induced pressure in the water and the motion of water particles. It beautifully supplements the concept of the spectrum. Initially, the book treats only open-water aspects of the linear wave theory, in other words, deep-water conditions without currents or a coast. This provides, together with the spectral description of the waves, an introduction to the energy balance of waves in oceanic waters. Sources and sinks are added to this balance, to represent the generation (by wind), the interaction amongst the waves themselves (wave–wave interactions) and the dissipation of the waves (by white-capping). Although several theories for these processes have been developed, the actual formulations in numerical wave models are still very much empirical and therefore relatively simple and descriptive. I will use these model formulations so that the reader will quickly become familiar with the basic ideas and results of these theories. This will satisfy many students of waves in oceanic waters. For those interested in waves in coastal waters, the book proceeds by adding the effects of sea-bottom topography, currents and a coast (shoaling, refraction, diffraction and reflection). The corresponding formulations of the generation, wave–wave interactions and dissipation in coastal waters are more diverse and empirical than those for oceanic waters and the presentation is consequently even more descriptive.

The text of the book provides an insight into basic theories and practical results, which will enable the reader to assess the importance of these in his or her field of engineering, be it coastal engineering, ocean engineering, offshore engineering or naval architecture. I have tried to balance the presentation of the material in a manner that will, I hope, be attractive to the practical engineer rather than the theoretically

---

[1] Reading a brief history of wave research may also be interesting (e.g., Phillips, 1981; Tucker and Pitt, 2001).

minded scientist. I am well aware that some basic knowledge that is required to understand certain parts of the text has sunken deep into the recesses of the reader's memory (statistics is a notorious example). In such cases, the required information is briefly reviewed in separate notes and appendices, which are intended as prompts rather than as true introductions. I hope that the scientifically minded reader may find the book sufficiently intriguing that it will lead him to more fundamental and advanced books (for instance Geernaert and Plant, 1990; Goda, 2000; Janssen, 2004; Komen *et al.*, 1994; Lavrenov, 2003; LeBlond and Mysak, 1978; Phillips, 1977; Sawaragi, 1995; Svendsen, 2006; and Young, 1999).

1.3 Physical aspects and scales

If the word 'waves'[2] is taken to mean 'vertical motions of the ocean surface',[3] then wind-generated gravity waves are only one type amongst a variety that occur in the oceans and along the shores of the world. All these waves can be ordered in terms of their period or wave length (see Fig. 1.1). The longest waves are *trans-tidal waves*, which are generated by low-frequency fluctuations in the Earth's crust and atmosphere. *Tides*, which are slightly shorter waves, are generated by the interaction between the oceans on the one hand and the Moon and the Sun on the other. Their periods range from a few hours to somewhat more than a day and their wave lengths accordingly vary between a few hundred and a few thousand kilometres. This is (very) roughly the scale of ocean basins such as the Pacific Ocean and the Northern Atlantic Ocean and of shelf seas such as the North Sea and the Gulf of Mexico. Although tides may be called waves, they should not be confused with 'tidal waves', which is actually a misnomer for tsunamis (see below).

The wave length and period of *storm surges* are generally slightly shorter than those of tides. A storm surge is the large-scale elevation of the ocean surface in a severe storm, generated by the (low) atmospheric pressure and the high wind speeds in the storm. The space and time scales of a storm surge are therefore roughly equal to those of the generating storm (typically a few hundred kilometres and one or two days). When a storm surge approaches the coast, the water piles up and may cause severe flooding (e.g., the flooding of New Orleans by hurricane Katrina in August of 2005, or the annual flooding of Bangladesh by

---

[2] Waves are basically disturbances of the equilibrium state in any given body of material, which propagate through that body over distances and times much larger than the characteristic wave lengths and periods of the disturbances.

[3] Waves *beneath* the ocean surface, for instance at the interface between two layers of water with different densities, are called 'internal waves'. They will not be considered in this book.

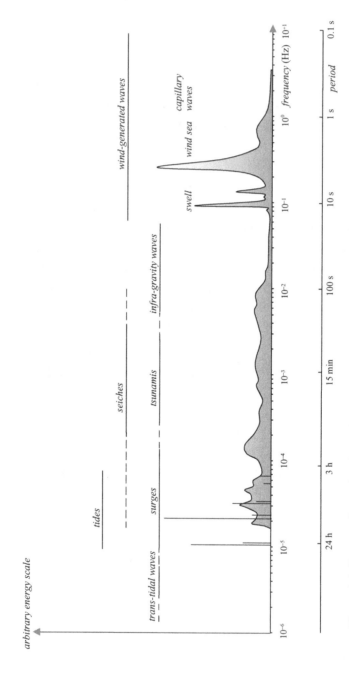

Figure 1.1 Frequencies and periods of the vertical motions of the ocean surface (after Munk, 1950).

cyclones[4]). The next, somewhat smaller scale of waves is that of *tsunamis*. These are waves that are generated by a submarine 'land' slide or earthquake. They have a bad reputation, since they are difficult to predict and barely noticeable in the open ocean (due to their low amplitude there) but they wreak havoc on unsuspecting coastal regions as they increase their amplitude considerably on approaching the coast (the Christmas tsunami of 2004 in the Indian Ocean being the worst in living memory). The waves at the next scale are even more difficult to predict. These are standing waves, called *seiches*, with a frequency equal to the resonance frequency of the basin in which they occur (in harbours and bays or even at sea, for instance in the Adriatic Sea). In a harbour, the amplitude of a seiche may be large enough (1 m is no exception) to flood low-lying areas of the harbour, break anchor lines and otherwise disrupt harbour activities. These waves are usually generated by waves from the open sea, the source of which is not well understood (although some, at least, are generated by storms). Next is the scale of *infra-gravity waves*. These waves are generated by groups of wind-generated waves, for instance in the surf zone at the beach, where these waves are called surf beat, with periods of typically a few minutes. The period of the next category, *wind-generated waves*, is shorter than 30 s. When dominated by gravity (periods longer than 1/4 s), they are called *surface gravity waves* (the subject of this book). While they are being generated by the local wind, they are irregular and short-crested, and called *wind sea*. When they leave the generation area, they take on a regular and long-crested appearance and are called *swell* (the beautiful swell on a tropical beach is generated in a distant storm). Waves with periods shorter than 1/4 s (wave lengths shorter than about 10 cm), are affected by surface tension and are called *capillary waves*.

The above types of waves are defined in terms of their wave period or wave length. Wind-generated surface gravity waves are thus characterised by their period of 1/4–30 s and corresponding wave length of 0.1–1500 m (in deep water). For describing the variation in space and time of *these* waves, other scales are used: the scales at which the processes of their generation, propagation and dissipation take place.

(1) On small scales, of the order of a *dozen or fewer* wave periods or wave lengths (however loosely defined), in other words, dimensions of about 10–100 s and 10–1000 m in real life (e.g., the dimension of the surf zone or a small harbour), waves can be described in great detail with theoretical models (details down to small fractions of the period or wave length). In these models, the basic hydrodynamic laws can be used to estimate

---

[4] Hurricanes occur in many parts of the world under different names. For the Atlantic Ocean and the eastern Pacific Ocean the term *hurricane* is used, whereas for the western Pacific Ocean, the term *typhoon* is used. In the Indian Ocean the term *cyclone* is used. A *tornado* is something entirely different. It denotes the much smaller atmospheric phenomenon of a relatively small but severe whirlwind (a diameter of a few hundred metres or less, whereas the scale of a hurricane is hundreds of kilometres with an eye of about 25 km) with a vertical axis extending from the clouds to the ground, usually occurring in thunderstorms, with much higher wind speeds and a much lower atmospheric pressure in the centre than in hurricanes.

the motion of the water surface, the velocity of the water particles and the wave-induced pressure in the water at any time and place in the water body, e.g., to compute the impact of a breaking wave on an offshore structure. Nothing in these models is left to chance; the Newtonian laws of mechanics control everything. In other words, in this approach the description and modelling of the waves are fully deterministic. Rapid variations in the evolution of the waves can be computed, e.g., waves breaking in the surf zone at the beach. Since this approach provides details with a resolution that is a small fraction of the wave length or period, it is called the *phase-resolving* approach.

(2) On a somewhat larger scale, of the order of a *hundred* wave periods or wave lengths, in other words, dimensions of about 100–1000 s and 100–10 000 m in real life, the above phase-resolving approach is not used. The reasons are as follows:

(a) the sheer amount of numbers needed to describe the waves would be overwhelming;

(b) details of the wind that generates the waves cannot be predicted at this scale and therefore the corresponding details of the waves cannot be predicted either;

(c) even if such details could be observed or calculated, they would be incidental to that particular observation or calculation and not relevant for any predicted situation; and

(d) the engineer does not require such details at this scale.

The description of ocean waves at this scale need therefore not be aimed at such details. Rather, such details should be ignored and the description should be aimed at characteristics that are relevant and predictable. This can be achieved by taking certain averages of the waves in space and time. This is the *phase-averaging* approach, in which statistical properties of the waves are defined and modelled. Meaningful averaging requires that, in some sense, the wave situation is constant within the averaging interval, i.e., the situation should be *homogeneous and stationary* in the space and time interval considered. If the waves are not too steep and the water is not too shallow, the physically and statistically most meaningful phase-averaged characteristic of the waves is the *wave spectrum*. This spectrum is based on the notion that the profile of ocean waves can be seen as the superposition of very many propagating harmonic waves, each with its own amplitude, frequency, wave length, direction and phase (the random-phase/amplitude model).

(3) Next are the three scales of coastal waters (of the order of *one* thousand wave lengths and periods), shelf seas (of the order of *ten* thousand wave lengths and periods) and oceans (of the order of a *hundred* thousand wave lengths and periods). In oceans and shelf seas, the time and space scales are generally determined by the travel time of the waves through the region, the spatial scale of the region itself and the scales of the wind and tides. In coastal waters, the space scale is also determined by coastal features such as beaches, bays and intricate topographical systems, such as tidal basins with barrier islands, channels and flats. For instance, a string of barrier islands may be 50–100 km long with a tidal basin behind it that is 10–20 km wide. The travel time to the mainland behind the islands is then typically only 15–30 min. In shelf seas and oceans, the space scale is determined by the size of the basin itself and by the space

scale of the weather systems. For instance, the North Sea is roughly 500 km wide and 1500 km long, while the weather systems there are only slightly smaller. The travel time across the North Sea for waves with period 10 s is typically 24 h, which is of the same order as the time scale of the storms there. The scale of the Pacific Ocean is roughly 10 000 km, and a 20-s swell takes about a week to travel that distance. All these scales are too large to use only *one* spectrum to characterise the waves. Instead, the spectrum under these conditions is seen as a function that varies in space and time. It can be forecast with numerical wave models, accounting for the generation, propagation and dissipation of the waves. The spectrum is thus determined in a *deterministic* manner from winds, tides and seabed topography. Note that we thus compute statistical characteristics of the waves (represented by the spectrum) in a deterministic manner.

(4) On a time scale of *dozens of years (or more)* the wave conditions can be characterised with long-term statistics (called wave climate) obtained from long-term wave observations or computer simulations. Acquiring a wave climate is basically limited to sorting and extrapolating a large number of such observations or simulations.

In summary: ocean waves are generally not observed and modelled in all their detail as they propagate across the ocean, into shelf seas and finally into coastal waters. Such details are generally not required and they are certainly beyond our capacity to observe and compute (except on a very small scale). The alternative is to consider the statistical characteristics of the waves. In advanced techniques of observing and modelling, these statistical characteristics are represented by the wave spectrum, which can be determined either from observations or with computer simulations based on wind, tides and seabed topography.

1.4 The structure of the book

The structure of the book follows roughly the above sequence of the various aspects of ocean waves, i.e., from *observing* ocean waves with instruments to *predicting* waves with computer models:

CHAPTER 1   INTRODUCTION
      The present, brief characterisation of this book and its contents.

CHAPTER 2   OBSERVATION TECHNIQUES
      The phenomenon of ocean waves is introduced by describing techniques to observe waves with *in situ* instruments or *remote-sensing* instruments. *In situ* instruments float on the ocean surface (buoys and ships), pierce the water surface (e.g., wave poles) or are mounted under water (e.g., pressure transducers). Remote-sensing

instruments, with their lenses or antennas, are usually located high above the oceans (e.g., laser or radar in airplanes and satellites).

CHAPTER 3   DESCRIPTION OF OCEAN WAVES

Having introduced the techniques used to observe the apparent chaos of ocean waves in the previous chapter, the techniques to describe this phenomenon are introduced. The basic concept for this is the *random-phase/amplitude* model. It leads to the definition of the variance density *spectrum*. Interpreted as the energy density spectrum, this spectrum provides the basis for modelling the physical aspects of the waves.

CHAPTER 4   STATISTICS

All *short-term* statistical characteristics of the waves can be expressed in terms of the spectrum (within the linear approach of the random-phase/amplitude model). Here, 'short-term' should be interpreted as the time during which the wave condition is, statistically speaking, stationary. This property of the spectrum is exploited to estimate, theoretically, important statistical parameters such as the significant wave height and the maximum individual wave height within a given duration (e.g., a storm). *Long-term* wave statistics can be arrived at only by collecting observations or by computing many wave conditions from archived wind data. Extrapolating such long-term statistical information to estimate extreme conditions, for instance to determine design conditions of an offshore structure, was, until recently, more an empirical art than a well-founded science.

CHAPTER 5   LINEAR WAVE THEORY (OCEANIC WATERS)

The linear theory of surface gravity waves is the basis for deriving the physical characteristics of wind-generated waves. This linear approach beautifully supplements the concept of the wave spectrum which assumes linear waves. The theory, as treated in this chapter for oceanic waters, addresses only local characteristics such as wave-induced orbital motions, wave-induced pressure fluctuations in the water and wave energy, together with such aspects as phase velocity and the propagation of wave energy. Only the simplest conditions are considered: the water has a constant depth, there are no obstacles, currents or coastlines and the wave amplitude is constant in space and time. The theory, being linear, ignores the effect of wind, dissipation and other nonlinear effects (these are treated in Chapters 6 and 8).

CHAPTER 6   WAVES IN OCEANIC WATERS

The concept of the wave spectrum, combined with the linear wave theory for oceanic waters, is the basis for describing the propagation of the waves on an oceanic scale with the spectral energy balance. Obviously, such modelling requires additional information on the generation of the waves (by wind), their dissipation (by white-capping) and other nonlinear effects (quadruplet wave–wave interactions).

CHAPTER 7   LINEAR WAVE THEORY (COASTAL WATERS)

In this chapter, the linear wave theory is continued for the more complex conditions of coastal waters with variable water depth, currents, obstacles, coastlines and rapidly varying wave amplitudes (compared with oceanic conditions). The corresponding phenomena of shoaling, refraction, diffraction, reflection, radiation stresses and wave-induced set-up are introduced.

CHAPTER 8   WAVES IN COASTAL WATERS

The modelling of waves in coastal waters, including the surf zone, is considerably more challenging than that in oceanic waters, not only because the propagation of the waves is more complicated, but also because the processes of generation, dissipation and nonlinear wave–wave interactions increase in number and complexity. The processes that dominate in oceanic waters are slightly modified in coastal waters but, more importantly, the processes of bottom friction, surf-breaking and triad wave–wave interactions are added.

CHAPTER 9   THE SWAN WAVE MODEL

To illustrate one application of the concepts and theories that are presented in this book, and to provide SWAN users with background information, the formulations and techniques of the third-generation SWAN model for waves in coastal waters are given in this final chapter.

2

Observation techniques

2.1 Key concepts

- *Visual* observations are often the only source of wave information available to the engineer. Sometimes measurements made with *instruments* are available.
- Measurement techniques can be divided into *in situ* techniques (instruments deployed in the water) and *remote-sensing* techniques (instruments deployed at some distance above the water).
- The most common *in situ* instruments are wave buoys and wave poles. Other *in situ* instruments are inverted echo-sounders, pressure transducers and current meters. These instruments need to be mounted on some structure at sea.
- The most common remote-sensing technique is radar, which is based on actively irradiating the sea surface with electro-magnetic energy and detecting the corresponding reflection. Radar may be deployed from the coast (e.g., with a receiving station in the dunes), from fixed platforms (e.g., oil-production platforms) or from moving platforms at relatively low altitude (airplanes) or high altitude (satellites).
- Radar can be used to obtain images of the sea surface, but it can also be used as a distance meter or as a surface-roughness meter.
- Each measurement technique has its own peculiarities as regards operational performance, accuracy, maintenance, cost and reliability.
- The most common result of a wave measurement is a time record of the sea-surface elevation at a fixed (horizontal) location.

2.2 Introduction

Waves are not only observed by surfers, swimmers or tourists from the beach. Experienced crew members onboard voluntary observing ships (VOS; or voluntary observing fleet, VOF), too, observe the waves and report wave height, period and direction daily to meteorological institutions around the world. Scientists and engineers too are watching waves. They want to quantify what they see; they want to record every detail of the moving sea surface to study and eventually predict waves. They therefore need to record the up-and-down motion of the surface, as a function of time (see Fig. 2.1), or as a function of horizontal co-ordinates (see Fig. 2.2).

Such detail is not available in *visual* observations but visual observations of the *height* of waves are fairly reliable if carried out by experienced observers who follow specific instructions (this is not true for the wave period) but they have their own peculiarities. For instance, ships try to avoid heavy weather and storms

10

surface elevation (m)

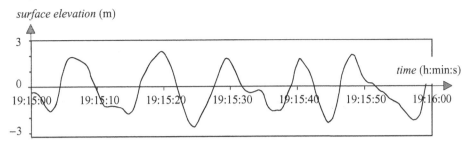

Figure 2.1 The up-and-down motion of the sea surface in a storm, as experienced by a buoy, i.e., the sea-surface elevation at one location as a function of time.

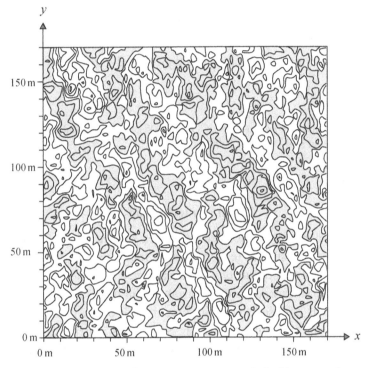

Figure 2.2 A bird's eye view of ocean waves, as recorded with stereo-photography with cameras looking down from two helicopters, i.e., the sea-surface elevation as a function of horizontal co-ordinates at one moment in time (the contour line interval is 0.20 m, shaded areas are below mean sea level; from the files of the author, see Holthuijsen, 1983a, 1983b).

and such conditions are therefore not properly represented in the statistics of wave observations from ships. Moreover, not all observers are qualified and their subjective assessments of wave conditions may well underestimate or overestimate the true wave conditions (e.g., high waves seem more impressive at night than during daytime). Still, visual observations should be treasured, because they are often the only source of information (albeit that measurements from satellites are emerging as an alternative source on a global scale).

To avoid the inherent problems of visual observations, one usually prefers measurements made with instruments. These are objective and seem to have little or no bias. That is generally true, but instruments have their own peculiarities too. The two most important are (a) limitations of the basic principle of the instrument (e.g., a buoy floating at the sea surface may swerve around or capsize in a very steep wave) and (b) sensitivity to the aggressive marine environment (e.g., mechanical impacts, marine fouling and corrosion[1]). The latter is certainly true for *in situ* techniques based on instruments positioned *in* the water. The alternative of *remote sensing*, which relies on instruments that are positioned *above* the water, is generally not sensitive to the marine environment but it may be sensitive to the atmospheric environment (e.g., rain, clouds, water vapour). This chapter treats the various observation techniques briefly, with references for further reading.

Literature:
Aage *et al*. (1998), Allender *et al*. (1989), COST (2005), Earle and Malahoff (1977), Tucker and Pitt (2001), Wyatt and Prandle (1999).

2.3 *In situ* techniques

An *in situ* instrument may be located *at* the sea surface (e.g., a floating surface buoy), or *below* the sea surface (e.g., a pressure transducer mounted on a frame at the sea bottom), or it may be *surface-piercing* (e.g., a wire mounted on a platform from above the sea surface, extending to some point below the sea surface). Most of these instruments are used to acquire time records of the up-and-down motion of the surface at one (horizontal) location. Sometimes a pier, extending from the beach across the surf zone, is used (e.g., the Field Research Facility of the U.S. Army Engineer Research & Development Center in Duck, North Carolina, USA or the Hazaki Oceanographical Research Facility of the Port and Airport Research Institute near Kashima, Japan) or a movable sled pulled along the seabed is used to acquire wave data along a transect or in a small area.

Literature:
Nakamura and Katoh (1992), Sallenger *et al*. (1983)

---

[1] To illustrate another type of marine aggression: greedy bounty hunters will 'salvage' a wave-recording buoy from the sea and sell it for scrap metal or collect the lost-and-found reward. Little do these vandals know that the buoy motions are continuously monitored (the buoy records tell revealing stories).

2.3.1 Wave buoys

One obvious way of measuring waves is to follow the three-dimensional motion of the water particles at the sea surface. This can be done with a buoy that closely follows the motion of these water particles by floating at the surface.[2] The most common technique for such a buoy is to measure its vertical *acceleration* with an onboard accelerometer (supplemented with an artificial horizon to define the vertical). The buoy also moves horizontally, but only over a small distance (roughly equal to the wave height), which is usually ignored. By integrating the vertical acceleration twice, the vertical motion of the buoy (the heave motion, see Note 2A) and thus of the sea-surface elevation is obtained as a function of time. Owing to the simultaneous horizontal motion of the buoy, the waves in the record tend to look more symmetrical (around the mean sea level) than they actually are. In reality, the crests are slightly sharper than measured and the troughs are slightly flatter. In addition, a buoy has a finite mass and size, causing the buoy generally to underestimate short waves and to resonate at its natural frequency (the eigenfrequency; thus overestimating waves near this frequency). For instance, the diameter of the NDBC[3] buoys in the USA, which usually carry a large array of meteorological sensors, may be as large as 10 m, whereas the diameter of the WAVERIDER buoy[4] (of Datawell, the Netherlands, which is the most commonly used buoy, see Fig. 2.3) is less than 1 m. In addition, a spherical buoy tends to avoid the steep parts of waves, circling around the crests of steep waves and thus avoiding maxima in the surface elevation. A buoy with a *flat* hull (e.g., a disc-shape) may even capsize in a steep wave. Some of these effects are known and can be corrected for in the analysis of the wave records. In spite of these shortcomings, buoys perform well in general.

The buoys are usually provided with radio communication to send their signals to a land- or platform-based receiving station. These links used to be based on ultra-high-frequency (UHF) radio (line-of-sight range $\approx$ 20 km) but new buoys are now often supplemented with satellite communication and position detection by the Global Positioning System (GPS, based on triangulation between dedicated satellites). As a matter of fact, GPS has become so accurate that, with some additional facilities, it can be used to measure waves: the Doppler shift of the satellite signal provides the velocity of the buoy. The accuracy can be enhanced by including a nearby fixed station in the GPS measurement (this mode is called 'differential

---

[2] Sometimes a ship is used as a wave-measuring 'buoy': measure its vertical motion and supplement this with a shipborne wave recorder (Haine, 1980; Tucker, 1956).

[3] National Data Buoy Centre of NOAA (National Oceanic and Atmospheric Administration, USA).

[4] To illustrate still another type of marine aggression: excited gun-toting crew members of a passing ship may use a WAVERIDER buoy as an interesting shooting target (it is after all a bright yellow circle moving up and down on the waves). This is, of course, anecdotal, but bullet scars on the buoy can be rather impressive (I have seen the evidence; whoever said that wave research is a safe occupation?).

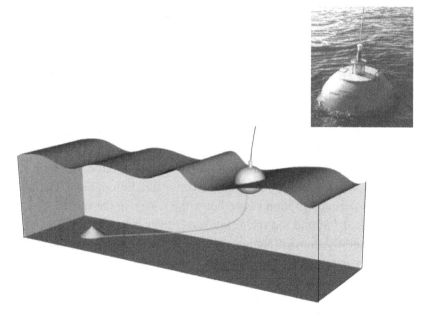

Figure 2.3 The WAVERIDER buoy at sea. The buoy measures its own vertical acceleration to estimate the sea-surface motion (photo courtesy of Datawell,
Haarlem, the Netherlands).

GPS' or D-GPS). This provides a new approach to wave measurements that is
already being exploited by the SMART buoy of OCEANOR (Norway) and the
GPS-WAVERIDER of Datawell.

The above heave buoys do not provide *directional* information. To obtain such
information, two other types of buoys have been developed. The first type measures
the *slope* of the sea surface. It is a relatively flat buoy (disc-shaped or doughnut-
shaped) and it measures, in addition to its heave, its own pitch-and-roll motion (see
Note 2A). This requires extra sensors (inclinometers) in the buoy to detect the tilt
of the buoy in two orthogonal directions and a sensor to monitor the direction to
geographic North. From these measurements, the mean wave direction and also the
degree of short-crestedness of the waves can be determined. A commercial version
of this buoy is the WAVEC buoy (WAve-VECtor, of Datawell). The second type of
buoy that can measure wave directions, measures its own *horizontal motion* (surge
and sway, see Note 2A). Similarly to the pitch-and-roll motion of the buoy, this
(horizontal) surge-and-sway motion of the buoy indicates the mean wave direction and the degree of short-crestedness. The DIRECTIONAL WAVERIDER (of
Datawell) is such a buoy. It uses the Earth's magnetic field to measure the surge
and sway. The SMART buoy and the GPS-WAVERIDER use GPS for the same

purpose. An even more sophisticated buoy is the cloverleaf buoy, which measures not only the surface elevation and its slope in two orthogonal directions but also the *curvature* of the surface in these two directions (the buoy actually consists of three pitch-and-roll buoys fixed to one another in a frame; it has only been used occasionally in scientific experiments).

NOTE 2A   The six degrees of freedom

The motion of a rigid body has six degrees of freedom: three translations and three rotations:

Translation:
surge = forward / backward
sway = left / right
heave = up / down

Rotation:
pitch = say 'yes'
roll = say 'so-so'
yaw = say 'no'

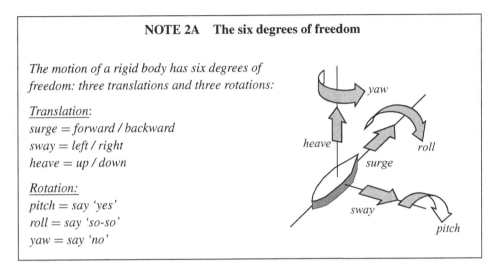

Literature:
Allender *et al.* (1989), de Vries *et al.* (2003), Holthuijsen and Herbers (1986), James (1986), Jeans *et al.* (2003a, b), Krogstad *et al.* (1997, 1999), Longuet-Higgins *et al.* (1963), Mitsuyasu *et al.* (1975, 1980), Nagai *et al.* (2004), Neumann and Pierson (1966), van der Vlugt *et al.* (1981).

2.3.2 Wave poles

When an offshore platform is available or purpose-built, a wire can be suspended vertically from that platform from above the water surface to a point somewhere beneath the water surface (see Fig. 2.4).[5] The vertical position of the water surface can then be measured as it moves along the wire (the instrument is called a 'wave pole' or a 'wave staff'). An obvious technique is to measure the length of the wire above the surface, e.g., by measuring the *electrical resistance* of this 'dry' part

---

[5] Sometimes a tall and slender buoy, floating *vertically* in the water, is used to provide a stable platform. The buoy is so long that it penetrates beneath the wave action, thus providing stability for sensors near the water surface. The generic name for such a buoy is a 'spar buoy' (e.g., Cavaleri, 1984; Tucker, 1982). Like many large buoys, spar buoys are also used to mount other instruments for oceanographic or meteorological observations. A famous example is the specially designed ship 'Flip' (Floating Instrument Platform) that floats horizontally to the required location where it is flipped vertically to provide the platform for observations (e.g., Fisher and Spiess, 1963; Snodgrass *et al.*, 1966).

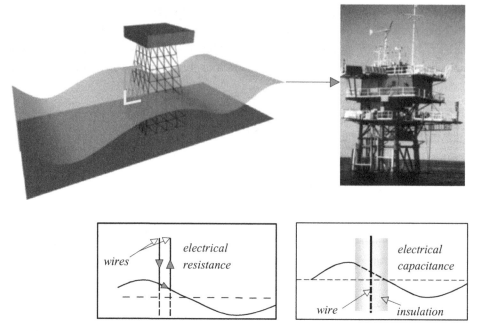

Figure 2.4 Two measurement techniques with a wave pole: electrical resistance and electrical capacitance (photo courtesy of the Institute of Marine Sciences, Venice, Italy).

of the wire. In practice two wires or one wire with a string of electrodes is used, which short-circuit at the water surface. Another technique is to measure the *electrical capacitance* of two parallel electric wires or of a single electric wire within an insulating rubber cord. It is also possible to send a high-frequency electrical signal down the wire, which will reflect at the water surface, again determining the position of the water surface. To illustrate that each *in situ* technique has its own peculiarities, it may be noted that the water surface, when moving down, tends to leave a thin film of water on the wire with a cusp-like edge between the wire and the sea surface. The dropping sea surface is therefore measured at a somewhat higher level than it would be in the absence of the wire. The error may occasionally be as large as several decimetres in rough seas, but normally the effect is relatively small and it introduces no problems.

Like a heave buoy, these wire techniques do not provide directional wave information. To obtain such information, one may use a *group* of vertical wires or poles. For instance, three poles at the corners of a small triangle can be used to estimate the slope of the surface (the triangle needs to be small compared with the lengths of the waves but not so small that measurement errors dominate; this set-up is called a *slope array*). The information is essentially the same as obtained with a pitch-and-roll buoy. When located at the corners of a larger triangle, the poles can be used to detect phase differences between the poles (this set-up is called a *phase array* with a size of the order of the wave length). For instance, if the crest of a (harmonic) wave passes through two poles simultaneously, the phase difference between these two poles is zero. A zero phase difference therefore indicates a wave direction normal to the fictitious line connecting the two poles. A third pole is needed to determine from which side the wave is approaching (left or right, in other words, there is a 180° ambiguity in the wave direction without the third pole). Any deviation from zero phase provides the wave direction relative to this reference direction. More advanced analysis techniques and more poles can provide more details of the directional character of the waves.

Literature:
Allender *et al.* (1989), Cavaleri (1979, 2000), Davis and Regier (1977), Donelan *et al.* (1985), Russell (1963), Young *et al.* (1996).

2.3.3 *Other* in situ *techniques*

The above buoys and poles are the most popular instruments used to observe waves. However, for many reasons (operational, financial etc.) one may want to use other techniques, which are less common but perfectly feasible in their own setting. Some are relatively well known. These are the inverted *echo-sounder*, the *pressure transducer* and the *current meter* (see Fig. 2.5). The inverted echo-sounder is an instrument, located at some depth beneath the sea surface, which measures the position of the water surface with a narrow, upward-looking sonic beam. A pressure transducer, located at some depth below the sea surface, can measure wave-induced pressure fluctuations. These fluctuations, combined with the linear wave theory (see Chapter 5), can be used to estimate wave characteristics. When deployed in a spatial pattern, a set of (at least three) inverted echo-sounders or pressure transducers can provide directional wave information. A current meter, mounted at some depth below the surface, measuring the wave-induced orbital motion, can also be used to estimate wave characteristics. With this instrument, directional information of the waves can be deduced without additional instrumentation, because the current is measured as a (horizontal) vector, i.e., with direction and magnitude. Sometimes, a combination of instruments is used (e.g., an inverted echo-sounder with an acoustic current meter or a set of inverted echo-sounders radiating at slightly different angles upwards from one under-water support; see Fig. 2.5).

A very refined instrument is the *wave-follower*, which consists of a small instrument package close to the water surface on a wave pole that moves up and down with the waves: the pole is carried vertically along a supporting structure by a small

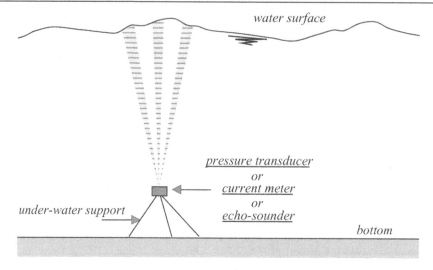

Figure 2.5 A pressure transducer, current meter or inverted echo-sounder mounted at the sea bottom (they may also be mounted at some depth on a platform piercing the water surface).

motor that is controlled by a wave sensor on the pole. It moves in such a way that the instrument package remains roughly at a fixed position above the sea surface. Sensors in the instrument package may then be used to measure the position of the sea surface more accurately or they may be used to measure other parameters, such as the air pressure just above the (moving) sea surface. It is a rather delicate set-up and it has been used only occasionally in scientific experiments.

Literature:
Bishop and Donelan (1987), Hashimoto *et al.* (1996), Hashimoto (1997), Hsiao and Shemdin (1983), Jeans *et al.* (2003a), Snyder *et al.* (1981), Takayama *et al.* (1994).

2.4 Remote-sensing techniques

Instruments that are mounted *above* the water surface on a fixed or moving platform are called *remote-sensing* instruments. The platform may be an observation tower at sea, a ship, an airplane or a satellite. Some instruments need not look downwards and these may therefore be located on land. The principle of these remote-sensing techniques is to receive reflections off the sea surface of visible or infra-red light or radar energy. The most important operational difference from *in situ* techniques is that large areas can be covered (nearly) instantaneously or in a short period of time, particularly if the platform is a satellite. However, remote sensing is often experimental and rather more expensive than *in situ* measurements. Then again,

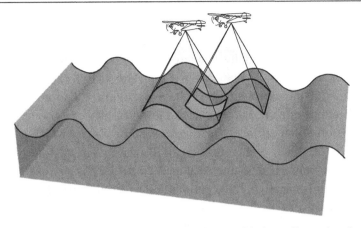

Figure 2.6 Stereo-photography from two airplanes with three-dimensional surface information where the two photographs overlap.

governments and international organisations often subsidise remote sensing and the costs of remote sensing are usually shared with many other users so that remote sensing may still be financially feasible for the individual user.

Literature:
Hwang *et al.* (1998).

2.4.1 Imaging techniques

Stereo-photography

Photography is an obvious technique to observe waves. With stereo-photography it is actually possible to obtain a three-dimensional image of the surface. It is an old and well-established technique for measuring terrestrial topography: a high-quality camera looking vertically down from an airplane takes photographs every few seconds of overlapping sections of the terrain below. The differences (*parallax*) in the overlapping photos can be converted into elevations, thus creating a three-dimensional image of the terrain. When this technique is applied to the moving sea surface, one camera is not enough because the surface itself would change between one photo and the next – if these photos were taken in sequence. For applications at sea, therefore, two *synchronised* cameras are required, usually operated from two airplanes flying in formation (see Fig. 2.6).

Literature:
Banner *et al.* (1989), Cote *et al.* (1960, also in Kinsman, 1965), Holthuijsen (1983a), Neumann and Pierson (1966).

Imaging and non-imaging radar

Conventional ship's radar is normally used to detect hard obstacles around a ship, i.e., obstacles that are potentially dangerous to the ship (marine radar, with its well-known screen, called the Plan Position Indicator or PPI, showing a scanning, map-like image of the surroundings). These radars are therefore always set to show the reflections off such hard surfaces. However, they can also be set to show the reflections off softer surfaces such as a beach or waves (which are normally considered to be 'clutter'). Such reflection off the waves is mostly due to *resonance* between the radar waves and features at the water surface (Bragg scatter). Since the radar wave length is usually in the centimetre range, only very short water waves reflect the radar waves (capillary waves, which are generated by wind, current or by breaking waves, but otherwise dominated by surface tension). These very short waves are modulated by longer waves (the waves that engineers are interested in) because, due to the orbital motion of the water particles in the longer waves, they are slightly shorter at the crest than in the troughs of these longer waves (see Section 5.4.4). The radar 'sees' this *modulation* and it is the modulation *pattern* that creates the image of the longer waves on the radar screen.

Radars that are based on the same principle have been built into airplanes and satellites to observe waves on a regional or oceanic scale. The problem for applications from high altitude is that the antenna needs to be very large in order to distinguish the individual longer waves in the modulation pattern.[6] However, by transmitting and receiving a properly programmed signal from the antenna (moving along the path of the airplane or satellite), such a large antenna can be simulated with a small one. Such radar with a programmed signal is called *synthetic aperture radar (SAR)*. The SAR images are realistic enough: everyone who sees such an image is convinced that it shows ocean waves. These images can be analysed, to obtain not the surface elevation itself[7] but statistical characteristics thereof in selected areas of limited size in the form of the two-dimensional wave-number spectrum (see Section 3.5.8). The data stream generated by a SAR is so large that the instrument cannot continuously send data to the receiving stations on Earth as the satellite orbits the Earth. It operates on request.

Other radar techniques are based on *non-imaging* returns of radar signals from the ocean surface (for instance, the frequency shift between the radiated and reflected signal). This can be exploited in various radar frequency bands, each providing operationally different (land-based or airborne) systems. One such radar can observe ocean waves from a long distance. This (low-frequency) radar is looking up, towards

---

[6] In general, for *any* antenna or lens; the larger the antenna or lens, the smaller the details that can be observed.
[7] One group of researchers (e.g., Borge *et al.*, 2004; Schulz-Stellenfleth and Lehner, 2004) claims to have retrieved the surface elevation of ocean waves from SAR images but I have not (yet) seen any validation of this.

the sky, with the radar energy reflecting off the ionosphere to the ocean surface and back. This can give a range of several thousand kilometres: sky-wave radar. Other, high-frequency (HF) radar can observe ocean waves at shorter ranges (up to just over the horizon): HF radar or ground-wave radar. Other non-imaging radar instruments are used as vertical distance meters (altimeters). These are treated below.

Literature:
Alpers *et al.* (1981), Borge *et al.* (1999), Georges and Harlan (1994), Hasselmann *et al.* (1985b), Hasselmann and Hasselmann (1991), Hasselmann *et al.* (1996), Heathershaw *et al.* (1980), Hessner *et al.* (2001), Kobayashi *et al.* (2001), Lehner *et al.* (2001), McLeish and Ross (1983), Schulz-Stellenfleth and Lehner (2004), Tomiyasu (1978), Wyatt and Ledgard (1996), Wyatt (1997, 2000), Wyatt *et al.* (1999), Young *et al.* (1985).

2.4.2 Altimetry

Laser altimetry

Another technique than photography that uses (visible or infra-red) light is the laser. As a distance meter, or rather, as an *altimeter*, a downward-looking laser can measure the vertical distance from the instrument to the sea surface rather accurately. It may be mounted on a fixed platform or in an airplane, but not on a satellite where its operation would be hindered too much by the weather. The deployment from an airplane has some special features, because the sea surface is measured along a line (the flight path of the airplane) and the airplane and the surface elevation *move* during the observation. Another technique by which to operate a laser altimeter from an airplane is to *scan* the sea surface with a moving laser beam (for instance, reflecting off a rotating mirror), along closely spaced lines at the sea surface, normal to the flight path *or* in a (forward-moving) circular pattern beneath the airplane. This technique provides a three-dimensional image of the sea surface, practically 'frozen' in time like in a stereo-photo (some distortions occur because the scanner needs time to build up the image and both the sea surface and airplane move in the time during which the scanner builds up the image). This system is called the airborne topographic mapper (ATM). These altimeter techniques are less cumbersome than stereo-photography but they share many of the operational problems (e.g., they both require a platform above the sea surface, airborne or not, and are weather-dependent).

Literature:
Allender *et al.* (1989), Hwang *et al.* (2000a), Ross *et al.* (1970), Schule *et al.* (1971).

Acoustic altimetry

Echo-sounders are not used only as *in situ* instruments (see Section 2.3.3) but also as remote-sensing instruments. When mounted above the water looking downwards, with a narrow beam, they can be used to measure the distance to the sea surface. This technique is operational at some sites in Japanese waters.

Literature:
Kuriyama (1994), Sasaki *et al.* (2005).

Radar altimetry

A narrow-beam *radar*, looking down at the sea surface, can also be used as an altimeter. If the radar is located near the water surface (at a fixed platform or in a low-flying airplane), the radar is accurate enough to measure the actual sea-surface elevation directly beneath the instrument. A variation of this technique is to scan the sea surface with the radar beam, in a manner almost identical to that of the laser-based airborne topographic mapper (ATM; see above). Such a system is called a *surface-contouring radar* or *scanning radar altimeter*.

From a larger distance, in particular from a *satellite*, the mode of operation of the radar altimeter is rather different. For such applications the (non-scanning) radar beam is pointing downwards to the sea surface, but its footprint (the spot at the sea surface that is 'illuminated' by the radar beam) is typically a few kilometres in diameter, which is too large to resolve individual waves. However, the radar signal that is reflected from the footprint to the satellite is somehow distorted by the presence of the waves in the footprint. This distortion can be used to estimate the *roughness* of the surface, which in turn can be converted into a characteristic wave height (the significant wave height, see Section 3.3.2). To explain this, consider a radar instrument transmitting a pulse of electromagnetic energy from the satellite to the sea surface (Fig. 2.7). This pulse, when originating from a sufficiently high altitude, arrives at the ocean surface as a (nearly) horizontal and flat front. When the water surface is horizontal and flat too, the reflection of the radar pulse is instantaneous and it is received by the satellite as a pulse. However, in the presence of waves, reflections occur first at the highest wave crests. This gives a weak onset of the reflection received by the satellite. As the radar front at the sea surface propagates further downwards, into the wave troughs, it meets more and more surface area and eventually it arrives at the bottom of the wave troughs. The reflection correspondingly builds up and dies down as it is received by the satellite. When the waves are very low, the distortion of the pulse is small and the return signal is short (narrow in time). If the waves are higher, the distortion is larger and the return signal broadens. This broadening is therefore a measure of the roughness

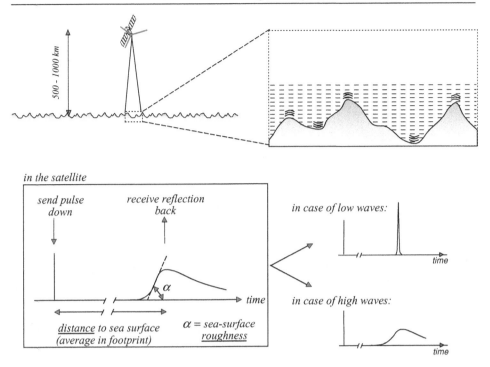

Figure 2.7 Radar altimetry from a satellite. Note that in this figure the time axes are interrupted to separate the time scale of the arrival of the radar return at the satellite (*distance* to mean sea surface) from the time scale of the shape of the radar return (giving sea-surface *roughness*).

of the sea surface and hence of some characteristic wave height in the footprint of the radar beam. In practice, the slope of the leading edge of the return signal (the angle α in Fig. 2.7) is used as a measure of the sea-surface roughness.

Literature:
Barrick (1968), Walsh *et al.* (1985, 1989).

3

Description of ocean waves

3.1 Key concepts

- The conventional *short-term* description of ocean waves requires statistical stationarity. A time record of actual ocean waves (the fluctuating sea-surface elevation as a function of time at one location) needs therefore to be as short as possible. However, characterising the waves with any reliability requires averaging over a duration that is as long as possible. The compromise at sea is a record length of 15–30 min. If the record is longer, it should be divided into such segments (possibly overlapping; each assumed to be stationary).
- The wave condition in a stationary record can be characterised with average wave parameters, such as the *significant wave height* and the *significant wave period*.
- The significant wave *height* is fairly well correlated with 'the' wave height as estimated visually by experienced observers. This is not true for the significant wave *period*.
- A more complete description of the wave condition is obtained by approximating the time record of the surface elevation as the sum of a large number of statistically independent, harmonic waves (wave components). This concept is called the *random-phase/amplitude model*.
- The random-phase/amplitude model leads to the concept of the one-dimensional *variance density spectrum*, which shows how the variance of the sea-surface elevation is distributed over the *frequencies* of the wave components that create the surface fluctuations.
- If the situation is stationary and the surface elevations are *Gaussian* distributed, the variance density spectrum provides a *complete statistical description* of the waves.
- The concept of the random-phase/amplitude model can be extended to the *three-dimensional*, *moving* sea surface, which is then seen as the sum of a large number of statistically independent, harmonic waves *propagating* in all directions across the sea surface. The corresponding two-dimensional variance density spectrum shows how the variance is distributed over the *frequencies and directions* of these harmonic wave components.
- The one-dimensional spectrum can be obtained from the two-dimensional spectrum by integration over all directions.
- The variance density spectrum provides also a description in a *physical* sense when multiplied with ρg (ρ is the density of water, g is the gravitational acceleration). The result is the *energy* density spectrum. It shows how the energy of the waves is distributed over the frequencies (and directions).
- The analysis of a time record of the sea-surface elevation, to obtain an estimate of the one-dimensional spectrum, is treated in Appendix C.

3.2 Introduction

The first step in describing wind waves is to consider the vertical motion of the sea surface at one horizontal position, for instance along a vertical pole at sea as addressed in the previous chapter. The ocean waves then manifest themselves as a surface moving up and down in time at that one location. It may sound odd, but it would be entirely legitimate to conclude from such motion, and from such motion alone, that the sea surface is a perfectly smooth, horizontal plane that moves vertically in a rather random manner. This of course is not the case, but we know that only because we have all seen real, three-dimensional, moving ocean waves. Such chaotic motion of ocean waves seems to defy any rational approach (see Fig. 2.1). In three dimensions, the situation looks even more problematic (see Fig. 2.2). However, a rational approach to describe this apparent chaos is entirely possible, as will be shown in this chapter.

To estimate wave conditions *visually*, even the most casual observer tends to concentrate his/her attention on the highest waves in the wave field. For instance, the estimates of the observers who report daily to the meteorological network of the World Meteorological Organisation (WMO) are based on '. . . the average height and period of 15 to 20 well defined, higher waves of a number of wave groups . . .' (guidelines of the WMO). These average wave characteristics are called the *significant wave height* and the *significant wave period*, denoted as H_s and T_s, respectively (or H_v and T_v to indicate that they have been estimated visually). The concept of the significant wave height and period is very useful in many situations. However, two wave parameters give only a limited description of the wave conditions. For instance, wave conditions may well be similar in the sense that the significant wave height and period are equal, but they may still be very different in detail: a mixed sea state of wind sea (short, irregular, locally generated waves) and swell (long, smooth waves, generated in a distant storm) may have the same significant wave height and period as a slightly higher wind sea without swell. To distinguish such conditions, more parameters are needed, for instance, a significant wave height and period for wind sea and swell separately. This is sometimes done and it may be adequate in some cases, but any small number of parameters would not, in general, completely characterise the wave conditions. For a complete description (in a statistical sense), another technique, the spectral technique, is required. It is based on the notion that the random motion of the sea surface can be treated as the summation of a large number of harmonic wave components.

3.3 Wave height and period

3.3.1 Waves

Before we can objectively define a wave height or period, we need to define more precisely what a *wave* is. This seems trivial but many people consider any elevation of the sea surface to be a wave. In the present context, this is not correct: we need to distinguish between the *surface elevation* and a *wave*. In a time record, the surface elevation is the instantaneous elevation of the sea surface (i.e., at any one moment

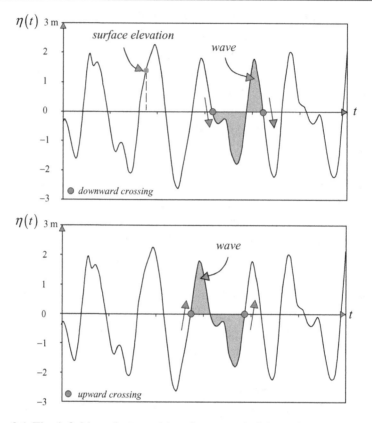

Figure 3.1 The definition of a 'wave' in a time record of the surface elevation with downward zero-crossings (upper panel) or upward zero-crossings (lower panel).

in time) relative to some reference level. In such a record, a wave is the *profile* of the surface elevation between two successive *downward* zero-crossings of the elevation (zero = mean of surface elevations, see Fig. 3.1). A surface elevation can be negative, whereas a wave cannot. Alternatives for defining a wave are possible, e.g., the profile between two successive *upward* zero-crossings (see Fig. 3.1).

If the surface elevation, denoted as $\eta(t)$, is seen as a Gaussian[1] process (see Appendix A), it does not matter whether the definition with downward zero-crossings or upward zero-crossings is used, because the statistical characteristics would be symmetrical. However, many prefer the definition with the downward crossings because in visual estimates the height of the crest relative to the *preceding* trough is normally considered to be the wave height. In addition, in a breaking

---

[1] Johann Carl Friedrich Gauss (1777–1855) was a German scientist with a wide range of interests. He had a passion for numbers and calculations (theory of numbers, algebra, analysis, geometry, probability and the theory of errors). He was also active in astronomy, celestial mechanics, surveying and geodesy. He made his fortune with shares in private companies.

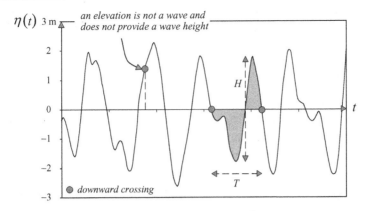

Figure 3.2 The definition of wave height and wave period in a time record of the surface elevation (the wave is defined with downward zero-crossings).

wave, the (steep) front, which is most relevant for the breaking process, is included in the definition with downward crossings (under such conditions, the waves are not symmetrical and the differences between the definition of a wave with zero-up or zero-down crossings becomes relevant). Characterising the waves in the wave record is based on averaging all of the individual wave heights and periods in the record. This requires the duration of the record to be short enough to be stationary but also long enough to obtain reasonably reliable averages. The commonly used compromise at sea is 15–30 min.

Literature:
Buckley *et al.* (1984), Goda (1986).

3.3.2 Wave height

It is natural to define the wave height H as the vertical distance between the highest and the lowest surface elevation in a wave (see Fig. 3.2). A wave will thus have only one wave height. In a wave record with N waves, the *mean* wave height $\overline{H}$ is then readily defined as

$$\overline{H} = \frac{1}{N} \sum_{i=1}^{N} H_i \tag{3.3.1}$$

where i is the sequence number of the wave in the record (i.e., $i = 1$ is the first wave in the record, $i = 2$ is the second wave, etc.).

Sometimes a quadratically weighted averaged value is used to define the root-mean-square wave height H_{rms}:

$$H_{rms} = \left(\frac{1}{N} \sum_{i=1}^{N} H_i^2 \right)^{1/2} \tag{3.3.2}$$

Such a measure of wave heights may be relevant for energy-related projects because the wave energy is proportional to the square of the wave height (see Section 5.5).

These characteristic wave heights $\overline{H}$ and H_{rms} seem to be rather obvious to define, but they are not very often used, probably because they bear little resemblance to the visually estimated wave height. Instead, another wave height, called the significant wave height H_s (just as in visual observations) is used. It is defined as the *mean of the highest one-third of waves* in the wave record:[2]

$$\boxed{significant\ wave\ height = \left| H_{1/3} = \frac{1}{N/3} \sum_{j=1}^{N/3} H_j \right.} \tag{3.3.3}$$

where j is *not* the sequence number in the record (i.e., sequence in time) but the rank number of the wave, based on wave *height* (i.e., $j = 1$ is the highest wave, $j = 2$ is the second-highest wave, etc.). This seems to be an odd way of defining a characteristic wave height but experiments have shown that the value of this wave height is close to the value of the visually estimated wave height (see Section 3.4). It is somewhat confusing that both the *visually estimated* characteristic wave height and this *measured* characteristic wave height are called the 'significant wave height'. To distinguish them from one another, the visually estimated significant wave height is therefore denoted here as H_v and the measured significant wave height (from a time record) as $H_{1/3}$ (pronounced as H-one-third). The significant wave height can also be estimated from the wave spectrum. It will be denoted as H_{m_0} (H-m-zero; see Section 4.2.2). Sometimes the mean of the highest one-tenth of waves is used to define $H_{1/10}$ (H-one-tenth),

$$H_{1/10} = \frac{1}{N/10} \sum_{j=1}^{N/10} H_j \tag{3.3.4}$$

but it has no obvious relation to the visually estimated significant wave height. It must be noted that, if the waves are not too steep and not in very shallow water, there is a (theoretically based) constant ratio between the various characteristic wave heights (e.g., $H_{rms} = \frac{1}{2}\sqrt{2}\,H_{1/3}$; see Section 4.2.2).

---

[2] If you are interested in history: this definition seems to have been introduced by Sverdrup and Munk (1946).

3.3.3 Wave period

It is equally natural to define the period T of a wave as the time interval between the start and the end of the wave (the interval between one zero-down crossing and the next; see Fig. 3.2). Since this wave period is defined with zero-crossings, it is called the zero-crossing period, T_0. The mean of this zero-crossing wave period, denoted as $\overline{T}_0$, is then defined, in analogy with the mean wave height $\overline{H}$, as

$$mean\ zero\text{-}crossing\ wave\ period = \overline{T}_0 = \frac{1}{N} \sum_{i=1}^{N} T_{0,i} \tag{3.3.5}$$

where i is the sequence number of the wave in the time record. In analogy with the significant wave height, the significant wave *period* T_s is defined as the mean period of the *highest* one-third of waves, $T_{1/3}$ (pronounced as T-one-third):

$$\boxed{significant\ wave\ period = T_{1/3} = \frac{1}{N/3} \sum_{j=1}^{N/3} T_{0,j}} \tag{3.3.6}$$

where, again, j is *not* the sequence number but the rank number of the wave, based on wave *height* (it is the same j as in the definition of the significant wave height, Eq. 3.3.3). To distinguish, in the notation, the visually estimated significant wave period from the measured significant wave period (obtained directly from the time record), the former is denoted as T_v and the latter as $T_{1/3}$. As with the wave heights, sometimes the mean of the highest one-tenth of waves is used to define $T_{1/10}$ (T-one-tenth):

$$T_{1/10} = \frac{1}{N/10} \sum_{j=1}^{N/10} T_{0,j} \tag{3.3.7}$$

Other characteristic wave periods are also used, but these are defined in terms of the wave spectrum. They will be treated in Chapter 4.

3.4 Visual observations and instrumental measurements

Wave measurements (i.e., recordings made with *instruments*) are routinely carried out at only a few locations in the world's oceans: mostly along the coasts of Europe, the USA, Canada and Japan (although satellite measurements are rapidly supplementing this on a worldwide scale). In most other places, the engineer has to find wave information from other sources. There are three alternatives (apart from starting a dedicated measurement campaign): visual observations, satellite measurements and computer simulations.

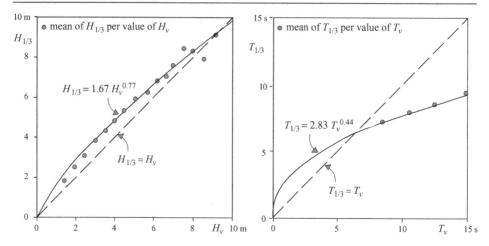

Figure 3.3 The relationship between the visually estimated significant wave height and period and the measured significant wave height and period (after Nordenstrøm, 1969). The standard deviation of the measured values is about 15% of the mean of the measurements at every value of H_v or T_v.

It is therefore of obvious interest to the engineer to know to what extent *visual* observations resemble estimates obtained from instrument measurements. To that end, one can carry out an experiment in which waves are visually estimated and simultaneously measured with instruments. The results of one such study are shown in Fig. 3.3. The agreement between the visually estimated significant wave height (H_v) and the measured significant wave height ($H_{1/3}$) is reasonable. The relationship can be represented by the best-fit power law for these data:

$$H_{1/3} = 1.67 H_v^{0.77} \qquad \text{(in m)} \tag{3.4.1}$$

so that $H_{1/3} \approx H_v$.

In contrast to this, the visually estimated significant wave *period* does *not* agree well with the instrumental measurement. The best-fit power-law relationship in the same study is

$$T_{1/3} = 2.83 T_v^{0.44} \qquad \text{(in s)} \tag{3.4.2}$$

so that $T_{1/3} \neq T_v$ (see Fig. 3.3).

Literature:
Battjes (1984).

3.5 The wave spectrum

3.5.1 Introduction

The aim of describing ocean waves with a spectrum is not so much to describe in detail one observation of the sea surface (i.e., one time record), but rather to describe the sea surface as a stochastic process, i.e., to characterise all possible observations (time records) that could have been made under the conditions of the actual observation. An observation is thus formally treated[3] as one realisation of a stochastic process (see Appendix A). Here, we base this treatment on the *random-phase/amplitude model*, which leads to the wave spectrum, which is the most important form in which ocean waves are described.[4]

The basic concept of the wave spectrum is simple, but its many aspects make it seem rather complicated. To distinguish the essence from these additional aspects, consider first a wave record, i.e., the surface elevation $\eta(t)$ at one location as a function of time, with duration D, obtained at sea with a wave buoy or a wave pole (see Fig. 3.4).

We can exactly reproduce that record as the sum of a large number of harmonic wave components (a Fourier series):

$$\eta(t) = \sum_{i=1}^{N} a_i \cos(2\pi f_i t + \alpha_i) \tag{3.5.1}$$

where a_i and α_i are the amplitude and phase, respectively,[5] of each frequency $f_i = i/D$ ($i = 1, 2, 3, \ldots$; the frequency interval is therefore $\Delta f = 1/D$). With a Fourier analysis, we can determine the values of the amplitude and phase for each frequency and this would give us the amplitude and phase spectrum for this record (see Fig. 3.4). By substituting these computed amplitudes and phases into Eq. (3.5.1), we exactly reproduce the record.

For most wave records, the phases turn out to have any value between 0 and 2π without any preference for any one value. Since this is almost always the case in deep

---

[3] The theory is taken from the description of noise (Tukey and Hamming, 1948) with some of the first applications to ocean waves by Barber and Ursell (1948) and Deacon (1949).

[4] For the alternative *wavelet* description, see the footnote in Appendix C.

[5] The Greek alphabet:

Notation		Name	Notation		Name	Notation		Name	Notation		Name
A	α	alpha	H	η	eta	N	ν	nu	T	τ	tau
B	β	beta	Θ	θ	theta	Ξ	ξ	xi	Υ	υ	upsilon
Γ	γ	gamma	I	ι	iota	O	o	omicron	Φ	ϕ, φ	phi
Δ	δ	delta	K	κ	kappa	Π	π	pi	X	χ	chi
E	ε	epsilon	Λ	λ	lambda	P	ρ	rho	Ψ	ψ	psi
Z	ζ	zeta	M	μ	mu	Σ	σ	sigma	Ω	ω	omega

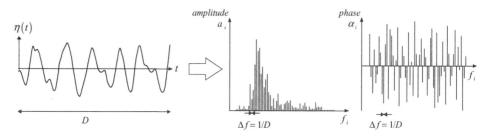

Figure 3.4 The observed surface elevation and its amplitude and phase spectrum.

water (not for very steep waves), we will ignore the phase spectrum (just keep this uniform distribution in mind and apply that knowledge when called for). Then, only the amplitude spectrum remains to characterise the wave record. If we were to repeat the experiment, i.e., measure the surface elevation again under statistically *identical* conditions (e.g., in an exact copy of the storm in which the first observation was made), the time record would be different and so would be the amplitude spectrum. To remove this sample character of the spectrum, we should repeat the experiment many times (M) and take the average over all these experiments, to find the *average* amplitude spectrum:

$$\bar{a}_i = \frac{1}{M} \sum_{m=1}^{M} a_{i,m} \qquad \text{for all frequencies } f_i \qquad (3.5.2)$$

where $a_{i,m}$ is the value of a_i in the experiment with sequence number m. For large values of M the value of $\bar{a}_i$ converges (approaches a constant value as we increase M), thus solving the sampling problem. However, it is more meaningful to distribute the *variance* of each wave component $\frac{1}{2}\overline{a_i^2}$ (see Note 3A). There are two reasons for this. First, the variance is a more relevant (statistical) quantity than the amplitude. For instance, the sum of the variances of the wave components is equal to the variance of the sum of the wave components (i.e., the random surface elevation).[6] In contrast to this, the sum of the amplitudes is not equal to the amplitude of the sum (there is no such thing as the amplitude of a random sea-surface elevation). Second, the linear theory for surface gravity waves (see Chapter 5) shows that the *energy* of the waves is proportional to the variance. This implies that, through the variance, a link is available to such physical properties as wave energy, but also wave-induced particle velocity and pressure variations. The variance spectrum $\frac{1}{2}\overline{a_i^2}$ is discrete, i.e., only the frequencies $f_i = i/D$ are present, whereas in fact all frequencies are present at sea. A first step to resolve this problem would be to distribute the variance

---

[6] Also, the square root of the variance is the standard deviation σ_η of the surface elevation, which can be seen as a vertical scale of the wave heights, for instance, the significant wave height $H_s \approx 4\sigma_\eta$ (see Section 4.2.2).

$\frac{1}{2}\overline{a_i^2}$ over the frequency interval $\Delta f = 1/D$, giving a variance *density* $\frac{1}{2}\overline{a_i^2}/\Delta f$ at each frequency (i.e., it is constant within the frequency band Δf). All frequencies would thus be represented because they have all been assigned a variance density. The variance is now distributed over all frequencies but its value still 'jumps' from one frequency band to the next (it is discontinuous). This is resolved by letting the frequency interval Δf approach zero ($\Delta f \rightarrow 0$). The definition of the variance density spectrum thus becomes

$$E(f) = \lim_{\Delta f \to 0} \frac{1}{\Delta f} \frac{1}{2}\overline{a^2} \quad \text{or} \quad E(f) = \lim_{\Delta f \to 0} \frac{1}{\Delta f} E\{\frac{1}{2}\underline{a}^2\} \tag{3.5.3}$$

(in the formal definition, to be treated below, the average $\frac{1}{2}\overline{a_i^2}$ will be replaced with the expected value $E\{\frac{1}{2}\underline{a}^2\}$ and the frequency band need not be the same for all frequencies). The underscore of $\underline{a}$ indicates that the amplitude will be treated as a random variable. This (one-dimensional) frequency spectrum $E(f)$ has been introduced here with only a brief explanation using the analysis of a measured time series and only to give the essence of the concept of the spectrum. It will be treated more extensively in the next sections.

3.5.2 The random-phase/amplitude model

The basic model for describing the moving surface elevation $\eta(t)$ is the random-phase/amplitude model, in which the surface elevation is considered to be the sum of a large number of harmonic waves, each with a constant amplitude and a phase randomly chosen for each realisation of the time record (for the concept of random variables and realisations, see Appendix A[7]):

$$\underline{\eta}(t) = \sum_{i=1}^{N} \underline{a}_i \cos(2\pi f_i t + \underline{\alpha}_i) \tag{3.5.4}$$

where N is a large number (of frequencies) and the underscores of amplitude $\underline{a}_i$ and phase $\underline{\alpha}_i$ indicate that these are now random variables (see Fig. 3.5).

The phases and amplitudes, being random variables, are fully characterised with their respective probability density functions. In this model, the phase at each frequency f_i is uniformly distributed between 0 and 2π (see Fig. 3.6):

$$p(\alpha_i) = \frac{1}{2\pi} \quad \text{for} \quad 0 < \alpha_i \leq 2\pi \tag{3.5.5}$$

---

[7] If you are not familiar with these concepts, you should read this appendix.

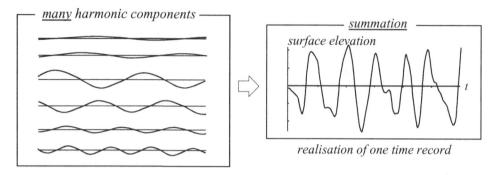

Figure 3.5 The summation of many harmonic waves, with constant but randomly chosen amplitudes and phases, creates a random sea surface.

and the amplitude $\underline{a}_i$ is at each frequency Rayleigh[8] distributed (with only one parameter μ_i varying over the frequencies; see Fig. 3.6):[9]

$$p(a_i) = \frac{\pi}{2} \frac{a_i}{\mu_i^2} \exp\left(-\frac{\pi a_i^2}{4\mu_i^2}\right) \qquad \text{for} \qquad a_i \geq 0 \qquad (3.5.6)$$

where μ_i is the expected value of the amplitude $\mu_i = E\{\underline{a}_i\}$ (see Appendix A for the notion of a mean value as an expected value). The exact values of the frequencies f_i in the summation of Eq. (3.5.4) are not important as long as (a) the frequencies are densely distributed along the frequency axis (i.e., the difference between two sequential frequencies f_i and f_{i+1} should be small compared with some characteristic wave frequency) and (b) they should be in the correct range (typically 0.05–1.0 Hz for waves at sea). Since $\mu_i = E\{\underline{a}_i\}$ is the only parameter in Eq. (3.5.6), the statistical characteristics of $\underline{a}_i$ are completely given by this one parameter (per frequency). The function that shows this mean amplitude along the frequency axis is called the amplitude spectrum $E\{\underline{a}_i\}$ (see Fig. 3.6).

For a given amplitude spectrum, a realisation of $\eta(t)$ can be created with Eq. (3.5.4) by drawing sample values of the amplitudes $\underline{a}_i$ and phases $\underline{\alpha}_i$ from their respective probability density functions, at each frequency separately and

---

[8] John William Strutt, Third Baron Rayleigh (1842–1919), was an English physicist whose work on gases (the discovery of argon) earned him the Nobel prize in 1904. He also worked in the field of acoustics, optics and wave propagation in fluids.

[9] The harmonic component may also be written as $\eta(t) = A_i \cos(2\pi f_i t) + B_i \sin(2\pi f_i t)$. In the random-phase/amplitude model, A_i and B_i would each be Gaussian distributed (with the same mean and standard deviation). Since from basic trigonometry $a_i = \sqrt{A_i^2 + B_i^2}$ and $\alpha_i = \arctan(-B_i/A_i)$, the amplitude a_i is Rayleigh distributed and the phase α_i is uniformly distributed. The distribution of the square of the amplitude, a_i^2 (to be considered later), is a χ^2-distribution with two degrees of freedom (which is an exponential distribution; see also the footnote in Appendix C, Section 3).

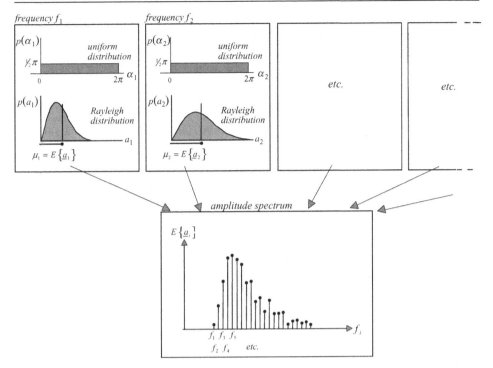

Figure 3.6 The random-phase/amplitude model: at every frequency there is one uniform distribution for the random phase and one Rayleigh distribution for the random amplitude (characterised by the expected value $E\{\underline{a}_i\}$). Top panels: for a series of frequencies, $f_i, i = 1, 2, 3, 4, 5$ etc. Bottom panel: the expected value of the amplitude as a function of frequency, i.e., the amplitude spectrum.

independently. A wave record at sea can be seen as one such realisation. For each new realisation of $\eta(t)$, the sample values of $\underline{a}_i$ and $\underline{\alpha}_i$ are again randomly drawn from these probability density functions. It is thus (hypothetically) possible to create a (large) set of realisations of the sea surface (this is called an *ensemble*).

Regarding the applicability of the random-phase/amplitude model to real ocean waves, the following remarks should be made.

• First, the random-phase/amplitude model generates a stationary (Gaussian) process. To use this approach for conditions at sea, which are never really stationary, a wave record needs to be divided into segments that are each deemed to be approximately stationary (a duration of 15–30 min is commonly used for wave records obtained at sea; these may be overlapping segments). In addition, at sea the wave components are not really independent from one another (as in the random-phase/amplitude model) because they interact to some degree. However, if the waves are not too steep and not in very shallow water, these interactions are weak and they can be ignored, leaving the random-phase/amplitude model in place as the basic model to describe ocean waves.

• Second, the random-phase/amplitude model is a summation of wave components at discrete frequencies f_i, whereas, in fact, a continuum of frequencies is present at sea. This aspect is the subject of the next section.

3.5.3 The variance density spectrum

The amplitude spectrum provides enough information to describe the sea-surface elevation realistically as a stationary, Gaussian process. However, for several reasons (see Section 3.5.1) it is more relevant to present the information in this spectrum in a different way: consider the variance $E\{\frac{1}{2} \underline{a}_i^2\}$ rather than the above-introduced expectation of the amplitude $E\{\underline{a}_i\}$. In other words, consider the *variance* spectrum instead of the *amplitude* spectrum (see Fig. 3.7 and Note 3A). This seems trivial and also enough to characterise the sea-surface elevation. However, both the amplitude and the variance spectrum are based on *discrete* frequencies, whereas Nature does not select such discrete frequencies. *All* frequencies are present at sea. The random-phase/amplitude model needs therefore to be modified. This is done by distributing the variance $E\{\frac{1}{2}\underline{a}_i^2\}$ over the frequency *interval* Δf_i at frequency f_i. The resulting variance *density* spectrum $E^*(f_i)$ is then[10]

$$E^*(f_i) = \frac{1}{\Delta f_i} E\{\tfrac{1}{2}\underline{a}_i^2\} \qquad \text{for all } f_i \tag{3.5.7}$$

and Δf_i is the interval between the frequencies. This spectrum is defined for all frequencies, but it still varies discontinuously from one frequency band to the next (see Fig. 3.7). A continuous version is obtained by having the width of the frequency band Δf_i approach zero (see Fig. 3.7):

$$E(f) = \lim_{\Delta f \to 0} \frac{1}{\Delta f} E\{\tfrac{1}{2}\underline{a}^2\} \tag{3.5.8}$$

> *This function $E(f)$ is called the variance density spectrum.*
> *It is the single most important concept in this book.*

The variance density spectrum gives a complete description of the surface elevation of ocean waves in a statistical sense, provided that the surface elevation can be seen as a stationary, Gaussian process. This implies that all statistical characteristics of the wave field can be expressed in terms of this spectrum (this is shown in Section 3.5.5).

---

[10] Note that the symbols for variance density $E^*(\, . \,)$ and $E(\, . \,)$ are different from the symbol for expected value $E\{ \, . \, \}$.

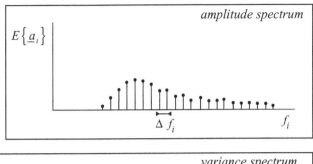

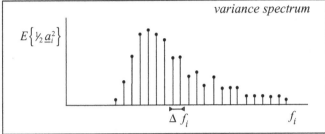

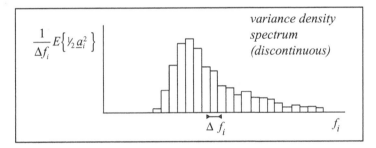

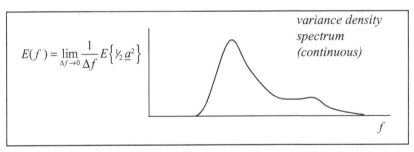

Figure 3.7 The transformation of the discrete amplitude spectrum of the random-phase/amplitude model to the continuous variance density spectrum.

The *dimension* and S.I. *unit* of the variance density $E(f)$ follow directly from its definition (Eq. 3.5.8): the dimension of the amplitude $\underline{a}$ is [length] and its S.I. unit is [m]; the dimension of the frequency band Δf is [time]$^{-1}$ and its S.I. unit is [s^{-1}], or rather [Hz]. The dimension of $E(f)$ is therefore [length^2/(1/time)] and its unit is either $[m^2 \ s]$ or $[m^2/Hz]$ (personally, I prefer the unit [Hz], because it shows better that frequencies are involved, rather than some time interval).

NOTE 3A   The variance of the sea-surface elevation

The variance of the surface elevation $\eta(t)$ is, by definition, the average of the squared surface elevation (relative to its mean) $\overline{\eta^2}$ (the overbar indicates time-averaging). For a *harmonic* wave with amplitude a, the variance is $\overline{\eta^2} = \frac{1}{2}a^2$.

In the random-phase/amplitude model for *random* ocean waves, a large number of harmonic waves is added and the variance of this sum, i.e., the random surface elevation $\eta(t)$, is equal to the sum of the individual variances ('the variance of the sum is the sum of the variances'):

$$variance = \overline{\eta^2} = E\{\underline{\eta}^2\} = \sum_{i=1}^{N} E\{\tfrac{1}{2}\underline{a}_i^2\} \qquad \text{for } E\{\underline{\eta}\} = 0$$

The square root of this variance is the standard deviation σ_η of the surface elevation, which can be seen as a vertical scale of the wave heights. For instance, the significant wave height $H_s \approx 4\sigma_\eta$ (see Section 4.2.2).

3.5.4 Interpretation of the variance density spectrum

The variance density spectrum was introduced in the previous section by transforming the discrete amplitude spectrum into a continuous distribution of the variance over frequencies. This spectrum shows how much Δvar a frequency *band* Δf contributes to the total variance (see Fig. 3.8):

$$\Delta var = \int_{\Delta f} E(f) df \tag{3.5.9}$$

It follows that the total variance $\overline{\eta^2}$ (see Note 3A) of the sea-surface elevation is the sum of the variances of all frequency bands Δf, or, for a continuous spectrum,[11]

$$total \ variance = \overline{\eta^2} = \int_0^\infty E(f) df \tag{3.5.10}$$

---

[11] Another way of finding the unit of $E(f)$ is to note that the S.I. unit of the total variance of the sea-surface elevation is $[m^2]$. Since the unit along the horizontal axis in Fig. 3.8 (frequency f) is [Hz], it follows that the unit along the vertical axis (variance density $E(f)$) should be $[m^2/Hz]$ to arrive at unit $[m^2]$ for the integral.

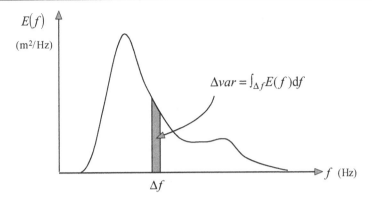

Figure 3.8 The interpretation of the variance density spectrum as the distribution of the total variance of the sea-surface elevation over frequencies.

Note that, in a *random* wave field, the contribution of a single frequency to the total variance is infinitely small, because the spectral bandwidth of a single frequency is zero: $\Delta f \to 0$ and its contribution $\Delta \mathrm{var} = E(f)\Delta f \to 0$. However, the spectrum of one *harmonic* wave, i.e., a wave with only one frequency, contains a finite energy. Its spectrum therefore consists of a delta function at that frequency (infinitely narrow and infinitely high, with an integral equal to the variance of the harmonic wave, see Fig. 3.9).

The variance density spectrum $E(f)$, showing how the variance of the sea-surface elevation is distributed over the frequencies, is rather difficult to conceive: a statistical characteristic (variance) is distributed over the frequencies of the harmonic components that make up the process. It may help if we multiply the spectrum by ρg. We then obtain the *energy* density spectrum (see below). This spectrum shows how the wave energy is distributed over the frequencies, which seems to be easier to comprehend.

The overall appearance of the waves can be inferred from the shape of the spectrum: the narrower the spectrum, the more regular the waves are. This is shown for three different wave conditions in Fig. 3.9. The narrowest spectrum corresponds to a harmonic wave. As indicated above, the spectrum then degenerates to a delta-function at one frequency. Distributing the variance over a slightly wider frequency band gives a slowly modulating harmonic wave because the components involved differ only slightly in frequencies and therefore get out of phase with one another only slowly, thus creating a fairly regular wave field. Distributing the wave variance over a wider frequency band gives a rather chaotic wave field (irregular waves), because the components in the time record get out of phase with one another rather quickly.

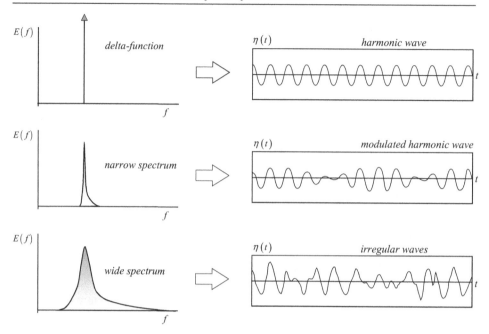

Figure 3.9 The (ir)regular character of the waves for three different widths of the spectrum.

As indicated above, the *energy* of the waves can be expressed in terms of the variance of the surface elevation because the energy of a harmonic wave (per unit horizontal ocean surface area) is equal to the mean-square elevation times the gravitational acceleration g and the density of water ρ (see the linear theory of surface gravity waves in Chapter 5), so the *total* energy (i.e. summed over all components; per unit horizontal ocean surface area) is

$$E_{total} = \rho g \overline{\eta^2} \qquad (3.5.11)$$

We can therefore multiply the variance density spectrum $E_{variance}(f) = E(f)$ by ρg and obtain the *energy* density spectrum as

$$\boxed{E_{energy}(f) = \rho g E_{variance}(f)} \qquad (3.5.12)$$

This close relationship leads to a rather inaccurate use of the word 'spectrum'. It refers to both the variance density spectrum and the energy density spectrum. Very often, the two terms are used indiscriminately, with the context indicating which of the two is actually meant. Just as the variance density spectrum is used to describe the *statistical* aspects of the waves, so can the energy density spectrum be used to describe the *physical* aspects of waves (within the limitations of the stationary, Gaussian model and the linear theory of surface gravity waves). In the above rationale of the spectrum, the wave components are assumed to be statistically independent. In other words, the wave components are assumed not to affect one another (they should behave as linear waves). This is usually quite realistic for wind-generated waves and it greatly simplifies the interpretation of the spectrum because the behaviour of linear waves is well understood.

3.5.5 Alternative definitions

The variance density spectrum can be defined in other ways than the one given above. The differences may relate to (a) the spectral *domain* or (b) the formal *definition*.

The spectral domain

The variance density has been defined in Section 3.5.4 above in terms of frequency $f = 1/T$ (where T is the period of the harmonic wave), but it can equally well be formulated in terms of *radian* frequency $\omega = 2\pi/T$. The corresponding spectrum $E(\omega)$ is then defined in the same manner as $E(f)$; the only difference being that $\cos(2\pi ft + \alpha)$ is replaced with $\cos(\omega t + \alpha)$. These spectra are obviously related: $E(\omega)$ can be expressed in terms of $E(f)$ and vice versa, but it must be borne in mind that the total variance $\overline{\eta^2}$ should be conserved in such transformations. In other words,

$$\overline{\eta^2} = \int_0^\infty E(\omega)d\omega = \int_0^\infty E(f)df \qquad (3.5.13)$$

which is readily achieved by taking

$$E(\omega)d\omega = E(f)df \qquad (3.5.14)$$

or

$$E(\omega) = E(f)\frac{df}{d\omega} = E(f)J \qquad (3.5.15)$$

where $J = df/d\omega$ is called the *Jacobian*.[12] In this case, that of transforming $E(f)$ into $E(\omega)$, it has the value $J = 1/(2\pi)$. Therefore, one must not only transform the frequencies f into ω, but also transform the density $E(f)$ into the density $E(\omega)$.

Formal definition

The variance density spectrum has been defined above in terms of the random-phase/amplitude model. An alternative definition of the spectrum, which has exactly the same interpretation in the sense that it shows the distribution of the total variance over the frequencies, is based on the Fourier transform of the *auto-covariance function* of the sea-surface elevation:

$$\tilde{E}(f) = \int_{-\infty}^{\infty} C(\tau)\cos(2\pi f \tau)d\tau \qquad \text{for } -\infty \leq f \leq \infty \qquad (3.5.16)$$

where the auto-covariance function $C(\tau)$ is defined as the average product of the elevations at moments t and $t + \tau$ (each relative to its mean; see Appendix A). For a stationary process, the value of t is not relevant (by definition all statistical characteristics are then constant in time) and the auto-covariance function depends only on the time *difference* τ:

$$C(\tau) = E\{\underline{\eta}(t)\,\underline{\eta}(t + \tau)\} \qquad \text{for } E\{\underline{\eta}(t)\} = 0 \qquad (3.5.17)$$

For a stationary wave condition, both $\tilde{E}(f)$ and $C(\tau)$ are even functions, i.e., $\tilde{E}(-f) = \tilde{E}(f)$ and $C(-\tau) = C(\tau)$. The variance density spectrum $E(f)$ is then defined as

$$E(f) = 2\tilde{E}(f) \qquad \text{for } f \geq 0 \qquad (3.5.18)$$

Since $C(\tau)$ contains all covariances of the joint probability density functions of $\underline{\eta}(t)$ and $\underline{\eta}(t + \tau)$ for any t and τ, it provides a complete description of the

---

[12] More generally, transforming one (density) function $F(x)$ into another $F(y)$, where y is a function of x, i.e., $y = f(x)$, under the condition that the integral is conserved, can be achieved with $F(x) = F(y)J$, in which $J = dy/dx$ is the (one-dimensional) Jacobian. This applies to all density functions of which the integral needs to be conserved in the transformation (i.e., not only spectra but also probability density functions). The transformation of a two-dimensional density function requires a two-dimensional Jacobian: $F(x_1, x_2) = F(y_1, y_2)\,J$, where the Jacobian J is the determinant of the matrix

$$\begin{vmatrix} \dfrac{\partial y_1}{\partial x_1} & \dfrac{\partial y_2}{\partial x_1} \\[2mm] \dfrac{\partial y_1}{\partial x_2} & \dfrac{\partial y_2}{\partial x_2} \end{vmatrix}$$

which is

$$J = \frac{\partial y_1}{\partial x_1}\frac{\partial y_2}{\partial x_2} - \frac{\partial y_2}{\partial x_1}\frac{\partial y_1}{\partial x_2}$$

process in a statistical sense (if the process is stationary and Gaussian and its mean is zero). It has been noted in Section 3.5.3 (without proof) that the variance density spectrum, *too*, provides such a complete description. This statement is based on the fact that the Fourier transform is reversible, so that, from Eq. (3.5.16): $C(\tau) = \int_{-\infty}^{\infty} \tilde{E}(f) \cos(2\pi f \tau) df$. This reversibility implies that the one function can be expressed in terms of the other without loss of information. The variance density spectrum therefore contains the same information as the auto-covariance function and it describes a stationary, Gaussian process as completely as the auto-covariance function does. It also follows that the total variance is $\overline{\eta^2} = C(0) = \int_{-\infty}^{\infty} \tilde{E}(f) df = \int_{0}^{\infty} E(f) df$, which indicates that the spectrum $E(f)$ gives the distribution of the total variance over the frequencies. Note that these *definitions* do not require the assumption of independent wave components (but it is usually quite realistic for wind-generated waves and it greatly simplifies the interpretation of the spectrum).

This definition of the variance density spectrum, which is based on the auto-covariance function, is not used very often because the corresponding computations (first, of the auto-covariance function and then of its Fourier transform) are rather inefficient compared with the calculation of the amplitudes of Eq. (3.5.8) directly from a wave record (with a technique called the Fast Fourier Transform, FFT; see Appendix C).

3.5.6 The frequency–direction spectrum

The above one-dimensional variance density spectrum characterises the stationary, Gaussian surface elevation as a function of time (at one geographic location). To describe the actual, *three-dimensional*, moving waves, the horizontal dimension has to be added. To that end we expand the random-phase/amplitude model by considering a harmonic wave that propagates in x, y-space, in direction θ relative to the positive x-axis (we use ω instead of f for the sake of brevity in the notation):

$$\eta(x, y, t) = a \cos(\omega t - kx \cos \theta - ky \sin \theta + \alpha) \qquad (3.5.19a)$$

or

$$\eta(x, y, t) = a \cos(\omega t - k_x x - k_y y + \alpha) \qquad (3.5.19b)$$

where the wave number $k = 2\pi/L$ (where L is the wave length of the harmonic wave), $k_x = k \cos \theta$, $k_y = k \sin \theta$ and θ is the direction of wave propagation (i.e., normal to the wave crest of each individual component). Analogously to the one-dimensional model, the corresponding three(!)-dimensional random-phase/

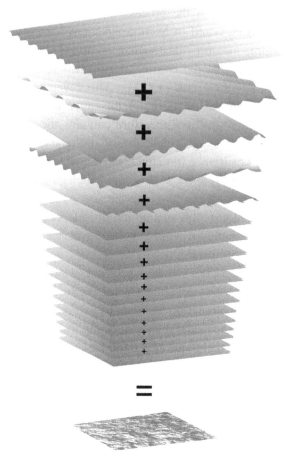

Figure 3.10 The random waves moving in time, i.e., the sum of a large number
of harmonic wave components, travelling across the ocean surface with different
periods, directions, amplitudes and phases (after Pierson *et al.*, 1955).

amplitude model (in x, y and t-space) is the sum of a large number of such *propagating* harmonic waves (see Fig. 3.10):

$$\underline{\eta}(x, y, t) = \sum_{i=1}^{N} \sum_{j=1}^{M} \underline{a}_{i,j} \cos(\omega_i t - k_i x \cos \theta_j - k_i y \sin \theta_j + \underline{\alpha}_{i,j}) \qquad (3.5.20)$$

Adding two dimensions to the original one-dimensional random-phase/
amplitude model (dimensions x and y, added to time t, or, equivalently, wave
number k and direction θ, added to frequency ω) would result in two more indices
in the summation of Eq. (3.5.20). However, the index for wave number k is equal

to the index for frequency ω because frequency and wave number are uniquely related (through the dispersion relationship of the linear theory for surface gravity waves; see Chapter 5: $\omega^2 = gk \tanh(kd)$, where d is the water depth). Every wave number k thus corresponds to one frequency ω and vice versa. The seemingly three-dimensional random-phase/amplitude model therefore reduces to a two-dimensional model in terms of frequency (or wave number) and direction. Each wave component is therefore indicated in Eq. (3.5.20) with only two indices: i (for the frequency or wave number) and j (for the direction).

As in the one-dimensional model, every individual wave component in this two-dimensional model has a random amplitude $\underline{a}_{i,j}$ (Rayleigh distributed) and a random phase $\underline{\alpha}_{i,j}$ (uniformly distributed). Furthermore, analogously to the definition of the one-dimensional spectrum, the exact values of the frequencies ω_i and the directions θ_j are not important as long as their interval is small (e.g. a small fraction of a characteristic wave period and a small fraction of $360°$, respectively), the frequencies are in the range of wind-generated waves and the directions cover the full circle. This two-dimensional random-phase/amplitude model represents a Gaussian process that is *stationary* in time and *homogeneous* in x, y-space: a spatial pattern of chaotically moving surface elevations, seen as the sum of many wave components propagating with various amplitudes, phases and frequencies (or wave lengths) in various directions across the ocean surface. The effect is a realistic representation of random, short-crested waves (see Fig. 3.10).

By using the same techniques as before, the discrete two-dimensional amplitude spectrum can be transformed into a continuous two-dimensional variance density spectrum so that, for all i and j (see Fig. 3.11),

$$E(\omega, \theta) = \lim_{\Delta\omega \to 0} \lim_{\Delta\theta \to 0} \frac{1}{\Delta\omega \, \Delta\theta} E\{\tfrac{1}{2}\underline{a}^2\} \tag{3.5.21}$$

or, in terms of frequency f,

$$E(f, \theta) = \lim_{\Delta f \to 0} \lim_{\Delta\theta \to 0} \frac{1}{\Delta f \, \Delta\theta} E\{\tfrac{1}{2}\underline{a}^2\} \tag{3.5.22}$$

Obviously (using the proper Jacobian; see Section 3.5.5),

$$E(\omega, \theta) = \frac{1}{2\pi} E(f, \theta) \tag{3.5.23}$$

The dimension and S.I. unit of $E(f, \theta)$ follow directly from its definition:[13] the dimension of the amplitude $\underline{a}$ is [length] and its S.I. unit is [m]. The dimension of

[13] Phillips (1977) defines a frequency–direction spectrum $\Psi_0(f, \theta)$ as $\Psi_0(f, \theta) = E(f, \theta)/f$. This is unusual and confusing and Phillips (1985) carefully points out the difference between $\Psi_0(f, \theta)$ and $E(f, \theta)$.

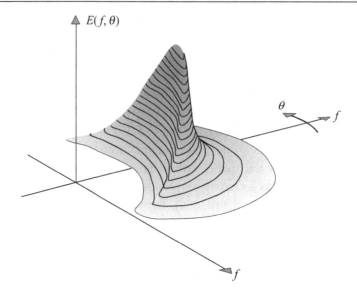

Figure 3.11 The two-dimensional spectrum of wind-generated waves (shown in polar co-ordinates).

the frequency band Δf is $[\text{time}]^{-1}$ and its S.I. unit is $[\text{s}^{-1}]$ or [Hz]. The direction band $\Delta \theta$ is dimensionless but its units are either radians or degrees. The dimension of $E(f, \theta)$ is therefore (from Eq. 3.5.22) $[\text{length}^2/(1/\text{time})]$ and its unit is $[\text{m}^2/\text{Hz/radian}]$ or $[\text{m}^2/\text{Hz/degree}]$.

The two-dimensional spectrum $E(f, \theta)$ shows how the variance of $\eta(x, y, t)$ is distributed over frequencies and directions just as the one-dimensional frequency spectrum shows how the variance is distributed over frequencies. The volume of $E(f, \theta)$ is therefore equal to the total variance $\overline{\eta^2}$ of the sea-surface elevation:

$$\overline{\eta^2} = \int_0^\infty \int_0^{2\pi} E(f, \theta) \mathrm{d}\theta \, \mathrm{d}f \tag{3.5.24}$$

The contribution of a spectral bin $(\Delta f, \Delta \theta)$ to the total variance is (see Fig. 3.12 and also Eq. 3.5.9)

$$\Delta var = \int_{\Delta f} \int_{\Delta \theta} E(f, \theta) \mathrm{d}\theta \, \mathrm{d}f \tag{3.5.25}$$

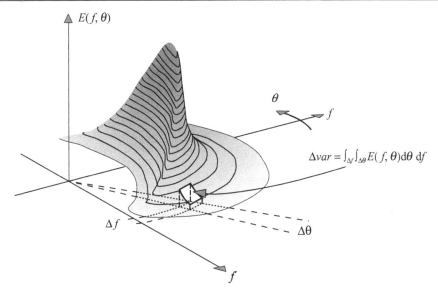

Figure 3.12 The contribution Δvar of a spectral bin (Δf, $\Delta\theta$) to the total variance of the waves.

The one-dimensional *frequency* spectrum $E(f)$, which does not contain any directional information, can be obtained from the frequency-direction spectrum $E(f, \theta)$ by removing all directional information by integration over all directions (per frequency):

$$E(f) = \int_0^{2\pi} E(f, \theta) \, \mathrm{d}\theta \qquad\qquad (3.5.26)$$

3.5.7 The spectrum at sea

Suppose that a storm in the Norwegian Sea generates swell travelling south into the North Sea (see Fig. 3.13). That swell will arrive one or two days later off the Dutch coast, where it may meet a young sea state being generated by a local breeze from westerly directions. The spectrum off the Dutch coast will then represent two wave systems: swell from the north and young wind sea from the west.

Swell is generally of a much lower frequency than a young wind sea, so in this case the two wave systems are well separated, both in frequency and in direction. Moreover, swell is rather regular and long-crested, so its spectrum is narrow (both in frequency and in direction; see Section 6.4.2 for an explanation of this). In contrast

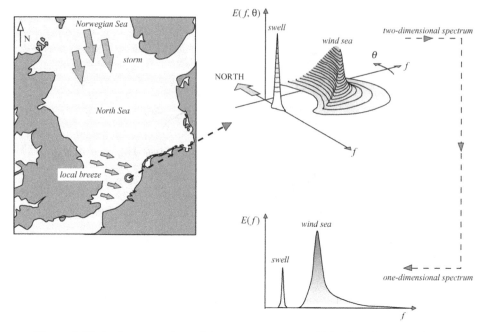

Figure 3.13 An interpretation of the wave spectrum off the Dutch coast when a northerly swell, generated by a storm off the Norwegian coast, meets a locally generated westerly wind sea.

to this, a young wind sea is irregular and short-crested and its spectrum is therefore much broader. The spectrum off the Dutch coast will therefore be rather distinctive in this situation: a narrow, low-frequency spectrum oriented in southerly directions (the direction of propagation) representing the swell and a much broader spectrum at higher frequencies oriented towards easterly directions, representing the locally generated wind sea. The one-dimensional spectrum obtained by integrating this two-dimensional spectrum over the directions is equally distinctive because the swell and the local wind sea are well separated in frequency.

3.5.8 Wave-number spectra

In the previous sections, the surface elevation was considered as a spatial, moving surface, i.e., as a function of space and time: $\eta(x, y, t)$. It can also be considered as a function of space alone, i.e., at *one moment* in time (a 'frozen' surface). The surface elevation can then be a function either of *two* spatial horizontal co-ordinates; $\eta(x, y)$, for instance, in a pair of stereo-photographs (see Fig. 2.2), or of *one* spatial (horizontal) co-ordinate only; $\eta(x)$, for instance in a photograph of the water surface seen through the glass side-wall of a wave flume. The description then requires either a one- or a two-dimensional wave-number spectrum.

The moving, spatial, sea surface $\eta(x, y, t)$ can also be described without assuming that the dispersion relationship of the linear wave theory is valid (as was done above). The description then requires a three-dimensional frequency–wave-number spectrum.

The one-dimensional wave-number spectrum

The rationale of the definition of the one-dimensional *wave-number* spectrum is identical to that of the one-dimensional *frequency* spectrum except that time t is replaced with the horizontal co-ordinate x and that radian frequency ω is correspondingly replaced with wave number k. The one-dimensional variance density spectrum in terms of wave number, $E(k)$, is then defined as

$$E(k) = \lim_{\Delta k \to 0} \frac{1}{\Delta k} E\{\tfrac{1}{2}\underline{a}^2\} \tag{3.5.27}$$

where Δk is the wave-number bandwidth. Since frequencies ω and wave numbers k are related through the dispersion relationship of the linear theory for surface gravity waves, the wave-number spectrum can be obtained from the frequency spectrum with

$$E(k) = E(\omega)\frac{d\omega}{dk} \tag{3.5.28}$$

where $d\omega/dk$ is the Jacobian needed to transform from radian frequency to wave-number domain (see Section 3.5.5; note that the value of this Jacobian happens to be equal to the velocity at which a group of waves propagates, $c_g = d\omega/dk$; see Section 5.4.3), so that

$$E(k) = c_g E(\omega) \tag{3.5.29}$$

All characteristics of this one-dimensional wave-number spectrum are identical or analogous to those of the frequency spectrum, e.g., the total wave variance is given by

$$\overline{\underline{\eta}^2} = \int_0^\infty E(k)dk \tag{3.5.30}$$

The two-dimensional wave-number spectrum

The harmonic wave components underlying the spectral description of the frozen sea surface, in *two* spatial dimensions x and y, can be written as

$$\eta(x, y) = a_{i,j} \cos(k_{x,i}x + k_{y,j}y + \alpha_{i,j}) \tag{3.5.31}$$

and the corresponding two-dimensional wave-number spectrum $E(k_x, k_y)$ is defined,[14] analogously to the above definitions of the various spectra, as

$$E(k_x, k_y) = \lim_{\Delta k_x \to 0} \lim_{\Delta k_y \to 0} \frac{1}{\Delta k_x \, \Delta k_y} E\{\tfrac{1}{2}\underline{a}^2\} \qquad (3.5.32)$$

where Δk_x and Δk_y are the spectral bandwidths. Obviously, $k_x = k \cos\theta$ and $k_y = k \sin\theta$, where $k = \sqrt{k_x^2 + k_y^2}$ and $\theta = \arctan(k_y/k_x)$, so that an equivalent two-dimensional spectrum can be defined as

$$E(k, \theta) = \lim_{\Delta k \to 0} \lim_{\Delta \theta \to 0} \frac{1}{\Delta k \Delta \theta} E\{\tfrac{1}{2}\underline{a}^2\} \qquad (3.5.33)$$

where Δk and $\Delta \theta$ are the spectral bandwidths. The two spectra are related by

$$E(k, \theta) = E(k_x, k_y)J \qquad (3.5.34)$$

where $J = k$ is the Jacobian used to transform the spectrum from the two-dimensional $\vec{k}$-domain to the two-dimensional k, θ-domain (see footnote in Section 3.5.5). The two-dimensional frequency–direction spectrum $E(\omega, \theta)$ is obtained from $E(k, \theta)$, simply with

$$E(\omega, \theta) = E(k, \theta)J \qquad (3.5.35)$$

where the Jacobian is $J = dk/d\omega = 1/c_g$ (see above). Similarly, we can find with $c = \omega/k$ that

$$E(k_x, k_y) = \frac{cc_g}{\omega} E(\omega, \theta) \qquad (3.5.36)$$

The one-dimensional wave-number spectrum is obtained from $E(k, \theta)$ by integrating over all directions:

$$E(k) = \int_0^{2\pi} E(k, \theta)d\theta \qquad (3.5.37)$$

The three-dimensional frequency–wave-number spectrum

In the previous sections, it was assumed that the dispersion relationship of the linear wave theory provides a unique relationship between the frequency ω and the wave number k. The spectra can thus be transformed from wave-number space to frequency space (and vice versa) and it allows the use of a two-dimensional rather than a three-dimensional spectrum to represent the moving, three-dimensional surface of ocean waves $\eta(x, y, t)$. If the dispersion relationship is *not* assumed a priori

---

[14] The spectrum $E(k_x, k_y)$ may also be written as $E(\vec{k}) = E(k_x, k_y)$, where $\vec{k} = (k_x, k_y)$ is called the wave-number vector.

(for instance to verify this relationship with observations), then these transformations cannot be made. The random sea surface should then be seen as the sum of a large number of propagating harmonic waves for which the frequencies and wave numbers are *independent*:

$$\underline{\eta}(x, y, t) = \sum_{i=1}^{N} \sum_{j=1}^{M} \sum_{l=1}^{P} \underline{a}_{i,j,l} \cos(\omega_i t - k_j x \cos\theta_l - k_j y \sin\theta_l + \underline{\alpha}_{i,j,l})$$

(3.5.38)

There are now three indices in this summation: the frequencies in this model have an index i that is different from the index of the wave number j. By using the same techniques as before, the three-dimensional variance density spectrum can now be defined as

$$E(\omega, k, \theta) = \lim_{\Delta\omega \to 0} \lim_{\Delta k \to 0} \lim_{\Delta\theta \to 0} \frac{1}{\Delta\omega \, \Delta k \, \Delta\theta} E\{\tfrac{1}{2}\underline{a}^2\}$$

(3.5.39)

The alternative frequency–wave-number *vector* spectrum $E(\omega, k_x, k_y) = E(\omega, \vec{k})$ can be transformed to this three-dimensional spectrum $E(\omega, k, \theta)$ with the proper Jacobian (see above) because the relationship between wave-number k and direction θ on the one hand and wave-number vector $\vec{k} = (k_x, k_y)$ on the other is retained (see Eq. 3.5.34). The definition of this spectrum $E(\omega, k_x, k_y)$ is based on the same model as used to define $E(\omega, k, \theta)$, with $k_j \cos\theta_l$ and $k_j \sin\theta_l$ replaced with $k_{x,j}$ and $k_{y,l}$.

If the waves *do* behave as linear waves, then the spectrum $E(\omega, k_x, k_y)$ collapses onto a curved plane in the spectral ω, k_x, k_y-space (the dispersion relationship). Deviations from the theoretical dispersion relationship (without ambient current) are then probably due to an ambient current. The current speed and direction can thus be inferred from observations of $E(\omega, k_x, k_y)$.

3.5.9 Spectrum acquisition

The techniques used to acquire the *one-* or *two*-dimensional spectrum are essentially the following:

measure the sea-surface elevation with *in situ* or remote-sensing techniques and analyse the records (see Appendix C), or

predict the spectrum with numerical wave models using wind, tide and seabed-topography information.

The *three*-dimensional spectrum is hardly ever acquired since it would involve measuring the sea surface as a three-dimensional, moving surface (but it is sometimes done e.g., with a stereo- or radar-film, involving techniques described in

Section 2.4.1). Operational spectral wave models always assume linear waves and predict the two-dimensional spectrum (usually $E(f, \theta)$). *In-situ* measurements (e.g., with a heave buoy), followed by an appropriate spectral analysis can provide reasonable estimates of the one-dimensional frequency spectrum $E(f)$ or $E(\omega)$. The mean direction and the directional spreading per frequency can also be determined if additionally some spatial characteristics of the waves are measured (e.g., with a pitch-and-roll buoy; see Section 6.3.4 for the concept of directional spreading). *Remote-sensing* and numerical wave models can estimate the full two-dimensional spectrum, usually the wave-number spectrum $E(k_x, k_y)$ (remote sensing) or $E(f, \theta)$ (wave models), from which the one-dimensional frequency spectrum $E(f)$ can readily be obtained.

3.6 Transfer functions and response spectra

In Section (3.5.5), we considered the transformation of a wave spectrum from frequency f-space to the radian-frequency ω-space. This involved the simultaneous transformation of f to ω and of $E(f)$ to $E(\omega)$, using a Jacobian. Such transformations concern the spectrum in various forms, without changing the random variable considered (the surface elevation). It is also possible to transform the spectrum of the surface elevation to the spectrum of some *other* wave variable, for instance, the wave-induced pressure in some point below the water surface. Such a transformation is not carried out with a Jacobian but with a transfer function. It requires that the relationship between the two variables can be treated as a linear system (a simple analytical expression suffices, but a complicated numerical model may also constitute a linear system). The word 'linear' is crucial here. It means the following: consider a system that,

on excitation with input $x(t)$, responds with output $X(t)$: $x(t) \rightarrow X(t)$ and

on excitation with input $y(t)$, responds with output $Y(t)$: $y(t) \rightarrow Y(t)$

then the system is called linear, if a linear combination of these excitations gives the corresponding linear combination of the responses:

$$\boxed{ax(t) + by(t) \rightarrow aX(t) + bY(t)} \qquad (3.6.1)$$

It shows that an amplification of the excitation gives an equal amplification of the response. It also shows that the responses to the excitations are *independent*: the response to excitation $x(t)$ is not affected by the response to excitation $y(t)$ and vice versa. The linear theory for surface gravity waves provides such a linear system. For instance, in this theory, the relationship between a sinusoidal wave surface elevation – the excitation $x(t)$ – and the corresponding sinusoidal water pressure at some point in the water – the response $X(t)$ – is a simple linear relationship in the above sense: if the amplitude of the wave is doubled, then the resulting pressure is doubled, independently of other excitations.

One characteristic of a time-constant, linear system is that, when the excitation is harmonic, with a given frequency, the response too is harmonic, with the same frequency:[15]

$$x(t) = \hat{x}\sin(2\pi ft + \alpha_x) \to \boxed{\begin{array}{c} \text{linear system} \\ \text{constant in time} \end{array}} \to X(t) = \hat{X}\sin(2\pi ft + \alpha_X)$$

The *response* $X(t) = \hat{X}\sin(2\pi ft + \alpha_X)$ differs from the *excitation* $x(t) = \hat{x}\sin(2\pi ft + \alpha_x)$ only in amplitude and phase. For a linear system therefore, the response can be described simply with the ratio of the amplitudes and the phase differences, which generally depend on the frequency of the excitation:

$$\boxed{\hat{R}(f) = \frac{\hat{X}(f)}{\hat{x}(f)}} \qquad \text{amplitude response function} \qquad (3.6.2)$$

and

$$R_\alpha(f) = \alpha_X(f) - \alpha_x(f) \qquad \text{phase response function} \qquad (3.6.3)$$

The *spectrum* of the response $E_X(f)$ is readily obtained as the product of the excitation spectrum $E_x(f)$ and the square of the amplitude response function $\hat{R}(f)$ (the square is used because the spectral density is a measure of the square of the amplitude):

$$E_X(f) = [\hat{R}(f)]^2 E_x(f) \qquad \text{response spectrum} \qquad (3.6.4)$$

The two functions $\hat{R}(f)$ and $R_\alpha(f)$ are called *frequency response functions* (they are the amplitude and phase response function, respectively). If the sea-surface elevation is stationary and Gaussian and the system is constant in time and linear, then the response too is stationary and Gaussian and as much characterised by its spectrum $E_X(f)$ as the excitation is characterised by its spectrum $E_x(f)$. In the linear theory of surface gravity waves, the response functions are simple linear analytical relationships, so the above transformation technique can then be used to transform spectra of one wave characteristic into that of another, e.g., to transform the surface-elevation spectrum into the spectrum of the wave-induced pressure.

If the response is direction-sensitive, two-dimensional response functions are required. For the amplitude this function is defined in the same way as the frequency

---

[15] In the random-phase/amplitude model, I used the cosine representation for the harmonic wave. Here and in many other places in the book I use the sine representation. This illustrates nicely that it is often immaterial whether the sine or cosine representation is used.

amplitude response function:

$$\hat{R}(f,\theta) = \frac{\hat{X}(f,\theta)}{\hat{x}(f,\theta)} \qquad \text{amplitude response function} \qquad (3.6.5)$$

where $\hat{R}(f,\theta)$ is the two-dimensional *frequency–direction amplitude response function*. The amplitudes are the amplitude of the *exciting* harmonic wave $\hat{x}(f,\theta)$ and the response wave $\hat{X}(f,\theta)$, respectively. The two-dimensional response spectrum is then readily calculated as

$$E_X(f,\theta) = [\hat{R}(f,\theta)]^2 E_x(f,\theta) \qquad (3.6.6)$$

This transformation seems obvious because it is totally analogous to the one-dimensional situation above. The exciting harmonic wave, being an ocean wave, is readily seen as a wave with a direction of propagation. However, the response is often not a wave propagating with a certain direction or the response has only one, fixed direction. For instance, the wave-induced motion of the water particles normal to a pipeline on the sea floor (exerting the wave-induced force normal to that pipeline), has a fixed direction by definition. Nonetheless, the amplitude of such a response (non-directional or fixed-directional) can be defined. It is simply the amplitude of the response when the system is excited by an ocean wave from a certain direction (even if the response itself has no direction or a fixed direction). In such cases the computation is carried out in two phases. First, the above two-dimensional response spectrum $E_X(f,\theta)$ is computed, and then the one-dimensional frequency response spectrum is calculated by integrating over all directions:

$$E_X(f) = \int_0^{2\pi} E_X(f,\theta)\mathrm{d}\theta \qquad (3.6.7)$$

The above introduction of response functions and response spectra has been given in the context of transforming the spectrum of the surface elevation into that of another wave variable. Such transformations are possible as long as the relationships involved are linear (in amplitude). The process can be inversed: from the spectrum of a (sub-surface) wave variable, the spectrum of the surface elevation can be obtained. This approach is exploited in wave measurements.

The notion of transforming the spectrum of the excitation to obtain the spectrum of the response is rather general; it applies to any linear system, for instance to compute stresses in or motions of *structures* in response to waves or wind. Here too, if the excitation is stationary and Gaussian, and the system is linear (which is generally true for small forces and motions) and constant in time, then the response (force or motion) too is stationary and Gaussian. The required response functions can sometimes be computed analytically but they can often also be measured, in actual structures in the field or in the laboratory. One technique is to excite the structure with one frequency after another, and measure the amplitude and phase response for all these frequencies separately. Another technique is to excite the structure with random waves and to divide the response spectrum by the excitation spectrum. More advanced techniques account for unrelated effects in such measurements (e.g., instrument errors, also called noise) but these require a more advanced spectral analysis: a cross-spectral analysis, which falls outside the scope of this book.

4

Statistics

4.1 Key concepts

Short-term statistics
- The theory describing short-term statistical characteristics of wind waves is based on the assumption that the surface elevation is a *stationary, Gaussian* process.
- For such a process, Rice (1944, 1945) has given an analytical expression for the mean frequency of level crossing in terms of the variance density spectrum.
- With this expression it can be shown that, for waves with a narrow spectrum, the crest height and the wave height are *Rayleigh* distributed with the zeroth-order moment m_0 of the wave spectrum as the only parameter. Observations have shown that this is also the case for waves with a broader spectrum.
- The *significant* wave *height* is readily estimated from the spectrum as $H_{m_0} = 4\sqrt{m_0}$. This is typically 5%–10% larger than the value of $H_{1/3}$ estimated directly from measured time series.
- Observations show that, for wind-sea spectra, the significant wave *period* is typically 5% shorter than the peak period of the spectrum.
- The *maximum individual wave height* in a given duration (under stationary conditions) is a random variable, with a corresponding probability density function that can be estimated from the wave spectrum and the duration. In most storms, this maximum individual wave height is about twice the significant wave height.
- The mean length of a wave *group*, in terms of the number of waves, can be estimated from the width of the variance density spectrum.

Long-term statistics
- Long-term wave statistics (relating to dozens of years or more) can be obtained from *observations* or from computer *simulations*. Some theoretical support to analyse such observations and simulations is provided by the extreme-value theory.
- Long-term *extreme* values of the significant wave height (the probability of exceedance or return periods) can be estimated with three approaches: the initial-distribution approach, the peak-over-threshold approach and the annual-maximum approach.
- The long-term distribution of the individual wave height can be obtained from a combination of short-term and long-term statistics.

4.2 Short-term statistics

In this chapter some statistical characteristics of wind waves are considered on a short-term and a long-term scale. The short-term characteristics are treated here, under the assumption that the surface elevation is a *stationary, Gaussian* process (see Appendix A). This is usually a reasonable assumption for the duration of a wave record (typically 15–30 min) but sometimes it is also assumed for the duration of a storm (typically 6–12 h), which is almost always an over-simplification. The

statistics relate (a) to cumulative effects of the waves, such as the fatigue in structures, and (b) to extreme values of individual waves, such as occur in survival
conditions that are used in the design of a structure. For these effects, some of
the most relevant quantities are the instantaneous surface elevation, crossings of
the surface elevation through certain levels, the crest heights and wave groups. For
extreme values, relevant quantities are the largest crest height or wave height within
a certain duration (e.g., a storm).

Literature:
Cartwright and Longuet-Higgins (1956), Huang *et al.* (1990b), Kimura (1981), Longuet-
Higgins (1957, 1975, 1984), Ochi (1998), Price and Bishop (1974), Soares (2003), Srokosz
(1990).

4.2.1 Instantaneous surface elevation

In the linear approximation of ocean waves (the random-phase/amplitude model),
the instantaneous sea-surface elevation as it appears at an *arbitrary moment* t_1
in time $\underline{\eta}(t) = \underline{\eta}(t_1)$, is *Gaussian* distributed. Assuming the mean to be zero, the
Gaussian probability density function can be written as

$$p(\eta) = \frac{1}{(2\pi m_0)^{1/2}} \exp\left(-\frac{\eta^2}{2m_0}\right) \qquad \text{for a zero-mean: } E\{\underline{\eta}\} = 0 \qquad (4.2.1)$$

where $m_0^{1/2}$ is the standard deviation σ of the surface elevation (see Note 4A).

NOTE 4A   The moments of the wave spectrum

When the random sea-surface elevation is treated as a stationary, Gaussian process,
then all its statistical characteristics are determined by the variance density spectrum
$E(f)$. These characteristics will be expressed in terms of the *moments* of that spectrum,
which are defined as

$$m_n = \int_0^\infty f^n E(f) df \qquad \text{for } n = \dots, -3, -2, -1, 0, 1, 2, 3, \dots$$

The moment m_n is called the '*n*th-order moment' of $E(f)$. For example, the variance
of the surface elevation $\overline{\eta^2}$ is equal to the zeroth-order moment:

$$\text{variance} = E\{\underline{\eta}^2\} = \int_0^\infty E(f) df = m_0 \qquad \text{for } \mu_\eta = E\{\underline{\eta}\} = 0$$

    An empirical confirmation that this is true would require obtaining a large set
of wave records at sea under statistically identical conditions (an ensemble; i.e., a
large number of identical storms) to obtain a large number of sample values of $\underline{\eta}$
at time t_1 to compare with the Gaussian distribution. This of course is impossible.

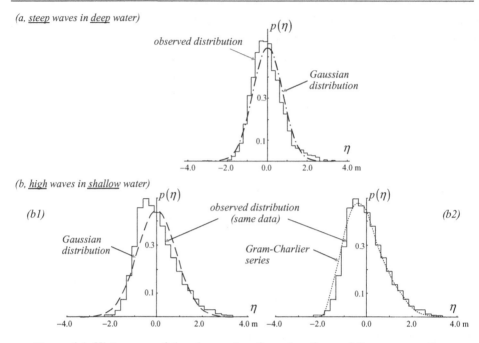

Figure 4.1 Histograms of the observed surface elevation and the corresponding Gaussian probability density functions. Panel a: steep waves in deep water (significant wave height 2.70 m, mean period 5.3 s, steepness 0.06, depth 70 m; data courtesy of FUGRO-OCEANOR Trondheim, Norway). Panels b1 and b2: high waves in shallow water (significant wave height 3.55 m, depth 8.8 m, significant wave height/depth ratio 0.44; after Ochi and Wang, 1984; same observations in panels b1 and b2). The Gram–Charlier series provides a better fit for the strongly nonlinear waves in shallow water.

Instead therefore, we consider the surface elevation as a function of time in one, stationary wave record (assuming that the process is ergodic, i.e., ensemble averages are equal to time averages, see Appendix A). The values of the surface elevation $\eta(t_j)$, $\eta(t_{j+1})$, $\eta(t_{j+2})$, ... then replace the ensemble values of the surface elevation $\eta(t_1)$ at time t_1. Usually, the agreement between the observed and theoretical probability density functions thus obtained is good, at least in open sea (deep water), but high crests are observed slightly more frequently than according to the Gaussian model and deep troughs slightly less frequently. For *steep* waves or in *shallow water* the discrepancies are larger because the waves are more nonlinear. Two such observations, with a fairly strong nonlinear character, one with rather steep waves, *steepness*[1] $= 0.06$, in deep water and one with a relatively large wave height in shallow water, $H_{m_0} = 0.44d$, where d is depth, are compared with the Gaussian distribution in Fig. 4.1.

---

[1] Steepness is defined here as a characteristic wave height divided by a characteristic wave length: $H_{m_0}/[g\overline{T}^2/(2\pi)]$, where $H_{m_0} = 4\sqrt{m_0}$ is an estimate of the significant wave height and the wave length is based on the mean wave period $\overline{T} = m_0/m_1$ (the zeroth- and first-order moments of the spectrum).

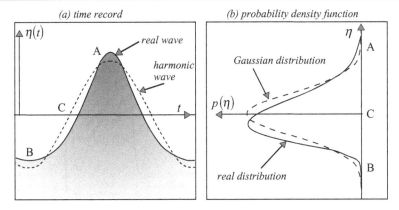

Figure 4.2 The analogy of the difference between a harmonic wave and a real (Stokes-type) wave (panel a) and the corresponding difference between a Gaussian distribution and the real (observed) probability density function of the surface elevation η at sea (panel b). Note that the axes in panel (b) are rotated $90°$ counter-clockwise relative to the usual orientation (to show η vertically, as in panel a).

The deviations are due to nonlinear processes that generally make the wave crests higher and sharper and the wave troughs shallower and flatter. The analogous differences between a harmonic wave and a nonlinear wave (e.g., a Stokes wave; see Section 5.6.2) are shown in Fig. 4.2.

It is obvious, because of the sharp and high *crests* in the nonlinear wave, that the surface elevation remains longer at the higher levels for the nonlinear wave than for the harmonic wave (near A in Fig. 4.2). Such a larger *fraction of time* corresponds to a larger *probability* that the surface elevation is located at these higher levels. This explains why the Gaussian model underestimates the observed high values of the random surface elevation. In contrast to this, for the large, negative values, the relatively shallow and flat *troughs* of the nonlinear wave (near B in Fig. 4.2) correspond to a *smaller* fraction of time than for the harmonic wave. Near the mean of the record ($\eta = 0$; near C in Fig. 4.2), the relatively steep *slope* of the surface of the real wave reduces the fraction of time of the elevation around the zero level and hence reduces the probability of occurrence in this interval. The total effect is a probability density function of the real surface elevation that is skewed to the negative values ('leaning' towards negative values; see Fig. 4.2, panel b) but its extremes, both positive and negative, are shifted upwards. This skewed character is also evident on visually inspecting a wave record on paper. It is immediately obvious (from the steeper crests and flatter troughs) whether the record is oriented upside-down or not. A probability density function that takes such skewness into account is Edgeworth's form of the type-A Gram–Charlier series (Longuet-Higgins, 1963; see Fig. 4.1 panel b2).

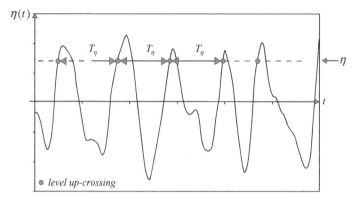

Figure 4.3 The up-crossings of the sea-surface elevation through level η and the corresponding time intervals T_η in a (statistically) stationary wave record.

Literature:
Cramer (1946), Edgeworth (1908), Longuet-Higgins (1963), Ochi and Wang (1984).

4.2.2 Wave height and period

The statistical characteristics of the wave *height*, which is probably the most important wave parameter for engineers, can be obtained theoretically from the wave spectrum. The derivation that is shown here is based on an expression due to Rice (1944, 1945, 1954), for the average *time interval* between *level crossings* in a stationary Gaussian process (see Fig. 4.3 for its application to the sea-surface elevation). The probability density function of the crest height and wave height can readily be derived from this expression if the spectrum is narrow.

Literature:
Cavanié *et al.* (1976), Longuet-Higgins (1952, 1983).

Wave period

The integral of the Gaussian probability density function gives the probability that $\eta(t)$ is below a certain level η, or the fraction of *time* that $\eta(t)$ is below that level. This fraction of time, by itself, does not give any information as to how *often* the surface elevation crosses that level or what the time *interval* between such crossings is (see Fig. 4.3).

The *average* of this time interval ($\overline{T}_\eta$ between successive up or down crossings through level η) can readily be expressed in terms of the spectrum as (Rice, 1944,

1945, 1954)

$$\overline{T}_\eta = \sqrt{\frac{m_0}{m_2}} \bigg/ \exp\left(-\frac{\eta^2}{2m_0}\right)$$ (4.2.2)

where m_0 and m_2 are the zeroth- and second-order moment (see Note 4A), respectively, of the variance density spectrum $E(f)$. The mean frequency of these level crossings $\overline{f}_\eta = \overline{T}_\eta^{-1}$ is correspondingly

$$\boxed{\overline{f}_\eta = \sqrt{\frac{m_2}{m_0}} \exp\left(-\frac{\eta^2}{2m_0}\right)}$$ (4.2.3)

A special case is the mean *zero*-crossing period $\overline{T}_0$ (see Section 3.3.3), which can be obtained from Eq. (4.2.2) with $\eta = 0$:

$$\overline{T}_0 = \sqrt{\frac{m_0}{m_2}}$$ (4.2.4)

It is sometimes denoted as $\overline{T}_0 = T_{m_{02}}$. The reciprocal of this, the mean zero-crossing frequency $\overline{f}_0 = \overline{T}_0^{-1}$, is obviously

$$\boxed{\overline{f}_0 = \sqrt{\frac{m_2}{m_0}}}$$ (4.2.5)

   Unfortunately, the value of m_2 (and therefore also the estimates of $\overline{T}_0$ and $\overline{f}_0$ with Eqs. 4.2.4 and 4.2.5) is sensitive to small errors or variations in the measurement or analysis technique. For instance, the integration interval used to compute m_0 and m_2 from the spectrum should strictly range from $f = 0$ to $f = \infty$, whereas in actual practice it is from 0 to some practical upper limit (e.g., the Nyquist frequency; see Appendix C). Moreover, in the definition of the moments of the spectrum, the energy density at high frequencies is enhanced, and more so the higher the order of the moment. This shows that the values of higher-order moments are rather sensitive to noise in the high-frequency range of the spectrum (where noise is usually relatively large). Similarly, if the wave record itself is used to estimate $\overline{T}_0$ directly (instead of using the moments of the spectrum), the definition says that *all* waves in the wave record should be included in the averaging procedure (see Section 3.3.3). In actual practice, a (low) threshold value is used (typically a few centimetres), to avoid including non-physical variations near the zero-level of the wave record (e.g., related to instrument noise). These considerations suggest that $\overline{T}_0$ is not always the most reliably estimated characteristic wave period. Another mean period is therefore sometimes used, which is less dependent on high-frequency noise. It is defined as the inverse of the mean frequency of the wave spectrum:

$$T_{m_{01}} = f_{mean}^{-1} = \left(\frac{m_1}{m_0}\right)^{-1}$$

(4.2.6)

Yet another characteristic wave period is the significant wave period $T_{1/3}$ (see Section 3.3.3). Like the mean period $T_{m_{01}}$, it is less dependent on high-frequency noise since it depends only on the higher waves. A theoretical expression for $T_{1/3}$ in terms of the spectrum is available but it is rather complicated and it will not be treated here (see Kitano *et al.*, 2001). The following relationships are empirical (i.e., based on observations or computer simulations, e.g., Goda, 1988a). For swell (or more precisely: waves with a narrow spectrum), $T_{1/3}$ is practically equal to the peak period of the spectrum (the inverse of the peak frequency):

$$T_{1/3} \approx T_{peak} \qquad \text{for swell}$$

(4.2.7)

For wind sea, this is not the case, but it has been found empirically (Goda, 1978) that, if a wind-sea spectrum is unimodal, the average period of the higher waves $\overline{T}_H$ is somewhat shorter than the inverse of the peak frequency f_{peak}:

$$\overline{T}_H \approx 0.95 T_{peak} \qquad \text{for wind sea and } H \geq 1.5\overline{H}$$

(4.2.8)

where $\overline{H}$ is the mean wave height. Since the significant wave period $T_{1/3}$ is taken from these higher waves:

$$T_{1/3} \approx 0.95 T_{peak} \qquad \text{for wind sea}$$

(4.2.9)

Crest height

It seems natural to define crests as maxima in the surface elevation and one might expect the corresponding heights η_{crest} to be always positive; after all, each crest is a maximum. If the spectrum is narrow, this is certainly true. For such a spectrum, the derivation of the statistical characteristics of the crest height is relatively simple. However, if the spectrum is wide, i.e., the waves are irregular, a crest height thus defined may well be negative (the definition allows local maxima to be counted; see Fig. 4.4). This difference shows that the width of the spectrum affects the statistics of crest heights: a narrow spectrum corresponds to positive crest heights only; a wide spectrum corresponds to positive and negative crest heights (if the definition allows local maxima to be counted).

For waves with a *narrow* spectrum, the total number of crests is equal to the number of up-crossings through the zero level. The number of high *crests*, i.e., those *above* a positive level η, is equal to the number of *up-crossings* through that

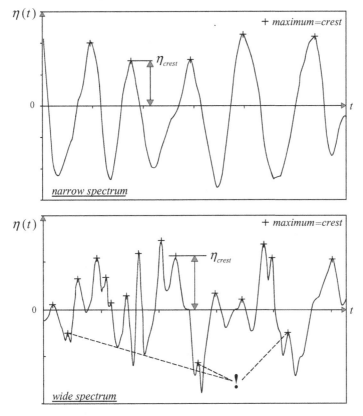

Figure 4.4 The exclusively positive crest heights under wave conditions with a narrow spectrum and the positive and negative crest heights under wave conditions with a wide spectrum.

level (see Fig. 4.5). For a duration D, the relative number of crests with height $\underline{\eta}_{crest} > \eta$ can then be estimated, from these numbers, as

$$\frac{\textit{number of crests with } (\underline{\eta}_{crest} > \eta) \textit{ in duration } D}{\textit{total number of crests in duration } D} = \frac{D/\overline{T}_\eta}{D/\overline{T}_0} = \frac{\overline{f}_\eta}{\overline{f}_0} \qquad (4.2.10)$$

Interpreting this relative number (fraction) as the probability of $\underline{\eta}_{crest}$ exceeding the level η and substituting the expressions for $\overline{f}_\eta$ and $\overline{f}_0$, Eqs. (4.2.3) and (4.2.5), into the right-hand side of Eq. (4.2.10) gives

$$\Pr\{\underline{\eta}_{crest} > \eta\} = \frac{\overline{f}_\eta}{\overline{f}_0} = \frac{\sqrt{\frac{m_2}{m_0}} \exp\left(-\frac{\eta^2}{2m_0}\right)}{\sqrt{\frac{m_2}{m_0}}} = \exp\left(-\frac{\eta^2}{2m_0}\right) \qquad (4.2.11)$$

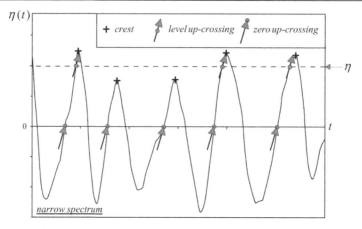

Figure 4.5 For waves with a narrow spectrum, the probability of a crest exceeding a certain level is equal to the relative number of crests above that level.

so the cumulative distribution function $\Pr\{\underline{\eta}_{crest} < \eta\} = 1 - \Pr\{\underline{\eta}_{crest} > \eta\}$ is

$$P_{\underline{\eta}_{crest}}(\eta) = \Pr\{\underline{\eta}_{crest} \le \eta\} = 1 - \exp\left(-\frac{\eta^2}{2m_0}\right) \qquad (4.2.12)$$

The probability *density* function of $\underline{\eta}_{crest}$ is obtained as the derivative of $P_{\underline{\eta}_{crest}}(\eta)$ (see Appendix A):

$$p_{\underline{\eta}_{crest}}(\eta) = \frac{\eta}{m_0} \exp\left(-\frac{\eta^2}{2m_0}\right) \qquad (4.2.13)$$

which is shown in Fig. 4.6 (see Note 4B for the notation in Eqs. 4.2.12 and 4.2.13). These functions are of the Rayleigh type (i.e., the independent variable η in the cumulative distribution function occurs to the second power in the exponent). A Rayleigh distribution has only one parameter, which in *this* case happens to be the zeroth-order moment m_0 of the variance density *spectrum* (and not the zeroth-order moment of the Rayleigh distribution or any other function!), which is equal to the variance of $\eta(t)$. Since all statistical characteristics of the crest heights are determined by this distribution, they can all be expressed in terms of this moment m_0 alone (provided that the spectrum is narrow). For instance, the *mean* and *standard deviation* of the crest height for waves with such a narrow spectrum are

$$\mu_{crest} = E\{\underline{\eta}_{crest}\} = \sqrt{\frac{\pi}{2}}\sqrt{m_0}$$

$$\qquad (4.2.14)$$

$$\sigma_{\eta crest} = \sqrt{E\{\underline{\eta}_{crest}^2\} - E^2\{\underline{\eta}_{crest}\}} = \sqrt{2 - \frac{\pi}{2}}\sqrt{m_0}$$

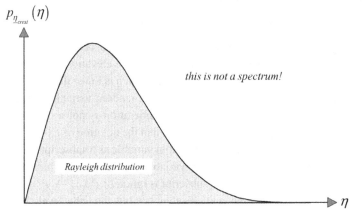

Figure 4.6 The Rayleigh probability density function of the crest height η_{crest} for wave conditions with a narrow spectrum.

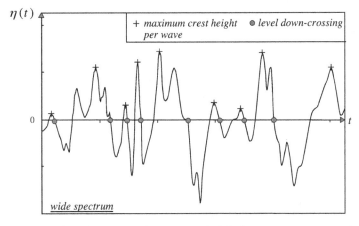

Figure 4.7 There is only one maximum crest height between two successive zero down-crossings, even for waves with a wide spectrum.

For waves with a *wide* spectrum, i.e., with an irregular appearance of the sea-surface elevation (see Note 4C for how to quantify the width of the spectrum), the probability density function is not readily derived. However, if we consider the maximum crest height *per wave* $\hat{\underline{\eta}}_{crest}$ (i.e., the maximum elevation between two consecutive zero up crossings; see Fig. 4.7; note the ^ in the notation), then *observations* have shown that the distribution function of this maximum is practically a Rayleigh distribution (at least for values of $\hat{\underline{\eta}}_{crest}$ that are not too low, e.g., $\hat{\underline{\eta}}_{crest} \geq \sqrt{m_0}$).

NOTE 4B   The notation in probability functions

The probability functions that we consider here relate to certain random variables, e.g., the cumulative distribution function of the surface elevation $\underline{\eta}$. This distribution function gives the probability that the surface elevation $\underline{\eta}$ is lower than some given level η: $P(\eta) = \mathrm{Pr}\{\underline{\eta} < \eta\}$. Note that in this notation, the random variable $\underline{\eta}$ is underscored whereas the level η is not (the level under consideration is not a random variable). Strictly speaking we should make that distinction in the notation of $P(\eta)$. This function should therefore be written as $P_{\underline{\eta}}(\eta)$: the random variable as a subscript and the level as the argument. However, when the random variable and the level are indicated with the same symbol (as in this example), the subscript is ignored: $P_{\underline{\eta}}(\eta) \rightarrow P(\eta)$. Sometimes the random variable is indicated with another symbol than the level. For instance, the crest height, $\underline{\eta}_{crest}$, may or might not exceed a certain level η. In such cases, we must retain the distinction, hence the notation $P_{\underline{\eta}_{crest}}(\eta)$ for the cumulative distribution function of $\underline{\eta}_{crest}$.

NOTE 4C   Spectral width parameters

The probability density functions of the crest height, defined as a maximum in the wave record, and of the wave height depend on the spectral width, which can be quantified with a parameter ε defined by Cartwright and Longuet-Higgins (1956) as

$$\varepsilon = \left(1 - \frac{m_2^2}{m_0 m_4}\right)^{1/2}$$

For $\varepsilon \rightarrow 0$ (i.e., a very narrow spectrum, for which $m_0 m_4 \rightarrow m_2^2$), $p_{\underline{\eta}_{crest}}(\eta)$ and $p(H)$ approach the Rayleigh distribution, whereas for $\varepsilon \rightarrow 1$ (i.e., a very wide spectrum), the distribution approaches a Gaussian distribution (a very irregular appearance of the waves with as many positive as negative crest heights). Note that this is a *theoretical* result for a spectrum with an *arbitrary* shape.

The spectrum of ocean waves very often has a tail with a shape given by $\alpha g^2 f^{-5}$ (where α is a constant and g is gravitational acceleration; see Section 6.3.3). The value of the fourth-order moment m_4, which is required to estimate ε, is then dominated by the upper limit of the integration to determine m_4 (usually the Nyquist frequency, see Appendix C). This, of course, is undesirable. Moreover, m_4 is a high-order moment of the spectrum and its estimation is therefore rather sensitive to noise in the spectrum at high frequencies (or the presence of nonlinear effects, which also distort the high-frequency tail of the spectrum). In actual practice, the value of ε therefore depends not only on the shape of the spectrum, but also, and to a high degree, on the high-frequency cut-off, errors and nonlinear distortions in the high-frequency part of the spectrum. The parameter ε should therefore be used with great care. An alternative spectral width parameter ν is given by Longuet-Higgins (1975):

$$\nu = \left(\frac{m_0 m_2}{m_1^2} - 1\right)^{1/2} = \left(\frac{T_{m\,02}^2}{T_{m\,01}^2} - 1\right)^{1/2}$$

which suffers to a lesser degree from the same problem. Another spectral width parameter, due to Battjes and van Vledder (1984), is defined as

$$\kappa^2 = \frac{1}{m_0^2}\left\{\left[\int_0^\infty E(f)\cos\left(\frac{2\pi f}{\bar{f}_0}\right)df\right]^2 + \left[\int_0^\infty E(f)\sin\left(\frac{2\pi f}{\bar{f}_0}\right)df\right]^2\right\}$$

$$\text{with } \bar{f}_0 = \sqrt{m_2/m_0}$$

That this parameter κ represents a spectral width is not so obvious. When the two integrals are interpreted as weighted surface areas of the spectrum (see illustration below), it is readily seen that the value of the first integral decreases as the width of the spectrum increases (the positive lobe of the cosine wave below the spectral peak dominates the value of this integral). The value of the second integral is zero for symmetrical spectra (symmetrical around $\bar{f}_0$) and deviates further from zero the more asymmetrical the spectrum.

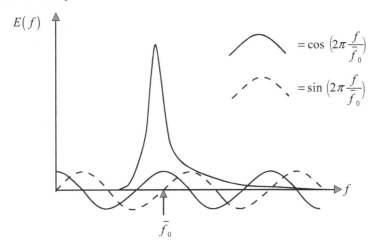

The weight functions in the two integrals in the definition of the spectral width/ groupiness parameter κ.

The above spectral-width parameters also control the groupiness character of the waves (see Section 4.2.3). In this role, κ is superior in several respects to ε and ν (see van Vledder, 1992).

Literature:
Forristall (2000), Rayleigh (1880).

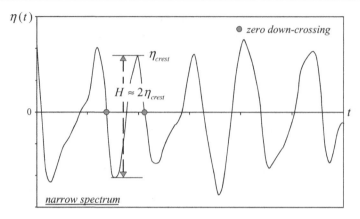

Figure 4.8 For wave conditions with a narrow spectrum, the wave height H is approximately equal to twice the crest height.

Wave height

For waves with a narrow spectrum *in deep water*, the height of the *wave* is practically equal to twice the height of the *crest*: $\underline{H} \approx 2\underline{\eta}_{crest}$ (the appearance of the sea-surface elevation is fairly regular; see Fig. 4.8). The probability density function of $\underline{H}$ is then readily determined from the probability density function of $\underline{\eta}_{crest}$ with a simple transformation, using a Jacobian as explained in Section 3.5.5:

$$p(H) = p_{\underline{\eta}_{crest}}(\eta)\frac{d\eta_{crest}}{dH} \tag{4.2.15}$$

so that, with Eq. (4.2.13), the probability density function of the wave height $\underline{H}$ is

$$p(H) = \frac{\eta_{crest}}{m_0} \exp\left(-\frac{\eta_{crest}^2}{2m_0}\right)\frac{d\eta_{crest}}{dH} \tag{4.2.16}$$

and with $\eta_{crest} = \frac{1}{2}H$ this is

$$\boxed{p(H) = \frac{H}{4m_0} \exp\left(-\frac{H^2}{8m_0}\right)} \tag{4.2.17}$$

which is also a Rayleigh distribution (see Fig. 4.9). The cumulative distribution function of $\underline{H}$ can be obtained by integrating the above probability density function. It can also be obtained from the cumulative distribution function of $\underline{\eta}_{crest}$ (Eq. 4.2.12; which is not a density function and the transformation does not require a Jacobian). Just substituting $\underline{\eta}_{crest} = \underline{H}/2$ into this distribution gives the cumulative

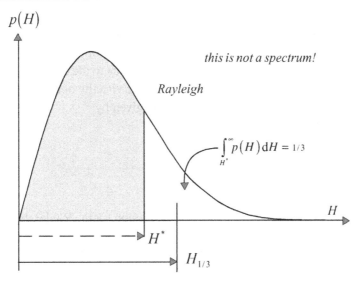

Figure 4.9 The significant wave height in the Rayleigh probability density function.

distribution function for the individual wave height $\underline{H}$:

$$\Pr\{\underline{H} \leq H\} = 1 - \exp\left(-\frac{H^2}{8m_0}\right) \tag{4.2.18}$$

All statistical characteristics of $\underline{H}$ follow from the Rayleigh distribution, e.g., the mean and the root-mean-square (*rms*) value of the wave height are

$$\overline{H} = E\{\underline{H}\} = \sqrt{2\pi m_0} \tag{4.2.19}$$

and

$$H_{rms} = E\{\underline{H}^2\}^{1/2} = \sqrt{8m_0} \tag{4.2.20}$$

As indicated in Section 3.3.2, the *significant wave height* is defined as the mean value of the highest one-third of wave heights. This fraction of the waves can be identified in the Rayleigh distribution, so the significant wave height can be determined from that distribution. The wave heights that are involved in this definition are located in the highest third of the Rayleigh distribution, i.e., where $H > H^*$, with H^* defined by (see Fig. 4.9):[2]

$$\int_{H^*}^{\infty} p(H)\mathrm{d}H = \frac{1}{3} \tag{4.2.21}$$

---

[2] Since $H = \sqrt{\ln(Q^{-1})}H_{rms}$ (where $Q = \Pr\{\underline{H} > H\}$ from Eqs. 4.2.18 and 4.2.20), it follows that $H^* = \sqrt{\ln(1/3)^{-1}}H_{rms} \approx 1.048H_{rms}$.

The *mean* value of these wave heights is by definition the significant wave height. It can be determined as an expected value, i.e., with the zeroth- and first-order moments of the highest third of the distribution. We will denote this estimate of the significant wave height as H_{m_0} (which is analogous to the notation of the mean zero-crossing period T_{m02}), to distinguish it from the visually obtained estimate and that obtained directly from the time record. It is given by

$$H_{m_0} = E\{\underline{H}\}_{H \geq H^*} = \frac{\int_{H^*}^{\infty} H p(H) dH}{\int_{H^*}^{\infty} p(H) dH} \tag{4.2.22}$$

Substituting Eq. (4.2.21) and the analytical expression for the Rayleigh distribution gives the following result:

$$H_{m_0} = 4.004 \ldots \sqrt{m_0} \tag{4.2.23}$$

or, for all practical purposes,

$$\boxed{H_{m_0} \approx 4\sqrt{m_0}} \qquad \text{deep water} \tag{4.2.24}$$

where m_0 is the zeroth-order moment of the variance density spectrum $E(f)$.

> ***This estimate of the significant wave height***
> ***is the second most important concept in this book.***

The significant wave height H_{m_0} can thus be estimated from the spectrum, which in turn can be obtained from a time series of the sea-surface elevation (see Appendix C) or from wind information with a numerical wave-prediction model (see Chapters 6 and 8).

Substituting this expression back into the Rayleigh distribution shows that 13.5% of the wave heights exceed this value. For the Rayleigh distribution, the ratio between H_{m_0} and other characteristic wave heights is fixed, for instance:

$$\overline{H} = E\{\underline{H}\} = \sqrt{\frac{\pi}{8}} H_{m_0} \tag{4.2.25}$$

and

$$H_{rms} = \frac{1}{2}\sqrt{2} H_{m_0} \tag{4.2.26}$$

Observations have shown that wave heights in deep water are indeed almost Rayleigh distributed (if the waves are not too steep). This is illustrated in Fig. 4.10

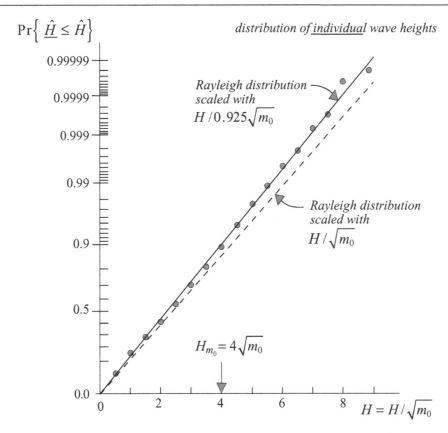

Figure 4.10 The *short-term* distribution function of observed *individual* wave heights H, normalised with the standard deviation of the surface elevation $\sqrt{m_0}$, from five hurricanes in the Gulf of Mexico (data from Forristall, 1978), with Rayleigh scales and the suggestion of Longuet-Higgins (1980) that one should re-scale the Rayleigh distribution for these observations with a factor of 0.925.

with a cumulative distribution of observed wave heights from five hurricanes in the Gulf of Mexico, plotted with Rayleigh scales.[3] The clustering of the observations around a straight line in Fig. 4.10 shows that these observations are indeed nearly Rayleigh distributed (i.e., the *shape* of the observed distribution is close to the *shape* of the Rayleigh distribution).

Although these and other observations confirm the applicability of the *shape* of the Rayleigh distribution, they show that the waves are somewhat smaller than those predicted with $H_{m_0} = 4\sqrt{m_0}$. There are several reasons for this. One is that, in the above theoretical derivation, a narrow spectrum was assumed and it was

---

[3] Plot $\{-\ln[1 - P(x)]\}^{1/2}$ against x.

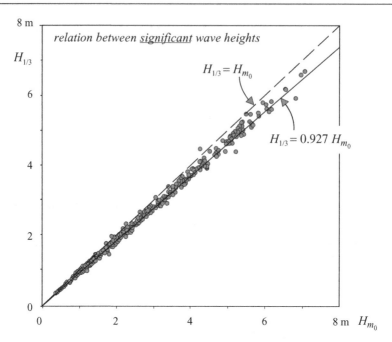

Figure 4.11 The significant wave height $H_{1/3}$ estimated directly with a zero-crossing analysis from the time records of the waves, compared with the theoretical estimate $H_{m_0} = 4\sqrt{m_0}$ from the spectrum of each record. Data are from location K13 in the southern North Sea (53.13° N, 03.13° E) during December 2003, courtesy of the Royal Netherlands Meteorological Institute. The best-fit linear approximation (least-squares fit) is close to the suggestion of Longuet-Higgins (1980) for the hurricane data of Forristall (1978), see Fig. 4.10.

assumed that $\underline{H} \approx 2\eta_{crest}$, which is not entirely correct. In addition, the surface elevation is not perfectly Gaussian distributed due to nonlinear processes such as wave breaking and nonlinear wave–wave interactions. The consequence is that the significant wave height estimated from a zero-crossing analysis, $H_{1/3}$, is 5%–10% lower than the significant wave height estimated from the spectrum with $H_{m_0} = 4\sqrt{m_0}$. Longuet-Higgins (1980) suggests $H_{1/3} = 0.925 H_{m_0}$ for the observations of Fig. 4.10. This is in close agreement with the results from other observations (see Fig. 4.11). Numerical simulations based on a linear superposition of independent wave components give slightly higher values than these findings from the field ($H_{1/3} = 0.95 H_{m_0}$; Goda, 1988a). A formulation of the Rayleigh distribution that is independent of such a discrepancy would be the following self-scaling formulation (substitute Eq. 4.2.24 into Eq. 4.2.18):

$$\Pr\{\underline{H} < H\} = 1 - \exp\left[-2\left(\frac{H}{H_{1/3}}\right)^2\right] \tag{4.2.27}$$

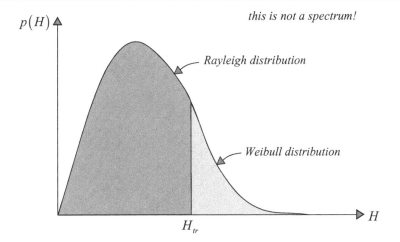

Figure 4.12 The probability density function of the wave height in shallow water according to Battjes and Groenendijk (2000).

The corresponding probability density function can be obtained by differentiating this cumulative distribution (or by substituting Eq. 4.2.24 into Eq. 4.2.17):

$$p(H) = \frac{4H}{H_{1/3}^2} \exp\left[-2\left(\frac{H}{H_{1/3}}\right)^2\right] \qquad (4.2.28)$$

Since the *shape* of the distribution function is not altered by this substitution, all relationships between characteristic values of the wave height $\underline{H}$ are unaffected and therefore still valid for real ocean waves (to the extent that the Rayleigh distribution applies).

In *shallow water* the distribution of the wave height deviates from the distribution in deep water, due to the effects of nonlinear phenomena, of which the most extreme example is wave breaking, particularly in the surf zone. Generally accepted theoretical derivations, such as those that lead to the Rayleigh distribution in deep water, are not available for shallow water. It appears that, in spite of this, the Rayleigh distribution fits the observations of waves in shallow water reasonably well. However, a closer inspection reveals that the distribution is affected at the higher values of the wave heights. This has led Battjes and Groenendijk (2000) to replace the tail of the Rayleigh distribution with the tail of the more general Weibull distribution (the Rayleigh distribution is a special case of the Weibull distribution):

$$\Pr\{\underline{H} < H\} = 1 - \exp\left[-2\left(\frac{H}{H_{ch,i}}\right)^{k_i}\right] \qquad \text{all individual waves in shallow water}$$

$$(4.2.29)$$

For wave heights lower than a transition wave height H_{tr}, the index $i = 1$, the characteristic wave height is $H_{ch,1}$ and the power in the exponent is k_1 (see Fig. 4.12).

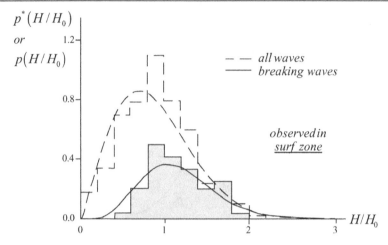

Figure 4.13 Histograms and fitted distributions (Eqs. 4.2.30 and 4.2.31) of the observed normalised wave height H/H_0 in the surf zone of individual breaking waves $p^*(H/H_0)$ and of all individual waves $p(H/H_0)$ (H_0 is the offshore rms wave height). The function $p^*(H/H_0)$ is scaled such that $\int_0^\infty p^*(H/H_0)\mathrm{d}(H/H_0)$ equals the probability of breaking (after Thornton and Guza, 1983).

For $H > H_{tr}$, the index $i = 2$, the characteristic wave height is $H_{ch,2}$ and the power is k_2. The values for the powers suggested by Battjes and Groenendijk (2000) are $k_1 = 2$ (which makes the expression a Rayleigh distribution for $H \leq H_{tr}$) and $k_2 = 3.6$. The values of the characteristic wave heights follow from the values of H_{rms} and the transition wave height H_{tr}, which depend on the zeroth-order moment of the wave spectrum, the local water depth and the local bottom slope.

Sometimes the probability density function of *breaking* waves (rather than all waves) is required, for instance to estimate the dissipation of waves in the surf zone (see Section 8.4.5). Thornton and Guza (1983) fitted a Rayleigh distribution to their observation of breaking wave heights (see Fig. 4.13) but found that a Rayleigh distribution *weighted* with a function $W(H)$ fitted these observations better:

$$p^*_{H_{br}}(H) = \frac{2H}{H^2_{rms}} \exp\left[-\left(\frac{H}{H_{rms}}\right)^2\right]W(H) \qquad \text{for individual breaking waves in shallow water} \qquad (4.2.30)$$

with

$$W(H) = \left(\frac{H_{rms}}{\gamma d}\right)^n \left\{1 - \exp\left[-\left(\frac{H}{\gamma d}\right)^2\right]\right\} \qquad (4.2.31)$$

where $p^*_{H_{br}}(H)$ is a density function, the surface area of which is the *probability of breaking* (not unity; it should therefore not properly be called a probability density function; hence the asterisk * in the notation). Thornton and Guza (1983) suggested $\gamma \approx 0.42$ and $n = 4$ for their observations.

Literature (including the joint probability density function of wave height and period):
Ahn (2000), Arhan and Ezraty (1978), Baldock *et al.* (1998), Buccino and Calabrese (2002), Cai *et al.* (1992), Forristall (1978, 1984, 2000), Goda (1975, 1988a), Haring *et al.* (1976), Hughes and Borgman (1987), Klopman and Stive (1989), Longuet-Higgins (1980), Mendez *et al.* (2004), Shum and Melville (1984), Soares (2003), Srokosz and Challenor (1987), Srokosz (1988), Stansell (2005), Tayfun (1981, 1990, 2004), Thornton and Schaeffer (1978), Thornton and Guza (1983), Tucker and Pitt (2001), Weggel (1972).

4.2.3 Wave groups

For some engineering problems, the arrival of a series of high waves (a *wave group*) is of some importance, for instance, for the stability of a rubble-mound breakwater or the overtopping of a dyke by waves. A wave group can be defined more precisely as an uninterrupted sequence of waves with wave heights higher than an arbitrarily chosen, but usually high, threshold value H_{gr}. The *length* of such a wave group is, by definition, the number of waves ($\underline{N}$) in the group.

To derive the probability that the length of an arbitrarily chosen wave group is larger than a value N, imagine a wave group, starting at wave height H_i and ending at wave height H_j, in a long, (statistically) stationary time series of wave heights:

$$\ldots, \underline{H}_{i-1}, \underline{H}_i, \underline{H}_{i+1}, \underline{H}_{i+2}, \ldots, \underline{H}_{j-1}, \underline{H}_j, \underline{H}_{j+1}, \underline{H}_{j+2}, \ldots$$

$$\underleftrightarrow{\text{group}}$$

The wave group starts at wave height $\underline{H}_i$ because this wave height is larger than the threshold value H_{gr} *and* the preceding wave height is smaller than the threshold value $\underline{H}_{gr}$. The group ends at $\underline{H}_j$ because this wave is still higher than the threshold value but the following wave height is smaller than that value. In other words, the wave group starts when $\underline{H}_i > H_{gr}$ *and* $\underline{H}_{i-1} < H_{gr}$ and it ends when $\underline{H}_j > H_{gr}$ *and* $\underline{H}_{j+1} < H_{gr}$.

If $j = i$, that is, the first wave of the group is also the last wave of the group, the length of the group is obviously $\underline{N} = 1$. This is rather trivial but by definition this single wave is a group. The probability of this occurring is given by the probability that $\underline{H}_{i+1} < H_{gr}$ (given that a group has been encountered, i.e., given that $\underline{H}_i > H_{gr}$):

$$\Pr\{\underline{N} = 1\} = \Pr\{\underline{H}_{i+1} < H_{gr} \mid \underline{H}_i > H_{gr}\} \tag{4.2.32}$$

This is a conditional probability (the probability that $\underline{H}_{i+1} < H_{gr}$ may depend on the value of $\underline{H}_i$). If such dependence does *not* exist, in other words, the wave heights are statistically *independent* (this is not quite true but I will modify the results later), it does not matter what the value of H_i is and the probability in Eq. (4.2.32) may be written as

$$\Pr\{\underline{N} = 1\} = \Pr\{\underline{H}_{i+1} < H_{gr}\}$$
$$= 1 - Q_H \tag{4.2.33}$$

where $Q_H = \Pr\{\underline{H}_{i+1} \geq H_{gr}\}$ and i is the sequence number (in the time series) of the first wave of the group (remember, this is a probability, given that a group *has* been encountered; it does not relate to the probability that a group *will* be encountered).

If the second wave is also high, i.e., $\underline{H}_{i+1} > H_{gr}$, and the *next* wave is low, i.e., $\underline{H}_{i+2} < H_{gr}$, then the length of the group is $\underline{N} = 2$. The probability of this occurring is given (if the wave heights are statistically independent) by

$$\Pr\{\underline{N} = 2\} = \Pr\{\underline{H}_{i+1} \geq H \text{ and } \underline{H}_{i+2} < H\}$$
$$= \Pr\{\underline{H}_{i+1} \geq H\} \cdot \Pr\{\underline{H}_{i+2} < H\} = Q_H(1 - Q_H) \tag{4.2.34}$$

In the same way, the probability that the group length $\underline{N} = 3$ is given by

$$\Pr\{\underline{N} = 3\} = \Pr\{\underline{H}_{i+1} \geq H_{gr} \text{ and } \underline{H}_{i+2} \geq H_{gr} \text{ and } \underline{H}_{i+3} < H_{gr}\}$$
$$= \Pr\{\underline{H}_{i+1} \geq H_{gr}\} \cdot \Pr\{\underline{H}_{i+2} \geq H_{gr}\} \cdot \Pr\{\underline{H}_{i+3} < H_{gr}\}$$
$$= Q_H^2(1 - Q_H) \tag{4.2.35}$$

The probability $\underline{N} = N$ is similarly given by

$$\Pr\{\underline{N} = N\} = Q_H^{N-1}(1 - Q_H) \tag{4.2.36}$$

These estimates are based on the assumption that the wave heights are statistically independent. However, wave heights are to some degree correlated. A high wave is generally followed by another high wave and a low wave generally by another low wave (i.e., they are *not* independent). The probability Q_H must therefore be replaced with a probability that involves the effect of the preceding wave height. This is a conditional probability, which in the present context is the probability that $\underline{H}_{i+1} > H_{gr}$, under the condition that the preceding wave height $\underline{H}_i > H_{gr}$. The notation of this conditional probability is $\Pr\{\underline{H}_{i+1} > H_{gr}|\underline{H}_i > H_{gr}\} = R_H$. The probability of the group length being equal to $\underline{N} = N$ would then be, with the same rationale as above, but now with this correlation between wave heights taken into account (compare with Eq. 4.2.36),

$$\Pr\{\underline{N} = N\} = R_H^{N-1}(1 - R_H) \tag{4.2.37}$$

The mean length of a wave group $\overline{N}$, expressed in terms of the number of waves in the group, is then given by (see Kimura, 1980)

$$\overline{N} = \overline{l}_{high\ waves} = (1 - R_H)^{-1} \tag{4.2.38}$$

It is likewise possible to define a group of *low* waves (i.e., a sequence of wave heights lower than a certain threshold H_{gr}). The average length of a group of such low waves is

$$\overline{l}_{low\ waves} = \left(1 - R_H^*\right)^{-1} \tag{4.2.39}$$

where

$$R_H^* = \Pr\{\underline{H}_{i+1} < H_{gr} | \underline{H}_i < H_{gr}\} \tag{4.2.40}$$

It is remarkable that, for typical wind-sea conditions (with a JONSWAP spectrum, see Section 6.3.3), the mean *distance* between two consecutive groups of high waves, i.e., the length of a group of high waves plus the length of a group of low waves: $\overline{l} = \overline{l}_{high\ waves} + \overline{l}_{low\ waves}$, is approximately seven (for a threshold value H_{gr} between the mean wave height $\overline{H}$ and the significant wave height $H_{1/3}$). In spite of the scepticism of some wave researchers, this corresponds nicely to the old rule-of-thumb: 'every seventh wave is the highest' (e.g., Rudyard Kipling, in *The First Jungle Book*, quoting the white seal saying that the seventh wave always goes farthest up the beach; and Henri Charrière in *Papillon* escaping from Devil's Island by jumping off the cliffs into a seventh wave).

Even a casual observation shows that, under conditions with relatively *regular* waves (a narrow spectrum), the wave groups are relatively long, whereas under conditions with irregular waves (a wide spectrum), the wave groups are relatively short. The groupiness of waves and the probability R_H depend therefore on the width of the spectrum (expressed in terms of the spectral width parameter κ; see Note 4C).

Literature:
Battjes and van Vledder (1984), Elgar *et al.* (1984), Ewing (1973), Goda (2000), Johnson *et al.* (1978), Kimura (1980), Longuet-Higgins (1984), Rye (1974), Soares (2003), van Vledder (1992).

4.2.4 Extreme values

For many engineering problems it is important to understand the statistical characteristics of extremes, in particular of the *maximum surface elevation* $\underline{\eta}_{max}$ *within a certain duration D* (e.g. in a storm; see Fig. 4.14) or the maximum *wave* height $\underline{H}_{max}$ within that duration (we are still considering stationary conditions).

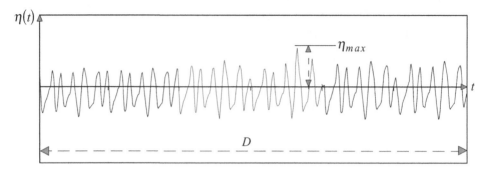

Figure 4.14 The maximum crest height in a duration D of stationary wave conditions.

The statistical characteristics of these maxima are fully described by the cumulative distribution functions of these maxima: for the maximum elevation $P_{\eta_{max}}(\eta)_D = \Pr\{\underline{\eta}_{max} < \eta\}_D$ and for the maximum wave height $P_{H_{max}}(H)_D = \Pr\{\underline{H}_{max} < H\}_D$. Note that the maximum *elevation* is equal to the maximum *crest height* within that duration. Also, the probability that the surface elevation η itself remains below level η within duration D is equal to the probability that the maximum elevation remains below that level, so that

$$\Pr\{\underline{\eta}_{max,crest} < \eta\}_D = \Pr\{\underline{\eta}_{max} < \eta\}_D = \Pr\{\underline{\eta} < \eta\}_D \qquad (4.2.41)$$

Extreme elevations

The rationale for arriving at the cumulative distribution function of the maximum crest height $\Pr\{\underline{\eta}_{max,crest} < \eta\}_D$ is the following. The probability that an *arbitrarily* chosen crest height *exceeds* a level η in a given sea state (i.e., *one* arbitrarily chosen crest height in a stationary sea state) is given by $\Pr\{\underline{\eta}_{crest} > \eta\}$. For the sake of brevity, I will denote this probability as $Q_{crest} = \Pr\{\underline{\eta}_{crest} > \eta\}$, which normally is taken to be a Rayleigh distribution. The probability that this arbitrarily chosen crest height does *not* exceed the level η is then $1 - Q_{crest}$. The probability that *two arbitrarily* chosen crest heights in the wave record do not exceed the level η is given by $(1 - Q_{crest})^2$ (if the crest heights are statistically independent, which is not entirely true, but the errors involved are acceptable, see below). For the same reason and under the same conditions: the probability that *all* crest heights in duration D do *not* exceed the level η is given by

$$\Pr\{\text{all } \underline{\eta}_{crest} \leq \eta\}_D = (1 - Q_{crest})^N \qquad (4.2.42)$$

where N is the total number of crests in duration D. If all crest heights are lower than the level η, then the maximum crest height and all elevations, including the maximum elevation, are lower than this level, so that[4]

$$
\begin{aligned}
\Pr\{\text{all } \underline{\eta}_{crest} < \eta\}_D &= \Pr\{\underline{\eta}_{max,crest} < \eta\}_D = \Pr\{\text{all } \underline{\eta} < \eta\}_D \\
&= \Pr\{\underline{\eta}_{max} < \eta\}_D = \Pr\{\underline{\eta} < \eta\}_D = (1 - Q_{crest})^N
\end{aligned}
$$

(4.2.43)

The probability that *one or more* crest heights will be *larger* than the level η within duration D is equal to the probability that *not all* crests heights are lower than level η (from Eq. 4.2.43):

$$
\Pr\{\underline{\eta}_{max,crest} > \eta\}_D = 1 - (1 - Q_{crest})^N
$$

(4.2.44)

(see Note 4D for an approximation). Note that the Gaussian distribution of the surface elevation $\underline{\eta}$ does not appear in this derivation but it is implicit if Q_{crest} is taken to be a Rayleigh distribution.

NOTE 4D   An approximation for $(1 - Q_{crest})^N$

If the number of crests is large (as in any duration of reasonable length, e.g., in a storm) ($N \gg 1$) and the probability of exceedance Q_{crest} is small (of the order of $1/N$), then $(1 - Q_{crest})^N \approx \exp(-NQ_{crest})$.

All that remains to be determined to compute the probability in Eq. (4.2.44), for a given value of Q_{crest}, is the total number of crests N within duration D. This can be done by noting that, for a narrow spectrum, the number of crests is equal to the number of upward or downward zero-crossings, which is determined by the mean zero-crossing frequency and the duration:

$$
N = \overline{f}_0 D = \sqrt{\frac{m_2}{m_0}} D
$$

(4.2.45)

The corresponding probability density function of the maximum crest height within duration D is the derivative of $\Pr\{\underline{\eta}_{max,crest} < \eta\}$:

$$
p_{\underline{\eta}_{max,crest}}(\eta) = \frac{d(1 - Q_{crest})^N}{d\eta}
$$

(4.2.46)

which is shown in Fig. 4.15. The maximum value of this probability density function is located at the *mode* of $\underline{\eta}_{max,crest}$ (which is interpreted as the most probable value

---

[4] Note that $\Pr\{\eta \le \eta\}_D = \Pr\{\text{all } \underline{\eta} \le \eta\}_D \ne \Pr\{\underline{\eta} \le \eta\}$; the first two give the probability of non-exceedance of *all* surface elevations within a duration D (the above distribution), whereas the last gives the probability of non-exceedance of *one* arbitrarily chosen surface elevation somewhere within duration D (the Gaussian distribution).

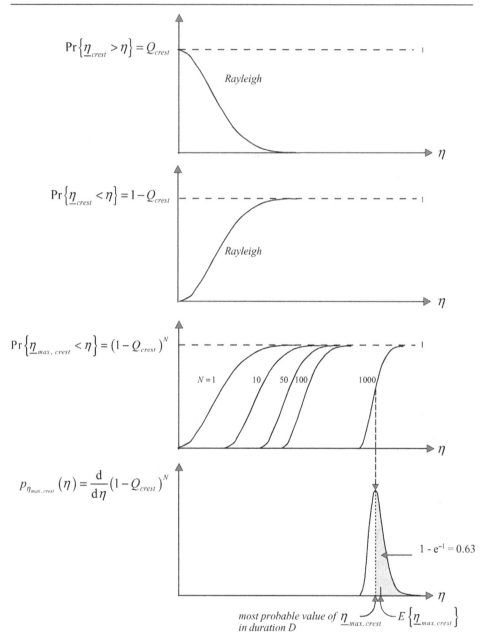

Figure 4.15 The distribution functions and probability density functions of $\underline{\eta}_{crest}$ and $\underline{\eta}_{max,crest}$.

Table 4.1. *The sensitivity of* $\mathrm{mod}(\underline{\eta}_{max,crest})$ *and* $E\{\underline{\eta}_{max,crest}\}$ *to errors in estimating the number of crests N*

N		$\sqrt{2\ln N}$	
2000	(real value)	3.90	(real value)
2500	(error relative to 2000 = 25%)	3.96	(error relative to 3.90 = 1.5%)
4000	(error relative to 2000 = 100%)	4.07	(error relative to 3.90 = 4.3%)

of the maximum crest height $\underline{\eta}_{max,crest}$). It can be shown that this most probable value is

$$\mathrm{mod}(\underline{\eta}_{max,crest}) \approx \sqrt{2\ln N}\sqrt{m_0} \tag{4.2.47}$$

The expected value (i.e., the mean) of the maximum crest height is slightly higher:

$$E\{\underline{\eta}_{max,crest}\} \approx \left(1 + \frac{0.29}{\ln N}\right)\sqrt{2\ln N}\sqrt{m_0} \tag{4.2.48}$$

Obviously, with more waves in a storm, the values of $\mathrm{mod}(\underline{\eta}_{max,crest})$ and $E\{\underline{\eta}_{max,crest}\}$ increase but Eqs. (4.2.47) and (4.2.48) show that the increase is only very slow because of the logarithm and the square root in these expressions. In other words, these values are rather insensitive to the value of N. An error in estimating the value of N therefore usually has no serious consequences for estimating $\mathrm{mod}(\underline{\eta}_{max,crest})$ and $E\{\underline{\eta}_{max,crest}\}$ (see Table 4.1). This illustrates that these results are also rather insensitive to the earlier condition that wave crests should be statistically independent.

It is generally not wise to use the value of $\mathrm{mod}(\underline{\eta}_{max,crest})$ to base a design of a structure at sea on, because the probability that the actually occurring maximum elevation exceeds $\mathrm{mod}(\underline{\eta}_{max,crest})$ is considerable; it is equal to 0.63. This implies that there is a probability of 0.63 that the actual maximum crest height within the duration will exceed its most probable value (but not by much, since the probability density function is rather narrow, see Fig. 4.15).

The agreement between observed values and the above theoretical estimate of the maximum crest height is very reasonable, as shown in Fig. 4.16 where the theoretical estimate (Eq. 4.2.48) is compared with observations (significant wave heights larger than 0.5 m).

Literature:
Cartwright and Longuet-Higgins (1956), Cartwright (1958).

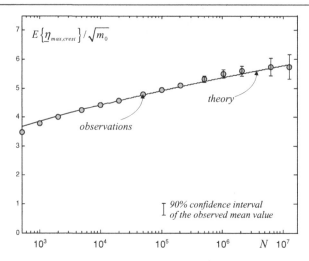

Figure 4.16 Measured and theoretically estimated values of the mean maximum crest height $E\{\underline{\eta}_{max,crest}\}$ of N waves (normalised with the standard deviation $\sqrt{m_0}$). These maxima were obtained from 10 years of observations with 4 buoys off the Mediterranean coast of Spain. With many thanks to Mercè Casas Prat who did the calculations (see Acknowledgements).

Extreme wave heights

The cumulative distribution function of the maximum individual *wave* height within a duration D can be derived with the same rationale as the above for the maximum *crest* height, with the following results:

$$\Pr\{\underline{H}_{max} \leq H\} = (1 - Q_H)^N \tag{4.2.49}$$

where $Q_H = \Pr\{\underline{H} > H\}$. The mode of $\underline{H}_{max}$, i.e., the most probable value of $\underline{H}_{max}$, is

$$\mathrm{mod}(\underline{H}_{max}) \approx H_{m_0}\sqrt{\frac{1}{2}\ln N} \tag{4.2.50}$$

The probability density function of $\underline{H}_{max}$ is just as narrow as the probability density function of $\underline{\eta}_{max,crest}$ (see Fig. 4.15), so for a given wave record $\underline{H}_{max} \approx \mathrm{mod}(\underline{H}_{max})$ (see Note 4E). Again, however (as with the most probable crest height), it is generally not wise to use $\mathrm{mod}(\underline{H}_{max})$ as the design wave height in engineering practice because the actually occurring maximum wave height has a probability of 0.63 of exceeding $\mathrm{mod}(\underline{H}_{max})$.

An easy rule-of-thumb based on Eq. (4.2.50) is that, in many storms, the maximum wave height is approximately equal to twice the significant wave height (see Note 4E):

$$\underline{H}_{max} \approx 2H_s \tag{4.2.51}$$

Very occasionally, very high waves occur, much higher than the above theory predicts. These waves are called 'freak' waves or 'rogue' waves and their origin is still a mystery (although some theoretical progress is being made; see Note 4E).

It is important to note that it is assumed in the above that the crest heights and wave heights are Rayleigh distributed. In water with a limited depth, this is not always the case and the statistical characters of the waves should be estimated on the basis of a slightly different distribution (see for instance Eq. 4.2.29), generally resulting in lower extremes.

Literature:
Borgman (1973), Cartwright and Longuet-Higgins (1956).

NOTE 4E   The maximum wave height and freak waves
(or rogue waves)

The fact that the probability density function of $\underline{H}_{max}$ is narrow is sometimes used to estimate $\underline{H}_{max}$ rapidly from the value of the significant wave height H_s (which is often the wave height predicted by a meteorological centre). If the duration of a storm is 6 h and the average zero-crossing wave period is about 10 s, then $N = 2160$ and it follows from Eq. (4.2.50) that

$$\underline{H}_{max} \approx \mathrm{mod}(\underline{H}_{max}) = 1.96 H_{m_0} \approx 2 H_s$$

Since this theoretical estimate of $\underline{H}_{max}$ is just as insensitive to the value of N as the theoretical estimate of $\mathrm{mod}(\underline{\eta}_{max,crest})$, this simple relationship between $\underline{H}_{max}$ and H_s is sometimes inverted by engineers to estimate the significant wave height H_s quickly from a wave record:

$$H_s \approx \frac{1}{2} \underline{H}_{max}$$

They need only look up the maximum wave height in the wave record and *presto*, they have estimated H_s! Tall tales, with a statement that 'the waves were . . . metres high!' often relate to the maximum wave height, so the significant wave height is often only half the stated value.

The above estimate of the maximum wave height in a storm of twice the significant wave height is only an estimate of the most probable maximum. In actual storms, the value will be somewhat higher or lower, but not by much because the probability density function of the maximum wave height is rather narrow. The occurrence of a very large wave height (larger than 2.5 times the significant wave height, say) is therefore exceptional. If such an exceptionally high wave occurs, one would expect a certain build-up towards such an event: the wave would probably be preceded by one or two or perhaps even three other high waves. However, that is not always the case. Sometimes, an extremely high wave occurs without any such warning. It appears out of nowhere, seemingly without any relation to the prevailing wave conditions. For a long time, tales of the sea about such monster waves crashing against ships were regarded as sailors' fantasy. However, extensive measurements at sea now available have revealed that at least some of these tales reflect actual facts. An example of a measurement of such a wave, with a *crest* height of 18.5 m when the significant *wave* height was 'only' 12 m, is given below.

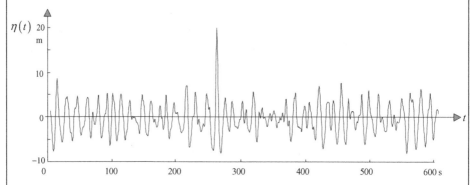

A freak wave measured (by laser altimeter) at the Draupner platform in the central North Sea on 1 January 1995. The crest height and the wave height of this wave were about 18.5 m and 26 m, respectively, whereas the significant wave height was 'only' about 12 m (courtesy of Statoil Norge AS; see for instance Haver and Andersen, 2000).

Such waves are called 'freak' waves or 'rogue' waves. The preceding trough is often very pronounced and it is sometimes referred to as 'a hole in the sea'. There is no consensus as to what exactly a freak wave is. I favour the description of survivors who have seen one and lived to tell the tale: an exceptionally high, steep breaking wave with an unusually long crest with an almost vertical front preceded by a deep trough. It seems to have a fairly stable form and it suddenly appears out of nowhere. This wave definitely stands out against its background. Here is an eyewitness account of such a wave: a single wave with a *crest* height of about 7 m in a situation in which the significant wave height was about 4 m. It is from Luigi Cavaleri, who operates an oceanographic observation tower in the Adriatic Sea just south of Venice (and helped me write this book; see the introduction):

We were on the tower during the night in the middle of a storm with about 4 m significant wave height. I was fixing an instrument at about 7 metres above the sea level, on a long horizontal extension of the platform. Suddenly I heard something like a train coming, and, looking in the dark, I could spot the whitish crest of a wave running against the tower at my height. There was nothing I could do. The crest passed barely below my feet before exploding against the structure. When I turned towards the tower, for a couple of seconds I could see only water. My colleagues, who were watching from an upper deck, were soaked. I was dry.

Focusing of wave energy by a meandering strong current (current refraction; see Section 7.3.5) has been suggested to explain accidents involving large waves in certain areas. A classical example is the Agulhas Current, off the southern tip of South Africa. However, energy focusing may lead to exceptionally high sea conditions in a certain area, but not necessarily to a single extremely high wave. A more promising and solid approach is based on theories that explain the instability of the sea surface under certain conditions. In a random sea we find sequences of larger and smaller waves. It may happen that, in a sequence of steep and fairly regular waves (a local inhomogeneity), a single large wave begins to extract energy from its neighbours, growing at their expense. As a matter of fact, one of the characteristics of a freak wave is that smaller waves and a deep trough precede the freak wave. This makes freak waves even more dangerous, since their occurrence surprises the crew of the unhappy ship that encounters such an event. A crew is generally not able to respond fast enough to avoid considerable damage, to the point of complete disaster (loss of the ship without trace). After a while, the wave dies down and returns its energy to its surroundings. Such a development of a single wave can be modelled numerically in great detail. However, it should be stressed that the *occurrence* of a freak wave is governed by statistics. There is no way we can predict where and when it will happen. At most we can estimate the probability of its appearance.

Literature:
Atkins (1977), Buckley (1983), Buckley *et al.* (1984), Draper (1965), Earle (1975), Godden (1977), Gumbley (1977), Günther and Rosenthal (2002), Haver and Andersen (2000), Janssen (2003), Lehner *et al.* (2001), Onorato *et al.* (2001, 2002, 2004), Osborne *et al.* (2000), Sand *et al.* (1990), Skourup *et al.* (1997), Stansell (2005).

4.3 Long-term statistics (wave climate)

In the previous sections, the statistical characteristics of the waves were considered for short-term, stationary conditions, usually for the duration of a wave record and sometimes for a storm. For *long-term* statistics, e.g., statistics over durations of a few dozen *years* or more, the conditions are not stationary and the problem of describing waves needs to be approached in an entirely different way. For these long time scales, it is not feasible to present the waves as a time series of the surface elevation itself. Instead, each stationary condition (with a duration of 15–30 min, say) is replaced with its values of the significant wave height, period and mean wave direction. This gives a long-term sequence of these values with a time interval of typically 3 h (see Fig. 4.17), which can be analysed to estimate the long-term statistical characteristics of the waves, for instance to obtain design conditions for marine structures. Usually the analysis is limited to the significant wave height, in particular its long-term distribution, and its return period. These long-term statistics can be estimated at a geographic location from (a) all available observations at that location (the initial-distribution approach), (b) the maximum value in storms at that location (the peak-over-threshold approach, or (c) the maximum value per year at that location (the annual-maximum approach).

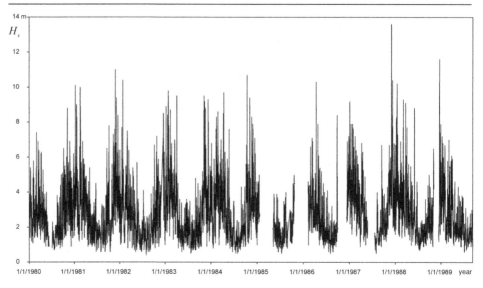

Figure 4.17 The significant wave height $H_{m_0} = 4\sqrt{m_0}$ over a ten-year period (1980–1989; NODC buoy 46005, position 131°W, 46°N, i.e., in the northern Pacific Ocean 600 km south-west of Seattle, data from the American National Oceanographic Data Center, downloaded from http://www.nodc.noaa.gov/BUOY/46005.html). Note the unusually high value in early 1988 and the gaps in other years. These data, supplemented with the mean zero-crossing period and with the years 1990–2003, will be used to illustrate various aspects of long-term statistics in this chapter.

A remarkable difference from the short-term statistics of waves is that there is no theoretical model for the basic long-term time series (e.g., of the significant wave height) such as a random-phase/amplitude model or a Gaussian model. However, long-term series can be analysed and interpreted using results of the extreme-value theory if certain fundamental conditions are fulfilled. The most important of these are that the values in the time series must be statistically *independent* from one another and they must be *identically distributed* (abbreviated to *i.i.d.*), i.e., each value should be an independent, random sample from one and the same population. These conditions can pose serious problems for real ocean waves because consecutive values in the time series, e.g., of the significant wave height, are usually *not* independent (they are correlated, i.e., a large value of the significant wave height is usually preceded and followed by another high value, at least when the time interval between the observations is less than a day or so).

To achieve statistical independence one should consider only values that are sufficiently far separated in time. This problem is often ignored, notably in the initial-distribution approach, to be described next. It can be solved by selecting values at large intervals, e.g., one value per storm, as in the peak-over-threshold approach, or one value per year, as in the annual-maximum approach, both to be described next. In addition, the values are often *not* identically distributed, because waves may have different sources. For instance, swell is generated in distant storms and wind sea is generated by local winds. For many oceanic locations, therefore, the wave climate should be separated into a swell climate and a wind-sea climate. Such distinction (which is also often ignored) might not be sufficient: each of these climates may have to be split again into two or more climates because swell may originate from different parts of the ocean, each with its own swell-generating weather patterns. For instance, swell off the Californian coast is generated in the northern hemisphere but also in the southern hemisphere with different weather climates. Wind sea may be generated by hurricanes in an area where the daily weather is dominated by trade winds, requiring different climate descriptions for the common, daily conditions and the extreme conditions of the hurricanes. In coastal waters, the situation is even more complicated, since the physical mechanisms that affect the waves may change as the significant wave height and period increase, due to the effect of the limited water depth (possibly imposing a maximum on the significant wave height due to depth-induced wave breaking). Such a physically imposed upper limit of the wave heights might not be noticeable in observed or numerically simulated wave heights because these may be too low, but such an upper limit would be very relevant when extrapolating to extreme conditions.

Literature:
Bauer and Staabs (1998), Castillo (1988), Coles (2001), Dacunha *et al.* (1984), Ewing *et al.* (1979), Goda and Kobune (1990), Goda (1992), Goda *et al.* (1993), Gorshkov (1986), Gumbel (1958), Hogben (1988, 1990a, 1990b), Kamphuis (2000), Leadbetter *et al.* (1983), Muir and El-Shaarawi (1986), Neu (1984), Peters *et al.* (1993), Petruaskas and Aagaard (1970), Repko *et al.* (2000), Soares (2003), van Vledder *et al.* (1993), WMO (1998).

4.3.1 *The initial-distribution approach*

Often, the first step in analysing the long-term time series of the significant wave height H_s, mean wave period $\overline{T}_0$ and mean wave direction $\overline{\theta}$ is to estimate the joint probability density function $p(H_s, \overline{T}_0, \overline{\theta})$, usually by sorting the observed values and presenting the results in two-dimensional histograms of H_s and $\overline{T}_0$ per directional sector, $\Delta\overline{\theta}$. The (actual or relative) *number* of observations is then presented (instead of the probability density) in bins of size ΔH_s, $\Delta\overline{T}_0$ (per directional sector, typically $\Delta\overline{\theta} = 30°$ or $45°$). These histograms may be given per season or per month (accumulated or averaged over a large number of years). By adding the numbers

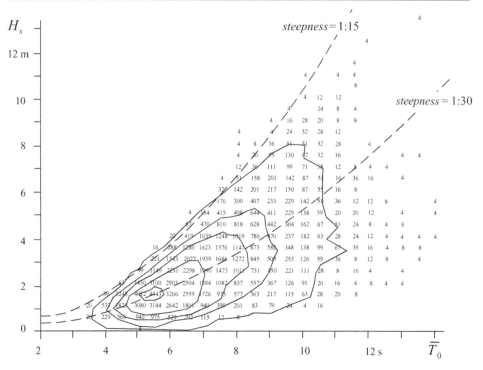

Figure 4.18 The histogram of the long-term, joint occurrence of significant wave height $H_{m_0} = 4\sqrt{m_0}$ and mean wave period $\overline{T}_0 = \sqrt{m_0/m_2}$ for the years 1980–2003 for NODC buoy 46005 of Fig. 4.17, representing the long-term joint probability density function $p(H_s, \overline{T}_0)$ (the frequency of occurrence is in units of 1 : 100 000). The dashed lines are lines of constant wave steepness $2\pi H_s/(g\overline{T}_0^2)$.

over the directional sectors one obtains the histogram for the significant wave height and period irrespective of direction, representing the joint distribution $p(H_s, \overline{T}_0)$.

Such a joint distribution is given in Fig. 4.18, which also shows that the observed wave steepness $2\pi H_s/(g\overline{T}_0^2)$ is limited to *steepness* $\leq 1 : 15$ approximately (this is a universal, physical limitation in deep water, imposed by wave breaking), while on average, in this example, *steepness* $\approx 1 : 30$ (it generally depends on the mix of swell and wind sea in the area). By summing the numbers for the mean period in the histogram for a given significant wave height *or* the numbers for the significant wave height for a given mean wave period, one obtains the histogram for either the significant wave height or the mean period separately, representing the probability density functions $p(H_s)$ and $p(\overline{T}_0)$, respectively (see Fig. 4.19).

For many applications, the histograms are adequate because only the statistics of the sorted values within the range of observed values are needed, e.g., to analyse

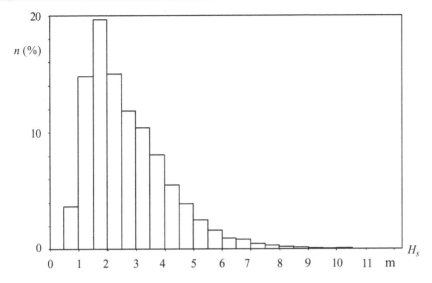

Figure 4.19 The histogram of the significant wave height for the years 1980–2003 for NODC buoy 46005 of Fig. 4.17 (n is the percentage of the total number of occurrences in the interval $\Delta H_s = 0.5$ m).

fatigue effects in a structure. However, extreme conditions usually fall outside the observed range and to estimate these one needs to *extrapolate* the observations (typically only for the significant wave height). This is usually done by fitting some curve through the histogram and extrapolating that curve to the desired low probability of occurrence. In the absence of any theory, the choice of the curve is entirely empirical: several candidate distributions (analytical expressions), each with several free parameters, are chosen, and the values of these parameters are estimated by fitting the candidate distributions to the data. The distribution that fits the data best is then used for the extrapolation. To facilitate *judging* such a fit by eye, it is convenient to use the cumulative distribution function $P(H_s) = \Pr\{\underline{H}_s \leq H_s\}$, rather than the probability density function $p(H_s)$, because, when plotted on paper with proper scales, the cumulative distribution function will appear as a straight line around which the data should cluster (if the candidate distribution fits the data; see for instance Fig. 4.20). Alternatively; objective goodness-of-fit tests are also available, e.g., the χ^2-test, the Kolmogorov–Smirnov test and the Anderson–Darling test.

The choice of the candidate distributions is rather arbitrary but past experience helps to limit the choice to only a few. A *two*-parameter distribution is the most convenient for a *fit by eye*, because a straight line on paper has only two free parameters (i.e., intercept and slope). However, it is obvious that a distribution

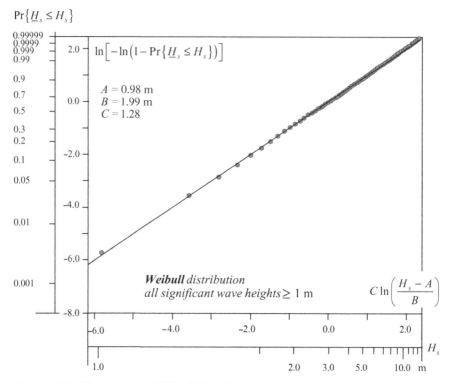

Figure 4.20 The long-term Weibull distribution of the significant wave height, for the years 1980–2003 for NODC buoy 46005 of Fig. 4.17 (note that more years are included than in Fig. 4.17; all values below 1.00 m, i.e., 2.3% of the data, have been removed). The straight line represents the best-fit candidate distribution (maximum likelihood). The position of the exceptional storm of early 1988 is not obvious (see Fig. 4.17).

with more free parameters would generally provide a better fit because it has more degrees of freedom. It is therefore advisable to consider also distributions with *three* free parameters. For an *objective fit* (see below), any number of parameters is permitted (within reason, and certainly considerably less than the number of data points). Two distributions that are widely used for the long-term distribution of the significant wave height are given in Note 4F: the log-normal distribution and the Weibull distribution.

To fit the candidate distributions to the observations requires that a probability of non-exceedance be assigned to each observed value. There are two procedures for this: (a), when the number of observations is large, one 'bins' the observations (i.e., one determines the numbers of observations falling within certain intervals, i.e., the 'bins' of a histogram; see Fig. 4.19); or (b), when the number of observations is small, one assigns a probability to each observation individually. In the *bin option*, the probability of $\underline{H}_{s,i}$ not exceeding the value $H_{s,i}$ (the lower limit of bin number i) is

$$\Pr\{\underline{H}_{s,i} < H_{s,i}\} = n_i/N \qquad (4.3.1)$$

where n_i is the number of observations lower than $H_{s,i}$ and N is the total number of observations.

When the number of observations is small, the intuitive estimate of the probability per *individual* observation would be $\Pr\{\underline{H}_{s,i} < H_{s,i}\} = 1 - i/N$ (where i is the ranking number of the observation; ranking the *highest* observation as $i = 1$; this

---

NOTE 4F   Long-term distributions for the significant wave height H_s

The *log-normal distribution* is given by

$$\Pr\{\underline{H}_s \leq H_s\} = \frac{1}{2\pi} \int_{-\infty}^{\frac{\ln H_s - A}{B}} \exp\left(-\frac{1}{2}x^2\right) dx$$

The *Weibull distribution* (this distribution is called the Weibull distribution for *minima*, although it is not used for minima here) is given by

$$\Pr\{\underline{H}_s \leq H_s\} = \begin{cases} 1 - \exp\left[-\left(\dfrac{H_s - A}{B}\right)^C\right] & \text{for } H_s > A \quad \text{and} \quad C > 0 \\ 0 & \text{for } H_s \leq A \end{cases}$$

The parameter A is a location parameter (the position of the distribution on the H_s-axis). In the Weibull distribution this parameter also represents the lower limit of the significant wave height (a permanent minimum background sea). The parameter B (>0) provides a normalisation (scaling), which determines the width of the distribution. The parameter C is a shape parameter. For $A = 0$, the Weibull distribution is called the two-parameter Weibull distribution. For $C = 1$, it reduces to an *exponential distribution* and for $C = 2$ to a *Rayleigh distribution*.

---

is called the 'plotting position'). Actually, statisticians tell us that, due to effects of sample variability of the observed values (slightly different results if the analysis were repeated with other samples from the same population), the plotting position depends on the distribution from which the observation is assumed to be taken, e.g., for the Weibull distribution, Goda (1988b, based on Petruaskas and Aagaard, 1970) recommends for the least-squares fitting technique

$$\Pr\{\underline{H}_{s,i} < H_{s,i}\} = 1 - \frac{i - \alpha}{N + \beta} \qquad \text{with } \alpha = 0.20 + 0.27/\sqrt{C} \text{ and}$$

$$\beta = 0.20 + 0.23/\sqrt{C} \qquad (4.3.2)$$

where C is the shape parameter of the distribution (see Note 4F).

Having thus established the probability values of the *observations*, one can then fit the various candidate distributions, subjectively by eye or objectively with a formal procedure, e.g., with the least-squares technique (i.e., minimise the sum of the squared differences between the observations and the candidate distribution), a maximum-likelihood technique (i.e., maximise the probability that the observations are taken from the candidate distribution), or a moments technique (i.e., compute the parameters of the candidate distribution such that the lower-order moments or the L-moments[5] of the observed distribution and of the candidate distribution are equal). The most primitive procedure seems to be the fit by eye: plot the values on paper with proper scales along the axes and, if the data belong to the distribution that corresponds to the scales, they should arrange themselves along a straight line (scatter will always remain because of sample variability). For the distributions considered in this chapter, such scales are readily constructed (except for the log-normal distribution) as *log* or *double-log* scales, depending on the distribution considered, e.g., plotting $y = \ln[-\ln(1 - \Pr\{\underline{H}_s < H_s\})]$ against $x = C \ln[(H_s - A)/B]$ gives a straight line[6] for the Weibull distribution (see Fig. 4.20;[7] this particular data set is also well approximated by the log-normal distribution when the seasonal variation in the significant wave height is removed, except for the high values of H_s, for which the fit remains poor; see Note 4G). The advantage of a fit by eye is that the engineer is able to favour the higher values in the data set at his professional discretion. The alternative, i.e., fitting the distribution with an objective procedure, is not as objective as it may sound because, without a theoretical basis, the *choice* of the technique is still subjective. In any case, an objective fit should always be inspected by eye to verify that the fit is reasonable, in particular for the high observed values in which engineers are usually most interested (results may unintentionally be seriously biased to the low values of the data, of which there are many). Assigning more importance to higher observed values can also be achieved in an objective fit, by properly weighting these higher values or by ignoring the lower values (which is called '*censoring*'; see Fig. 4.20 where the low values $H_s < 1$ m were removed).

---

[5] Moments that are based on the quantile function (see Appendix A and Pandey *et al.*, 2004).

[6] There are several methods by which to represent the data as straight lines. Statisticians and mathematicians tend to plot the observed values of H_s against the fitted value of H_s (at given values of non-exceedance of the fitted candidate distribution; for a perfect fit, this would give a straight line). These plots are called quantile–quantile plots (or, Q–Q plots; see Note 4G). The traditional technique in engineering is different: plot, in one figure, both the observed and the candidate probabilities along the vertical axis and H_s along the horizontal axis, using log or double-log scales (see Fig. 4.20). The candidate values always appear as a straight line (due to the scaling), whereas the observed values will appear as a straight line only in the case of a perfect fit. Such a figure can be used for estimating by eye the parameters of a two-parameter distribution (from the slope and intercept of a straight line that is fitted through the observed values).

[7] With many thanks to Sofia Caires (see the acknowledgements), who did all the calculations for the long-term statistics in this book.

NOTE 4G   Seasonal variation removed from the long-term distribution of the significant wave height (the initial-distribution approach)

The fit of the *log-normal* distribution to the observations of Fig. 4.20 is given in the left-hand illustration below. This fit is reasonable for the middle section of the observed significant wave heights (1.5 m < H_s < 7.5 m, say) but rather poor for the lower and higher sections. When the seasonal variation is removed from the time series of H_s (by scaling the values of H_s in the time series with a harmonic component with a period of 365 days, resulting in the *scaled* significant wave height H_s^*), then the fit of the log-normal distribution improves markedly for most of the observations, but it is still not good for the high section (ln H_s^* > 1.2):

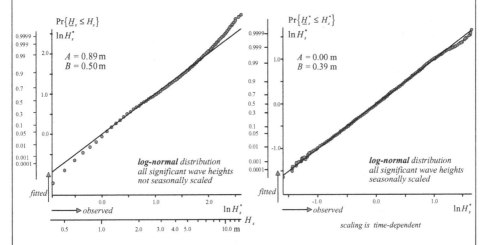

The long-term, log-normal distribution of the significant wave height (Q–Q plots), for the years 1980–2003 for NODC buoy 46005 of Fig. 4.17. Left-hand panel: no seasonal scaling; right-hand panel: seasonally scaled values. The straight lines represent the best-fit (maximum-likelihood) candidate distributions.

The extrapolation of the long-term distribution provides the probability that an (unobserved) high value of the significant wave height is exceeded. It does not indicate *when* such an event will happen. That of course is unpredictable (in the long-term), but with some extra information it is possible to determine *how often* it will happen. In many engineering design procedures this is expressed in terms of a *return period*, i.e., the average time interval between occurrences of an extreme significant wave height, or, better stated: the average time interval between successive up crossings of the significant wave height through a chosen level (see Fig. 4.21). This return period can be estimated from the long-term cumulative distribution function $P(H_s) = \mathrm{Pr}\{\underline{H}_s \leq H_s\}$, if also the average duration of exceedance per event is known (an event, corresponding to a storm, is defined as a series of consecutive

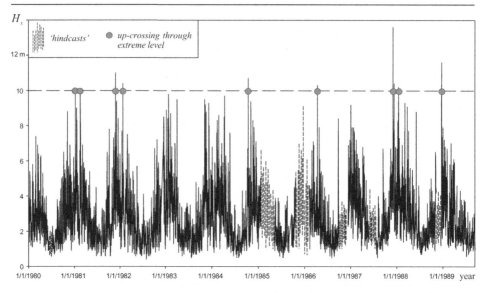

Figure 4.21 The up-crossing of the significant wave height through a high level (observations of Fig. 4.17, supplemented with an artists impression of hindcast results).

values of H_s that are all above a chosen level, preceded and followed by lower values; the duration of an event is also known as its 'persistence'). To obtain this estimate of the return period, consider a long period of N years during which the significant wave height crosses the chosen level n times (in the upward direction only).

The average time interval between these up crossings is, by definition, the return period $RP_{\underline{H}_s > H_s} \approx N/n$ years. For estimating the number of these up-crossings n, we first need to interpret the probability of exceedance $Pr\{\underline{H}_s > H_s\}$ as the fraction of time during which $\underline{H}_s > H_s$. For instance, if the probability of exceedance of the level 10 m is 0.001 83, then the total duration D during which the significant wave height exceeds this level of 10 m is $D = 16$ h per year on average (i.e., averaged over many years). If the average duration per *event* $(\overline{d}_{\underline{H}_s > 10\,m})$ is 8 h, then, obviously, the up crossing through the level of 10 m occurs twice per year (on average). This frequency of occurrence (number of occurrences per year) is apparently determined from $0.001\,83 \times 24 \times 365/8$ (year^{-1}), when $\overline{d}_{\underline{H}_s > 10\,m}$ is expressed in *hours*. The return period is the inverse of this frequency, $RP_{\underline{H}_s > 10\,m} = 8/(0.001\,83 \times 24 \times 365)$ $= 4371$ h, about half a year. Expressed analytically:

$$RP_{\underline{H}_s > H_s} = \frac{\overline{d}_{\underline{H}_s > H_s}}{Pr\{\underline{H}_s > H_s\} \times 24 \times 365}\ \text{year}$$

when $\overline{d}_{\underline{H}_s > H_s}$ is in hours, with the initial-distribution approach

$$(4.3.3)$$

or

$$\mathrm{RP}_{\underline{H}_s>H_s} = \frac{\overline{d}_{\underline{H}_s>H_s}}{\mathrm{Pr}\{\underline{H}_s > H_s\}} = \frac{\overline{d}_{\underline{H}_s>H_s}}{1 - P(H_s)} \quad (\text{unit of } \overline{d}_{\underline{H}_s>H_s})$$

$$\text{initial-distribution approach} \qquad (4.3.4)$$

where $P(H_s)$ is the cumulative distribution function: $P(H_s) = \mathrm{Pr}\{\underline{H}_s < H_s\}$. This estimation of the return period requires information about $\overline{d}_{\underline{H}_s>H_s}$, which can be obtained only from observed or simulated time series of H_s. Strangely enough, the return period is sometimes estimated as $\mathrm{RP}_{\underline{H}_s>H_s} = \Delta t_{H_s} / [1 - P(H_s)]$, where Δt_{H_s} is the time interval between the observations of H_s, typically 3 h, or even as $\mathrm{RP}_{\underline{H}_s>H_s} = 1/[1 - P(H_s)]$, which implies that $\overline{d}_{\underline{H}_s>H_s}$ would be unity (the dimension and unit of this estimate of $\mathrm{RP}_{\underline{H}_s>H_s}$ being mysterious because it has the same dimension and unit as $\overline{d}_{\underline{H}_s>H_s}$, which has been replaced by unity). Needless to say, the return period thus estimated is seriously wrong or even nonsensical.

The return period of *calms*, i.e., the periods during which $\underline{H}_s < H_s$, which is important for activities such as towing an offshore platform to its location of operation, is estimated similarly:

$$\mathrm{RP}_{\underline{H}_s<H_s} = \frac{\overline{d}_{\underline{H}_s<H_s}}{\mathrm{Pr}\{\underline{H}_s < H_s\}} = \frac{\overline{d}_{\underline{H}_s<H_s}}{P(H_s)} \quad (\text{unit of } \overline{d}_{\underline{H}_s<H_s})$$

$$\textit{calms} \text{ in the initial-distribution approach} \qquad (4.3.5)$$

Literature:
Battjes (1972a), Gerson (1975), Goda (1992, 1988b), Graham (1982), Gringorten (1963), Kuwashima and Hogben (1986), Ochi (1992), Salih *et al.* (1988), Tucker and Pitt (2001).

4.3.2 The peak-over-threshold approach

The statistics of extreme values of the significant wave height can also be estimated with another approach than the above. In the peak-over-threshold (POT) approach considered here, only the maximum value of H_s in each of a large number of *storms* is considered (see Fig. 4.22). A storm is defined here as an uninterrupted sequence of values of H_s all exceeding a certain, fairly high value (threshold value $H_{s,threshold}$), preceded and followed by a lower value.[8] The value to be chosen for this threshold depends very much on the local conditions. For severe climates, a threshold value of 5 m may be needed, whereas for calm climates, a value of 1 m may be better suited. The criterion is that a sufficient number of storms (preferably several dozen or more) can be identified in the long-term time record. For each such storm the *maximum* significant wave height is then identified as the highest (i.e., peak) value in that storm: $H_{s,peak}$ (the peak over threshold).

---

[8] Sometimes small gaps between such storms are ignored, to avoid breaking up a phenomenon that obviously is one storm, seen from a meteorological point of view.

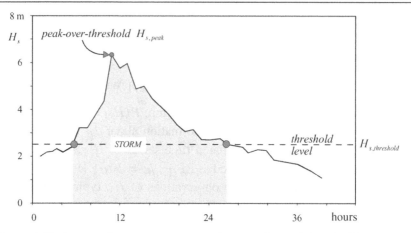

Figure 4.22 A storm between two successive crossings of the significant wave height through a threshold level.

The extreme-value theory (e.g., Castillo, 1988; Coles, 2001) tells us that the distribution of the maximum in such a sequence of values above a threshold is the generalised Pareto distribution (see Note 4H). In other words, the maximum significant wave height in a storm should be Pareto distributed (under certain conditions, e.g., the values must be independent and identically distributed (i.i.d.) and the threshold value must be relatively high; see Fig. 4.23).[9]

This POT approach has two important advantages over the initial-distribution approach treated in the previous section: (a) if the wave climate contains more than one distribution due to the occurrence of different physical regimes, selection of only the high values of the significant wave height tends to concentrate the analysis on the regime that dominates the (high) extremes; and (b) the storms are statistically independent events, providing a more solid theoretical base and simplifying the interpretation of the results of the analysis (e.g., estimating the sample errors involved).

Once the parameters of the distribution of $\underline{H}_{s,peak}$ have been determined by fitting the distribution to the data, an estimate of the return period $RP_{\underline{H}_{s,peak}}$ can be made. This return period is defined in analogy with the return period in the initial-distribution approach: it is the average time interval between *storms* during which $\underline{H}_{s,peak} > H_{s,peak}$. To introduce the estimation of this return period, consider, in a long-term time series, all storms for which $H_s > H_{s,threshold}$ (=4 m, say) and suppose that the peak value in these storms exceeds the level of 9 m with a probability of $\Pr\{\underline{H}_{s,peak} > 9\,\text{m}\}_{threshold=4\,m} = 0.005$. It then follows that one out of every 200

---

[9] Stated differently: the distribution of the maximum of a large set of *independent and identically distributed* (i.i.d.) random variables $\underline{x}_i$ larger than some threshold value $x_{threshold}$ is a generalised Pareto distribution (GPD), or, in mathematical terms, the (convergence) theorem is $max\{\underline{x}_1, \ldots, \underline{x}_n | \underline{x}_i > x_{threshold}\} \to$ GPD for $n \to \infty$.

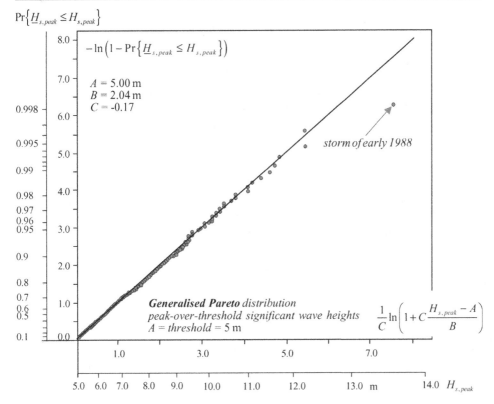

Figure 4.23 The long-term generalised Pareto distribution of the peak-over-threshold significant wave height for threshold value $A = H_{s,threshold} = 5\,\mathrm{m}$ (resulting in 24 storms per year on average), for the years 1980–2003 for NODC buoy 46005 of Fig. 4.17. The straight line represents the best-fit candidate distribution (maximum likelihood; $C = 0$ would correspond to an exponential distribution; the indicated value of $C = -0.17$, estimated with a standard deviation of 0.04, is so far from zero that the observed values are probably not taken from such an exponential distribution). The position of the exceptional storm of early 1988 is obvious.

storms has such a peak value (a *severe* storm). Suppose that the average interval between all storms (defined by $H_s > 4$ m) is $\Delta T_{storm} = 16$ weeks, then the average interval between *severe* storms (i.e., with $H_{s,peak} > 9$ m) is 200×16 weeks ≈ 60 years. Expressed analytically:

$$RP_{\underline{H}_{s,peak} > H_{s,peak}} = \frac{\Delta T_{storm}}{\mathrm{Pr}\{\underline{H}_{s,peak} > H_{s,peak}\}_{threshold}} = \frac{\Delta T_{storm}}{1 - P(H_{s,peak})_{threshold}}$$

(units of ΔT_{storm}) the POT approach (4.3.6)

where $P(H_{s,peak})_{threshold}$ is the cumulative distribution function: $P(H_{s,peak})_{threshold} = \mathrm{Pr}\{\underline{H}_{s,peak} < H_{s,peak}\}_{threshold}$. One advantage of the POT approach is that it has some theoretical basis – the extreme-value theory – and that only storms with the significant wave height higher than the threshold value need to be considered (thus reducing any numerical simulation efforts to these storms only). This approach therefore seems to be an attractive compromise between, on the one hand, the initial-distribution approach, in which a very large number of values of the significant wave height is used but which has no theoretical basis, and on the other hand, the annual-maximum approach (to be treated next) which also has the theoretical support of the extreme-value theory, but for which usually only a small number of observations is available (equal to the number of years in the long-term time record).

NOTE 4H Long-term distribution for the maximum significant wave height per storm (the peak-over-threshold approach)

The *generalised Pareto distribution* is given by

$$\Pr\{\underline{H}_{s,peak} \leq H_{s,peak}\}_{threshold} = 1 - \left(1 + C\frac{H_{s,peak} - A}{B}\right)^{-1/C}$$

$$\text{for} \quad \underline{H}_{s,peak} \geq A \qquad\qquad \text{if} \quad C > 0$$

$$\text{for} \quad A \leq \underline{H}_{s,peak} \leq A - B/C \quad \text{if} \quad C < 0$$

The parameter A is the threshold value $A = H_{s,threshold}$. The parameter B (>0) provides a normalisation (scaling) and the parameter C is a shape parameter. For $C \rightarrow 0$, the distribution reduces to a shifted *exponential* distribution:

$$\Pr\{\underline{H}_{s,peak} \leq H_{s,peak}\}_{threshold} = 1 - \exp\left(-\frac{H_{s,peak} - A}{B}\right) \qquad \text{for } \underline{H}_{s,peak} > A$$

Literature:
Caires and Sterl (2003, 2005), Ferreira and Soares (1998), Goda (1992), van Gelder and Vrijling (1999).

4.3.3 *The annual-maximum approach*

Occasionally, another approach, the annual-maximum approach, is used. Consider a population of random values (its distribution is called the parent distribution) from which a set of samples is arbitrarily drawn. The extreme-value theory tells us that (a), under fairly general conditions, the distribution of the *maximum* of that set is the generalised extreme-value (GEV) distribution;[10] and (b), if the parent distribution is a Weibull or log-normal distribution, the GEV distribution of this

---

[10] The distribution of the maximum of a large set of *independent and identically distributed* (i.i.d.) random variables is a generalised extreme-value distribution. Stated in mathematical terms: the convergence theorem is that $max\{x_1, \ldots, x_n\} \rightarrow$ GEV for $n \rightarrow \infty$.

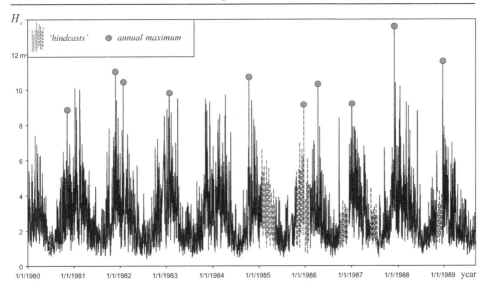

Figure 4.24 The annual maxima in a long-term time record of the significant wave height (observations as in Fig. 4.17, supplemented with an artist's impression of hindcast results).

maximum reduces to a Gumbel distribution[11] (e.g., Castillo, 1988; Coles, 2001; see Note 4I). To use these theoretical findings in a wave-climate analysis, consider the original population (the parent distribution) to be the significant wave height over many years and the set of samples to be *one year* of these significant wave heights. The maximum of this sample set is then the maximum significant wave height per year $H_{s,AM}$. A time series of N years thus gives N values of $H_{s,AM}$ (see Fig. 4.24). Since the parent distribution of the significant wave height is often close to a Weibull or log-normal distribution (see Section 4.3.2), it follows that $H_{s,AM}$ should be (nearly) Gumbel distributed (see Fig. 4.25). The parameters of this GEV distribution can be estimated from the observed values of $H_{s,AM}$ with any of the methods mentioned earlier in Section 4.3.2.

To introduce the estimation of the *return period* in this approach, consider a situation in which the probability of $H_{s,AM}$ exceeding the level of 7.5 m is 0.02. This exceedance then occurs (on average) twice every hundred samples. Since one sample corresponds to one year, the exceedance occurs twice in every hundred years, or once in every fifty years. The return period $\mathrm{RP}_{\underline{H}_{s,AM}>H_{s,AM}}$ can apparently be estimated with

$$\mathrm{RP}_{\underline{H}_{s,AM}>H_{s,AM}} = \frac{1}{\mathrm{Pr}\{\underline{H}_{s,AM} > H_{s,AM}\}} \quad \text{(year)} \qquad \text{annual-maximum approach}$$
$$(4.3.7)$$

---

[11] Statisticians say that, under these conditions, the Gumbel distribution is the *domain of attraction* both of the Weibull and of the log-normal distribution (e.g., Castillo, 1988, p. 120).

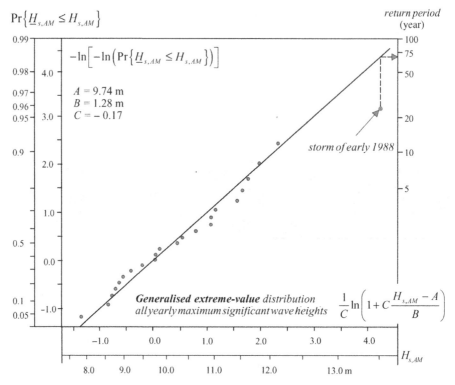

Figure 4.25 The long-term generalised-extreme-value (GEV) distribution of the annual maximum significant wave height, for the years 1980–2003 for NODC buoy 46005 of Fig. 4.17. The straight line represents the best-fit candidate distribution (maximum likelihood); $C = 0$ would correspond to a Gumbel distribution; the standard deviation of the estimated value of C is 0.12, so the value of $C = 0$ is located in the 95% confidence interval of $C = -0.17$. The position of the exceptional storm of early 1988 is obvious: the fitted distribution (i.e., the solid straight line) suggests that the value of 13.6 m in this storm would occur only once in 70 years (on average), as shown by the vertical dashed line.

or

$$RP_{\underline{H}_{s,AM} > H_{s,AM}} = \frac{1}{1 - P(H_{s,AM})} \text{ (years)} \qquad \text{annual-maximum approach}$$

(4.3.8)

where $P(H_{s,AM})$ is the cumulative distribution function $P(H_{s,AM}) = \Pr\{\underline{H}_{s,AM} < H_{s,AM}\}$. One advantage of this approach is that basic support is provided by the extreme-value theory and that only the highest value in a year needs to be considered (thus reducing any numerical simulation effort to only a few storms). A serious disadvantage is that generally not enough years of observations or hindcasts are available to estimate the parameters of the distribution and hence the return period with reasonable reliability.

Literature:
Goda *et al.* (1993), Goda (2000), Mathiesen *et al.* (1994).

NOTE 4I   Long-term distribution for the annual maximum significant wave height (the annual-maximum approach)

The *generalised extreme-value (GEV) distribution* is given by

$$\Pr\{\underline{H}_{s,AM} \leq H_{s,AM}\} = \exp\left[-\left(1 + C\frac{H_{s,AM} - A}{B}\right)^{-1/C}\right] \qquad \text{for } B > 0$$

For $C < 0$, this distribution is also known as the *Weibull* distribution (for maxima, with an upper bound $H_{s,AM} \leq A - B/C$), which is usually written as

$$\Pr\{\underline{H}_s \leq H_s\} = \exp\left[-\left(-\frac{H_s - A^*}{B^*}\right)^{C^*}\right]$$

with $A^* = A - B^*$, $B^* = -B/C$ and $C^* = -1/C$. For $C > 0$ this distribution is also known as the *Fréchet* distribution or *Fisher–Tippett II* distribution (with a lower bound $H_{s,AM} \geq A - B/C$). For $C \to 0$ (which is often the case at sea), the distribution reduces to the *Gumbel* distribution or *Fisher–Tippett I* distribution:

$$\Pr\{\underline{H}_{s,AM} \leq H_{s,AM}\} = \exp\left[-\exp\left(-\frac{H_{s,AM} - A}{B}\right)\right] \qquad \text{for } B > 0$$

The parameter A is a location parameter (the position of the distribution on the H_s-axis). The parameter B provides a normalisation (scaling) and the parameter C is a shape parameter.

4.3.4 Individual wave height

The *long-term* distribution of the *individual* wave height can be determined from a combination of the short-term statistics and the long-term statistics (the Battjes method, see Battjes, 1972a, and Tucker and Pitt, 2001). As shown in Section 4.2.2, the short-term distribution of the individual wave height (in deep water) is usually a Rayleigh distribution, with only the significant wave height as parameter. In the long term, the conditions are not stationary and the significant wave height varies with time. To determine the long-term statistical properties of the individual wave height we must therefore account for this (random) variation of the significant wave height.

Consider a duration D, during which the wave condition is stationary (the short term). The number of waves $N_{\underline{H}>H, stationary}$ with a wave height $\underline{H}$ that is higher than a certain level H, is then given by

$$N_{\underline{H}>H, stationary} = N_{total} \cdot \Pr\{\underline{H} > H\} \tag{4.3.9}$$

where the total number of individual waves $N_{total} = D/\overline{T}_0 = D\,\overline{T}_0^{-1}$ (in which $\overline{T}_0$ is the mean zero-crossing wave period) and the probability that the wave height $\underline{H}$ is higher than H is given by the Rayeigh distribution $\Pr\{\underline{H} > H\} = \exp[-2(H/H_s)^2]$, so that

$$N_{\underline{H}>H, stationary} = D\,\overline{T}_0^{-1} \exp[-2(H/H_s)^2] \tag{4.3.10}$$

If the wave conditions are *not* stationary, then the total number of waves during the duration D can be estimated with the long-term average (expected value) of $D\overline{T}_0^{-1}$

$$N_{total, non\text{-}stationary} = \int_0^\infty D\,\overline{T}_0^{-1}\, p(\overline{T}_0)\mathrm{d}\overline{T}_0 \tag{4.3.11}$$

The total number of waves with a wave height above H under these conditions $N_{\underline{H}>H, non\text{-}stationary}$ can be estimated with Eq. (4.3.9) as the average (expected value) of $N_{total} \cdot \Pr\{\underline{H} > H\}$ over that duration:

$$N_{\underline{H}>H, non\text{-}stationary} = E\{N_{total} \cdot \Pr\{\underline{H} > H\}\} \tag{4.3.12}$$

Since $N_{total} \cdot \Pr\{\underline{H} > H\}$ in this expression depends on the mean zero-crossing wave period $\overline{T}_0$ and on the significant wave height H_s, estimating $N_{\underline{H}>H, non\text{-}stationary}$ requires the long-term joint probability density function of H_s and $\overline{T}_0$ (compare with Eq. 4.3.10):

$$N_{\underline{H}>H, non\text{-}stationary} = \int_0^\infty \int_0^\infty D\,\overline{T}_0^{-1} \exp\left[-2(H/H_s)^2\right] P(H_s, \overline{T}_0)\mathrm{d}H_s\, \mathrm{d}\overline{T}_0 \tag{4.3.13}$$

The distribution of the individual wave height can then be determined as the relative number of high waves (from Eqs. 4.3.11 and 4.3.13):

$$\Pr\{\underline{H} > H\}_{non\text{-}stationary} = \frac{N_{\underline{H}>H, non\text{-}stationary}}{N_{total, non\text{-}stationary}}$$

$$= \frac{\displaystyle\int_0^\infty \int_0^\infty D\,\overline{T}_0^{-1} \exp[-2(H/H_s)^2] P(H_s, \overline{T}_0)\mathrm{d}H_s\, \mathrm{d}\overline{T}_0}{\displaystyle\int_0^\infty D\,\overline{T}_0^{-1}\, p(\overline{T}_0)\mathrm{d}\overline{T}_0} \tag{4.3.14}$$

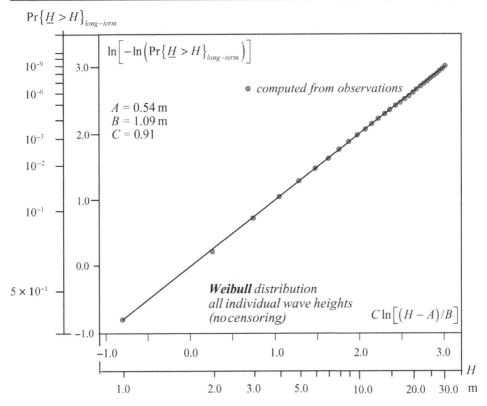

Figure 4.26 The *long-term* distribution of the *individual* wave height computed with the Battjes method, for the years 1980–2003 for NODC buoy 46005 of Figs. 4.17 and 4.18 (not censored) and plotted with Weibull scales. The straight line represents the best-fit candidate distribution (least-squares). The value of the shape parameter $C = 0.91$, reduces this Weibull distribution to nearly an exponential distribution (for which $C = 1$).

The duration itself is immaterial since it appears as a constant in the denominator and the numerator and this result can be applied to long-term situations:

$$\Pr\{\underline{H} > H\}_{long\text{-}term} = \frac{\int_0^\infty \int_0^\infty \overline{T}_0^{-1} \exp[-2(H/H_s)^2] P(H_s, \overline{T}_0) dH_s \, d\overline{T}_0}{\int_0^\infty \overline{T}_0^{-1} p(\overline{T}_0) d\overline{T}_0}$$

(4.3.15)

The distribution of the individual wave height H thus obtained from observed joint distributions $P(H_s, \overline{T}_0)$ turns out to be very close to a Weibull distribution (see Note 4F and Fig. 4.26) with the value of the shape parameter $C \approx 1$ (Battjes, 1972a), which reduces it to an exponential distribution.

NOTE 4J   Long-term distributions for the *individual* wave height H

The *Weibull distribution* (it differs from the one in Note 4F only in that H_s has been replaced with H) is given by

$$\Pr\{\underline{H} \le H\}_{long\text{-}term} \begin{cases} = 1 - \exp\left[-\left(\dfrac{H-A}{B}\right)^C\right] & \text{for} \quad H > A \quad \text{and} \quad C > 0 \\ = 0 & \text{for} \quad H \le A \end{cases}$$

The parameter A is a location parameter (the position of the distribution on the H-axis). In the Weibull distribution this parameter also represents the lower limit of the wave height (a permanent minimum background sea). The parameter B provides a normalisation (scaling), which determines the width of the distribution. The parameter C is a shape parameter. For $A = 0$, the Weibull distribution is called the two-parameter Weibull distribution. For $C = 1$, it reduces to an *exponential distribution* and for $C = 2$ to a *Rayleigh distribution*.

To estimate the long-term return period of the *individual* wave height, we interpret the probability of exceedance $\Pr\{\underline{H} > H\}_{long\text{-}term}$ as the fraction of waves higher than H. For instance, if the probability of exceeding the value of 20 m is 10^{-6}, then one out of 1 000 000 waves is higher than 20 m. The number of waves between such high waves would therefore be 1 000 000 on average. This number of waves can be converted to a time interval with the mean wave period that was used in the analysis, giving the return period $\mathrm{RP}_{\underline{H}>H}$ for the individual wave height:

$$\mathrm{RP}_{\underline{H}>H} = \frac{E\{\overline{T}_0\}}{\Pr\{\underline{H} > H\}_{long\text{-}term}} \quad \text{(unit of } \overline{T}_0\text{)}$$

long-term return period of individual wave height   (4.3.16)

where $E\{\overline{T}_0\} = \int_0^\infty \overline{T}_0 p(\overline{T}_0) d\overline{T}_0$ is the long-term average value of the mean zero-crossing period. Since the value of exceedance of 30 m (!) for the individual wave height in the data set of Fig. 4.26 is 10^{-9} (the NODC buoy 46005 in the North Pacific Ocean) and the average zero-crossing period for this data set is 6.8 s (see Fig. 4.18), we find with Eq. (4.3.16) that the return period of this extreme individual wave height is approximately 200 years. Similarly, an individual wave height of more than 25 m would occur about twice a year. Obviously, the location of this buoy is a most interesting place for a marine architect or ocean engineer.

Literature:
Battjes (1972a), Tucker and Pitt (2001).

4.3.5 Wave atlases

Long-term wave statistics are often based on wave observations carried out in the context of routine observation programmes of government agencies or private industry, using buoys, wave poles or ocean-looking satellites. Some of these observations are used only for operational purposes and are never stored, or are stored only for a short period of time. Other observation programmes are specifically aimed at acquiring and storing data to provide a basis for estimating long-term wave statistics. An alternative to these long-term observation programmes is provided by computer simulations. Such simulations (called 'hindcasts'; see Chapters 6 and 8) are based on *archived wind fields* that are available at national or international meteorological institutes. These hindcasts can often provide wave information over much longer periods than observations can, because meteorological (wind) archives are generally much older than wave archives. Moreover, such hindcast studies can be carried out within one or two years whereas an equivalent observation programme would require dozens of years.

The results of the statistical analysis, usually *histograms* that represent long-term probability distribution functions (see Fig. 4.18), are sometimes published as wave atlases. Some cover all the world's oceans; others cover only selected regions. Examples of such atlases are the following:

those based on (visual) observations:
- *Ocean Wave Statistics* (Hogben and Lumb, 1967)
- *Global Wave Statistics* (Hogben *et al.*, 1985)
- *Wind and Wave Climate in the Netherlands Sector of the North Sea between 53° and 54° North Latitude*(Bouws, 1978)

one based on satellite observations:
- *Atlas of the Oceans: Wind and Wave Climate* (Young and Holland, 1996, 1998),

those based on hindcasts:
- *Marine Climatic Atlas of the World* (U.S. Navy, 1974 etc.)
- *Navy Hindcast Spectral Ocean Wave Model Climatic Atlas: North Atlantic Ocean* (US Navy, 1983),
- *Statistical Database of Winds and Waves around Japan* (NMRI, 2005),
- *European Wave Energy Atlas* (WERATLAS, Pontes *et al.*, 1996)
- *Statistica delle onde estreme Mare Tirreno* (Tosi *et al.*, 1984)
- *Medatlas* (Stefanakos *et al.*, 2004a, 2004b)

and those based on mixed sources:
- *Wind and Wave Climate Atlas of Canada* (MacLaren Plansearch Limited, 1991)
- *World Wave Atlas* (Barstow, 1996)

Occasionally the actual time series of the significant wave height are published on the Internet (see Fig. 4.17 for an example).

5

Linear wave theory (oceanic waters)

5.1 Key concepts

- In this book, *oceanic* waters are deep waters (such that the waves are unaffected by the seabed) with straight or gently curving coastlines, without currents or obstacles such as islands, headlands and breakwaters. In anticipation of the treatment of waves in coastal waters (see Chapter 7), a limited but constant water depth (i.e., a horizontal bottom) is also considered here.
- The *linear theory* of surface gravity waves that is considered here applies, strictly speaking, only to water with idealised physical properties and motions and with *gravitation* as the only external force.
- This theory (also known as the *Airy wave theory*) is based on only two equations: a *mass* balance equation and a *momentum* balance equation. Both can be expressed in terms of an auxiliary function ϕ (the velocity potential function). This results in the Laplace and Bernoulli equations.
- For certain (linearised) kinematic boundary conditions, a propagating harmonic wave with *constant* and relatively *small* amplitude is one of the solutions of the Laplace equation. This wave is the basic component of the random-phase/amplitude model of Chapter 3.
  - For this harmonic wave, the Laplace equation provides expressions for the wave-induced *motions* of the water particles (velocity and path).
  - In combination with a dynamic boundary condition (free-wave condition), the Laplace equation also provides a relationship between wave period and wave length (the *dispersion relationship*). This in turn provides an expression for the propagation speed of the wave (the *phase speed*), which in turn provides an expression for the propagation speed of a group of waves (the *group velocity*).
- The Bernoulli equation, in linearised form, in combination with the above results of the Laplace equation, provides expressions for the wave-induced *pressure* in the water beneath the wave.
- Wave *energy* and its (horizontal) *transport* are nonlinear properties of the harmonic wave, which can be estimated with the above results of the linear wave theory. The propagation speed of wave energy turns out to be equal to the group velocity.
- Very short waves (wave length shorter than a few centimetres) are affected by surface tension (*capillary waves*).
- The linear theory should not be used for steep waves or waves in very shallow water. For these waves, nonlinear theories are available, such as the *Stokes wave theory, cnoidal wave theory* and the *stream-function theory*.
- The linear and nonlinear wave theories introduced here apply to waves with a constant profile in water with a constant depth (permanent waves). These are local theories in the sense that they can be used only to compute *local* wave characteristics (on the scale of one wave length or period). Computing the *evolution* of the waves over time or distance requires additional modelling (see Chapter 6 for oceanic waters and Chapters 7 and 8 for coastal waters).

5.2 Introduction

As shown in the previous chapters, *describing* random ocean waves is based on the notion of summing a large number of independent harmonic waves. *Understanding* random waves is correspondingly based on understanding these harmonic waves. This is possible with the linear theory for surface gravity waves, which describes in detail such harmonic waves. It is based on only two fundamental equations and some simple boundary conditions, describing certain kinematic and dynamic aspects of the waves. When these equations and boundary conditions are linearised, freely propagating, harmonic waves are solutions of these equations. This linear character implies that these waves do not affect one another while they travel together across the water surface, in perfect agreement with the basic assumption underlying the random-phase/amplitude model for random waves (see Chapter 3). The main requirement for the linear theory to apply is that the amplitudes of the waves are small, i.e., small compared with the wave length and small compared with the water depth. This is the *small-amplitude approximation*. This linear theory is also known as the *Airy wave theory*[1] (Airy, 1845) and the harmonic wave involved is therefore sometimes called the Airy wave. Nonlinear wave theories are briefly addressed in Section 5.6.

The linear theory of surface gravity waves has been the basic theory for ocean waves for about 150 years. It is presented in many books; usually with a scientific or engineering approach against the background of mathematics, physics, oceanography, ocean engineering or coastal engineering. The approach here is definitely that of an engineer. The theory is treated in two parts. The first part relates to waves in *oceanic* waters (this chapter), i.e., waters that are deep enough not to affect the waves, without currents or obstacles such as islands, headlands and breakwaters. In anticipation of the treatment of waves in coastal waters (Chapter 7), a finite water depth with a horizontal seabed is also considered.

Literature:
Barber (1969), CEM (2002), Crapper (1984), Dean and Dalrymple (1998), Dingemans (1997a, 1997b), Goda (2000), Herbich (1990), Kamphuis (2000), Kinsman (1965), Lamb (1932), Leblond and Mysak (1978), LeMéhauté (1976), Lighthill (1978), Massel (1996), Mei (1989), Mei *et al.* (2006), Phillips (1977), Rahman (1995), Sorensen (1993), Stoker (1957), Svendsen (2006), Tucker and Pitt (2001), Whitham (1974), Wiegel (1964), Young (1999).

5.3 Basic equations and boundary conditions

To develop the linear theory for surface gravity waves, the water is assumed to be an ideal fluid with only the Earth's *gravitation* inducing the forces that control the

---

[1] George Biddell Airy (1801–1892), was an English professor of mathematics and astronomy. He improved the theory of the orbital motions of Venus and the Moon. He also studied optics, e.g., astigmatism (the eye defect) and the rainbow.

motions of the water particles. Of course, water is not an ideal fluid, but we will tend to ignore this when applying the results of the theory to real ocean waves. This usually has no serious consequences (the results of the theory are surprisingly robust). However, in extreme situations some of the idealisations are violated too severely and the theory no longer applies (e.g., when waves are steep).

5.3.1 Idealisations of the water and its motions

As an ideal fluid, the water is assumed to be *incompressible*, to have a constant *density* (i.e., constant in space and time) and to have no *viscosity*. In addition, the water body must be *continuous*. The first of these conditions (incompressibility) seems to be reasonable, since the forces involved are so small that the corresponding compression of water can be ignored. With respect to the idealisation of constant density, the horizontal distances over which it normally varies in the ocean or in coastal waters (due to variations in temperature and salinity; usually over dozens of kilometres or more) are much larger than the scales at which the linear theory is applied (usually over a distance of only a few wave lengths; at least with constant parameters). Locally, the density and viscosity can therefore be considered to be constant (horizontally). The vertical variations are usually also ignored, but, in river estuaries, with salt water from the sea moving upriver during an incoming tide, beneath the fresh water moving downriver, the vertical variations may be relevant. In such cases the linear theory, as presented here, should be applied only with due caution. Time variations in density and viscosity are usually so slow that these too can be ignored. The effect of (viscosity-induced) *internal forces* is usually negligible for the wave lengths considered. The condition of continuity of the water seems to be a strange condition because water is normally quite continuous. However, water may contain discontinuities in the form of air bubbles. When this happens to any significant degree, e.g., when waves break, the linear theory does not apply (the waves would probably be too steep for the linear theory to apply anyway).

The next assumed idealisations relate to the *motion* of the water particles. Water particles may neither leave the *surface* nor penetrate the (fixed) *bottom*. A porous or moving bottom is therefore not admissible in the conventional linear wave theory (as treated here), but it is admissible in other versions of the linear wave theory (e.g., Dean and Dalrymple, 1984).

The water should be subjected to only one external force: *gravitation*. Wind-induced pressure is therefore excluded and wave generation by wind is not part of the linear wave theory (it will be treated separately in Chapter 6). Excluding another external force, *surface tension*, implies that the waves in the linear wave theory should be longer than a few centimetres, say. However, this force is readily included in the linear wave theory. Excluding another external force, the *Coriolis* force (acceleration), implies that the waves should be shorter than a few kilometres, say. The wave length of surface gravity waves is therefore limited to the range of a few centimetres to a few kilometres. Excluding *bottom friction* is not a serious limitation for the linear wave theory because its *local* effect (the generation of turbulence) is not transported into the main water body anyway.[2] The large-scale effect of bottom friction (energy dissipation) is treated in Chapter 8.

For the theory to be linear, certain aspects of the wave kinematics (*motion*) and wave dynamics (*forces*) need to be neglected. For this, it is enough that the amplitude of the waves is small, relative to the wave length and to the water depth. Normally the linearisation is introduced after the basic equations have been treated with the nonlinear terms still in place. This leads to some tedious reading, which can be avoided by introducing the linearisation much earlier. I have done this so as to present simpler equations and shorter explanations. For readers who are more inquisitive, I have given the conventional treatment in Appendix B.

The linear wave theory gives the *harmonic wave* as its most interesting result. The corresponding analytical expressions for the particle velocities and wave-induced pressure in the water are found with an elegant mathematical technique that uses a rather abstract mathematical concept. It is the *velocity potential function*, which is a scalar function representing the particle velocities in the water. The use of this function requires the motion of the water particles to be *irrotational* (the particles may not rotate around their own axis). The concepts of rotation (for fluid motions also called vorticity) and the velocity potential function are explained in Appendix B. The assumption that the motion of the water particles is irrotational is reasonable because, in the present context, vorticity can be generated only by turbulence at the bottom, and, as indicated above, this turbulence does not penetrate very far into the main water body. The expressions for particle velocities and wave-induced pressure are used to find expressions for other wave characteristics, such as phase speed and wave energy.

5.3.2 Balance equations

As usual in fluid mechanics, we will consider balance equations as the basis for the linear wave theory. Here, only two are needed: a *mass* balance equation and a *momentum* balance equation. Since the derivation for these equations is nearly identical, I will give only one general derivation for the balance equation of an

---

[2] Turbulence (in the absence of wave breaking) is continuously generated and dissipated near the bottom by the wave-induced motion of the water particles (if the water is sufficiently shallow). This turbulence cannot travel very far from the bottom because the (thin) layer of turbulence near the bottom that the wave-induced velocities have built up in a quarter of the wave cycle is destroyed in the following quarter of the cycle. It is built up again during the next quarter of the cycle, when the water motion turns back, but it is destroyed again when the cycle is completed.

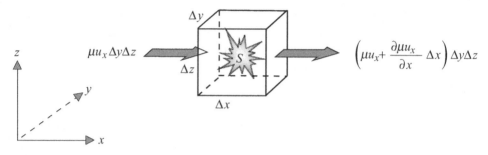

Figure 5.1 A property μ being transported by the water in the x-direction, through
a volume $\Delta x \Delta y \Delta z$ in a three-dimensional situation.

arbitrary property μ, which we can subsequently interpret as either mass density
or momentum density.

Consider a body of fluid in x, y, z-space (the orthogonal x- and y-axes form the
horizontal plane and the z-axis is directed vertically upwards; see Fig. 5.1). The fluid
transports some arbitrary, conservative property[3] through a volume $\Delta x \Delta y \Delta z$ (it
could be a scalar property, such as heat, or a vector property, such as the momentum
of the water itself). The property will be indicated by its density μ (i.e., per unit
volume).

The derivation of the balance equation essentially involves balancing the local
storage of the property μ in volume $\Delta x \Delta y \Delta z$ against the sum of inflow, outflow
and local production over a time interval Δt:

storage of μ during time interval Δt

　　 = net import of μ during time interval Δt

　　　　 + local production of μ during time interval Δt　　　　　　　　(5.3.1)

The storage term on the left-hand side is equal to the quantity of μ at the *end* of the
time interval, minus the quantity of μ at the *start* of the time interval:

storage of μ during time interval Δt

　　 = quantity at end of interval − quantity at start of interval

$$= \left(\mu \, \Delta x \, \Delta y \, \Delta z + \frac{\partial(\mu \, \Delta x \, \Delta y \, \Delta z)}{\partial t} \, \Delta t \right) - \mu \, \Delta x \, \Delta y \, \Delta z$$

$$= \frac{\partial \mu}{\partial t} \, \Delta x \, \Delta y \, \Delta z \, \Delta t \qquad\qquad\qquad\qquad (5.3.2)$$

The first term on the right-hand side of the balance equation Eq. (5.3.1) is the *net
import* of μ (during interval Δt). For the x-direction it is equal to the import in the
x-direction, through the left-hand side of the volume (with surface area $= \Delta y \, \Delta z$),

---

[3] A property that can change in quantity only by external factors.

minus the export in the x-direction, through the right-hand side of the volume. Assuming that μ is transported with the velocity of the water particles $\vec{u} = (u_x, u_y, u_z)$, the net import in the x-direction, with velocity component u_x, can then be written as

net import of μ in the x-direction during time interval Δt

$\qquad = import - export$

$\qquad = \mu u_x \, \Delta y \, \Delta z \, \Delta t - \left(\mu u_x + \dfrac{\partial \mu \, u_x}{\partial x} \Delta x \right) \Delta y \, \Delta z \, \Delta t$

$\qquad = -\dfrac{\partial \mu \, u_x}{\partial x} \Delta x \, \Delta y \, \Delta z \, \Delta t \qquad\qquad\qquad\qquad (5.3.3)$

The net imports in the y- and z-directions during interval Δt are similarly

net import of μ in the y-direction during time interval Δt

$\qquad = -\dfrac{\partial \mu \, u_y}{\partial y} \Delta x \, \Delta y \, \Delta z \, \Delta t \qquad\qquad\qquad\qquad (5.3.4)$

net import of μ in the z-direction during time interval Δt

$\qquad = -\dfrac{\partial \mu \, u_z}{\partial z} \Delta x \, \Delta y \, \Delta z \, \Delta t \qquad\qquad\qquad\qquad (5.3.5)$

The second term on the right-hand side of the balance equation Eq. (5.3.1) is the local production of μ in the volume, during interval Δt:

local production of μ during time interval Δt

$\qquad = S \, \Delta x \, \Delta y \, \Delta z \, \Delta t \qquad\qquad\qquad\qquad\qquad\qquad (5.3.6)$

where S is the production of μ per unit time, per unit volume. Substituting Eqs. (5.3.2)–(5.3.6) into Eq. (5.3.1) gives

$$\dfrac{\partial \mu}{\partial t} \Delta x \, \Delta y \, \Delta z \, \Delta t = -\dfrac{\partial \mu \, u_x}{\partial x} \Delta x \, \Delta y \, \Delta z \, \Delta t - \dfrac{\partial \mu \, u_y}{\partial y} \Delta x \, \Delta y \, \Delta z \, \Delta t$$

$$-\dfrac{\partial \mu \, u_z}{\partial z} \Delta x \, \Delta y \, \Delta z \, \Delta t + S \, \Delta x \, \Delta y \, \Delta z \, \Delta t \qquad (5.3.7)$$

Dividing all terms by $\Delta x \, \Delta y \, \Delta z \, \Delta t$ and moving the transport terms to the left-hand side gives the balance equation for μ *per unit volume, per unit time*:

$$\dfrac{\partial \mu}{\partial t} + \dfrac{\partial \mu \, u_x}{\partial x} + \dfrac{\partial \mu \, u_y}{\partial y} + \dfrac{\partial \mu \, u_z}{\partial z} = S \qquad\qquad\qquad (5.3.8)$$

The first term on the left-hand side represents the *local rate of change* of μ. The three following terms represent the effect of transportation and are called the *advective* terms (or sometimes *convective* terms, but convection usually refers to vertical transport only, at least in oceanography and meteorology). Lastly, the term on the right-hand side is called the *source* term, (or sometimes the *sink* term when it is negative). It represents the generation (or dissipation) of μ (per unit volume per unit time).

Mass balance and continuity equations

If we take the mass density of water as $\mu = \rho\,(\approx 1025\mathrm{kg/m^3}$ for sea water) and substitute this into Eq. (5.3.8), we obtain the *mass balance equation*:

$$\frac{\partial \rho}{\partial t} + \frac{\partial \rho u_x}{\partial x} + \frac{\partial \rho u_y}{\partial y} + \frac{\partial \rho u_z}{\partial z} = S_\rho \qquad (5.3.9)$$

Since we assume the mass density to be constant (i.e., all derivatives of ρ are zero) and we assume that there is no production of water (i.e., $S_\rho = 0$) this equation reduces to the following equation, which is known as the *continuity equation*:

$$\boxed{\frac{\partial u_x}{\partial x} + \frac{\partial u_y}{\partial y} + \frac{\partial u_z}{\partial z} = 0} \qquad \text{continuity equation} \qquad (5.3.10)$$

The continuity equation is thus derived from the mass balance equation, but mass or mass density *as such* has disappeared from the equation. It is a *linear* equation in terms of the water particle velocities u_x, u_y and u_z.

Momentum balance

If we want to obtain the *momentum* balance equation, we take μ as the momentum density of the water, which by definition is the mass density of water times the velocity of the water particles (a vector quantity), $\mu = \rho \vec{u} = (\rho u_x, \rho u_y, \rho u_z)$. This balance equation is therefore a vector equation. When it is written in terms of components, we need three component equations (one for each component of the vector). By substituting $\mu = \rho u_x$ into Eq. (5.3.8), we find the balance equation for the *x-component*:

$$\frac{\partial (\rho u_x)}{\partial t} + \frac{\partial u_x(\rho u_x)}{\partial x} + \frac{\partial u_y(\rho u_x)}{\partial y} + \frac{\partial u_z(\rho u_x)}{\partial z} = S_x \qquad (5.3.11)$$

where S_x is the production of momentum in the *x*-direction. Such production of momentum per unit time is by definition a force acting on the volume (i.e., the force per unit volume). (Remember that the second law of mechanics of Newton[4] states

---

[4] Sir Isaac Newton (1642–1727) was an English mathematician and physicist. He studied the refraction of light by a glass prism and discovered the gravitational force (and much more). It was he, who, as the story goes, upon seeing an apple fall from a tree, concluded that the motion of the apple and the motion of the Moon are governed by the same force.

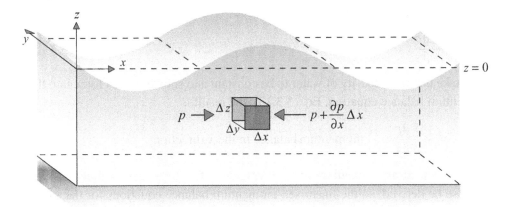

Figure 5.2 The horizontal pressure gradient in the water beneath a wave.

that $K\,\mathrm{d}t = \mathrm{d}(mv)$, or $K = \mathrm{d}(mv)/\mathrm{d}t$, where K is the force, m is the mass and v is the velocity of the body so that mv is momentum, and, consequently, *force is the rate of change of momentum.*) Equation (5.3.11) may therefore also be written as

$$\frac{\partial(\rho u_x)}{\partial t} + \frac{\partial u_x(\rho u_x)}{\partial x} + \frac{\partial u_y(\rho u_x)}{\partial y} + \frac{\partial u_z(\rho u_x)}{\partial z} = F_x \qquad (5.3.12)$$

where F_x is the body force in the x-direction per unit volume. The advective terms (the second, third and fourth terms on the left-hand side of Eq. 5.3.12) contain the velocities in quadratic combinations (nonlinear terms). They should therefore be removed to make the theory linear, so that the momentum balance equation Eq. (5.3.12) reduces to the *linearised* momentum balance equation:[5]

$$\frac{\partial(\rho u_x)}{\partial t} = F_x \qquad (5.3.13)$$

For the situation considered here, this horizontal force F_x is due solely to the horizontal pressure gradient $\partial p/\partial x$ in the water (see Fig. 5.2). The pressure is due to gravitation (see Eq. 5.4.33), which conforms to the condition that gravitation should be the only external force. The total horizontal force on the volume $\Delta x\,\Delta y\,\Delta z$ is equal to the pressure-induced force on the left-hand side of the volume *minus* the pressure-induced force on the right-hand side:

$$p\,\Delta y\,\Delta z - \left(p + \frac{\partial p}{\partial x}\,\Delta x\right)\Delta y\,\Delta z \qquad (5.3.14)$$

---

[5] In conventional presentations of the linear wave theory, the narrative continues from here while retaining these nonlinear terms (resulting in an equation that is called the Bernoulli equation for unsteady fluid motion). This conventional approach, in which these terms are removed at a later stage, is given in Appendix B.

Per unit volume this is (divide by $\Delta x \, \Delta y \, \Delta z$)

$$F_x = -\frac{\partial p}{\partial x} \qquad\qquad\qquad (5.3.15)$$

If we take the mass density of water to be constant and substitute this force into the momentum balance equation, Eq. (5.3.13), the result is

$$\frac{\partial u_x}{\partial t} = -\frac{1}{\rho}\frac{\partial p}{\partial x} \qquad \text{momentum balance in the } x\text{-direction} \qquad (5.3.16)$$

The corresponding momentum balance equations for the y- and z-directions can likewise be derived, so the linearised momentum balance equations for the x-, y- and z-directions are, respectively,

$$\boxed{\begin{aligned} \frac{\partial u_x}{\partial t} &= -\frac{1}{\rho}\frac{\partial p}{\partial x} \\[2mm] \frac{\partial u_y}{\partial t} &= -\frac{1}{\rho}\frac{\partial p}{\partial y} \\[2mm] \frac{\partial u_z}{\partial t} &= -\frac{1}{\rho}\frac{\partial p}{\partial z} - g \end{aligned}} \qquad \begin{array}{l}\text{linearised momentum balance}\\ \text{equations for the } x\text{-}, y\text{- and } z\text{-direction}\end{array} \qquad (5.3.17)$$

Note that the equation for the z-direction contains the term $-g$ and the other two equations do not. The reason is obvious: the weight of the volume, $\rho g \, \Delta x \, \Delta y \, \Delta z$, should be added as an external force in the z-direction (which was defined earlier as positive upwards so g appears with a minus sign).

5.3.3 Boundary conditions

To find expressions for such aspects as the propagation speed of the wave and the wave-induced pressure in the water, we must solve the above continuity equation and momentum balance equations for specific boundary conditions. These boundary conditions are of a kinematic nature (related to the motions of the water particles) and of a dynamic nature (related to forces acting on the water particles).

   The *lateral* boundaries i.e., at the up-wave side and down-wave side of the x-domain, will be controlled by the assumption that the wave is periodic with infinitely long crests in the y-direction. This reduces the wave to be described by this theory to a periodic two-dimensional wave (i.e., there exist only variations in the x- and z-directions; there is no variation in the y-direction). The remaining boundaries to be considered are the water surface and the bottom.

   At the water surface, the kinematic boundary condition is that particles may not leave the surface. In other words, the velocity of the water *particle* normal to the surface is equal to the speed of the *surface* in that direction. In the linearised approach, this is expressed as (for the nonlinear version see Appendix B)

$$u_z = \frac{\partial \eta}{\partial t} \qquad \text{at } z = 0 \tag{5.3.18}$$

where η is the surface elevation, measured vertically upwards from $z = 0$ (located in the still-water level; see Fig. 5.2). At the bottom, the kinematic boundary condition is that particles may not penetrate the (fixed, horizontal) bottom:

$$u_z = 0 \qquad \text{at } z = -d \tag{5.3.19}$$

To ensure that the wave is a free wave,[6] i.e., subject only to gravity, the (atmospheric) pressure at the water surface is constant (we will take it to be zero). This is the dynamic surface boundary condition:

$$p = 0 \qquad \text{at } z = 0 \tag{5.3.20}$$

5.3.4 The velocity potential function

Finding analytical solutions for the above balance equations and boundary conditions seems a daunting task, but mathematicians have found an elegant way to approach this problem. It requires the use of a rather abstract function, the velocity potential function $\phi = \phi(x, y, z, t)$, which is defined as a function of which the *spatial* derivatives are equal to the velocities of the water particles:

$$\phi(x, y, z, t) \quad defined \ such \ that \quad u_x = \frac{\partial \phi}{\partial x}, \ u_y = \frac{\partial \phi}{\partial y} \ and \ u_z = \frac{\partial \phi}{\partial z}$$

$$\tag{5.3.21}$$

but which can exist only if the motion of the water particles is irrotational (see Appendix B). If this is the case (and it generally is; see Section 5.3.1), we can write the continuity equation Eq. (5.3.10) in terms of this function ϕ by substituting the spatial derivatives of Eq. (5.3.21) into Eq. (5.3.10), giving

$$\frac{\partial^2 \phi}{\partial x^2} + \frac{\partial^2 \phi}{\partial y^2} + \frac{\partial^2 \phi}{\partial z^2} = 0 \qquad \text{from the continuity equation} \tag{5.3.22}$$

This equation is called the *Laplace equation*. Remember that it is derived from the mass balance equation, but mass or mass density *as such* has disappeared from the equation.

---

[6] In contrast to 'forced' water waves, which are affected by other external forces; for instance, waves generated by a corrugated metal sheet moving horizontally in the water surface.

The kinematic boundary conditions at the surface and at the bottom can also be expressed in terms of the velocity potential function (just substitute the spatial derivatives of Eq. 5.3.21 into Eqs. 5.3.18 and 5.3.19):

$$\boxed{\frac{\partial \phi}{\partial z} = \frac{\partial \eta}{\partial t}} \qquad \text{at } z = 0 \tag{5.3.23}$$

$$\boxed{\frac{\partial \phi}{\partial z} = 0} \qquad \text{at } z = -d \tag{5.3.24}$$

The three momentum balance equations Eq. (5.3.17) can also be expressed in terms of ϕ by substituting the spatial derivatives of Eq. (5.3.21) into these equations. For the momentum in the x-direction, the result is

$$\frac{\partial}{\partial t}\left(\frac{\partial \phi}{\partial x}\right) = -\frac{1}{\rho}\frac{\partial p}{\partial x} \tag{5.3.25}$$

Changing the order of differentiation and moving the term on the right-hand side to the left-hand side allows us to write this equation as

$$\frac{\partial}{\partial x}\left(\frac{\partial \phi}{\partial t}\right) + \frac{\partial}{\partial x}\left(\frac{p}{\rho}\right) = 0 \qquad \rightarrow \qquad \frac{\partial}{\partial x}\left(\frac{\partial \phi}{\partial t} + \frac{p}{\rho}\right) = 0 \tag{5.3.26}$$

We can add a term gz between the brackets without altering the meaning of the equation because this term would disappear when the derivative in the x-direction is taken. The other two momentum balance equations can be treated likewise (except that for the momentum balance in the z-direction the term gz does not disappear on taking the derivative in the z-direction; it represents gravitation, as in Eq. 5.3.17), with the result that

$$\frac{\partial}{\partial x}\left(\frac{\partial \phi}{\partial t} + \frac{p}{\rho} + gz\right) = 0$$

$$\frac{\partial}{\partial y}\left(\frac{\partial \phi}{\partial t} + \frac{p}{\rho} + gz\right) = 0 \tag{5.3.27}$$

$$\frac{\partial}{\partial z}\left(\frac{\partial \phi}{\partial t} + \frac{p}{\rho} + gz\right) = 0$$

The sum of terms between brackets appears in all three equations, expressing that this sum is not a function of x, y or z. It can therefore be only an (arbitrary) function of time t: $f(t)$, for which we take the simplest possible, $f(t) = 0$, so that

$$\boxed{\frac{\partial \phi}{\partial t} + \frac{p}{\rho} + gz = 0} \qquad \textit{from the momentum balance equations} \tag{5.3.28}$$

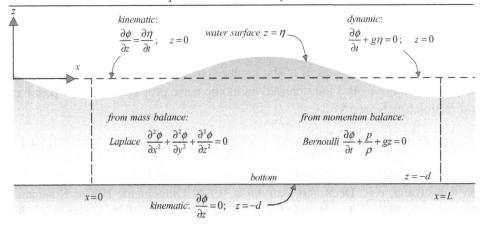

Figure 5.3 The (linearised) basic equations and boundary conditions for the linear wave theory, in terms of the velocity potential.

This is the *linearised Bernoulli[7] equation* for unsteady flow (for the nonlinear version, see Appendix B)

The dynamic surface boundary condition can, like the kinematic surface boundary condition, also be expressed in terms of the velocity potential. Taking the linearised Bernoulli equation at the surface $z = \eta$ (but in the linear approximation at $z = 0$), with $p = 0$ (see Eq. 5.3.20), gives

$$\boxed{\frac{\partial \phi}{\partial t} + g\eta = 0} \qquad \text{at } z = 0 \tag{5.3.29}$$

(note that we take the surface elevation η into account in the equation, but apply the boundary condition at $z = 0$). The above equations and boundary conditions are summarised in Fig. 5.3.

The *Laplace* equation and the *kinematic* boundary conditions will be used in the following to obtain the solution for the velocity potential and hence all *kinematic* aspects of the waves. The momentum balance equations and the dynamic boundary conditions are not required for this! The linearised *Bernoulli* equation and the linearised dynamic boundary condition will subsequently be used, in combination with the solution for the velocity potential, to obtain the expressions for some dynamic aspects of the waves.

---

[7] Daniel Bernoulli (1700–1782) was a Swiss scientist (born in the Netherlands) who started his career in medicine but, shortly after receiving his doctorate, published his first mathematical work applied to fluid dynamics. He won the Grand Prize of the Paris Academy ten times, for work on tides (jointly with Euler), magnetism, measuring time at sea, ocean currents, forces on ships and the pitch and roll of ships. See Kinsman (1965, p. 104) for a very brief but interesting family history.

5.4 Propagating harmonic wave

5.4.1 Introduction

One of the *analytical solutions* of the Laplace equation with the above kinematic boundary conditions is a long-crested harmonic wave propagating in the positive x-direction (see Note 5A):[8]

$$\boxed{\eta(x, t) = a \sin(\omega t - kx)} \tag{5.4.1}$$

with the following velocity potential function (see textbooks that give more details on this subject, e.g., Dean and Dalrymple, 1998; Dingemans, 1997a; Kinsman, 1965; Lamb, 1932; Leblond and Mysak, 1978; LeMéhauté, 1976; Lighthill, 1978; Massel, 1996; Mei, 1989; Mei *et al.*, 2006; Phillips, 1977):

$$\boxed{\phi = \hat{\phi} \cos(\omega t - kx) \qquad \text{with} \qquad \hat{\phi} = \frac{\omega a}{k} \frac{\cosh[k(d + z)]}{\sinh(kd)}} \tag{5.4.2}$$

This is the first time that the harmonic wave appears in the linear wave theory. Remember that the linear wave theory is based on the *small-amplitude* approximation (see Section 5.2), i.e., the amplitude of the wave should be small compared with the wave length and the water depth ($ak \ll 2\pi$ and $a \ll d$, respectively).

NOTE 5A   A propagating harmonic wave

The solution to the Laplace equation shown here is a cylindrical wave with constant wave height, propagating in the positive x-direction (i.e., a wave without variations normal to the direction of propagation and therefore having infinitely long crests). We will take the following *sine* wave as representing this wave (we might equally well have chosen a *cosine* wave; it would have made no difference in the results, except that everywhere *sin* is replaced with *cos*, which implies a 90° phase difference, which is immaterial here):

$$\eta(x, t) = \frac{H}{2} \sin\left(\frac{2\pi}{T} t - \frac{2\pi}{L} x\right)$$

where H is the wave height, T is the wave period and L is the wave length (see illustration below). It is usually more convenient to express the wave in terms of *amplitude* $a = H/2$, *radian frequency* $\omega = 2\pi/T$ and *wave number* $k = 2\pi/L$, so that the

---

[8] In the random-phase/amplitude model, I used the cosine representation for the harmonic wave, but, as I said earlier, in Section 3.6, it is often immaterial whether the sine or cosine representation is used.

propagating harmonic wave can be written as

$$\eta(x,t) = a\sin(\omega t - kx)$$

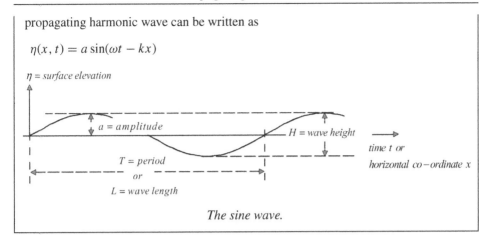

η = surface elevation

a = amplitude

H = wave height

time t or

horizontal co–ordinate x

T = period

or

L = wave length

The sine wave.

That this wave is a *propagating* wave (in the positive x-direction) is readily shown as follows. The propagation of a wave is best 'seen' by an observer riding at a crest of the wave. His forward speed, i.e. the speed of a fixed position in the moving surface *profile*, is by definition the forward speed of the wave, i.e., where the *phase* of the wave remains constant, or, expressed differently, where the time derivative of the phase $\omega t - kx$ is zero:

$$\frac{\partial(\omega t - kx)}{\partial t} = 0 \quad\text{or}\quad \frac{\partial(\omega t)}{\partial t} - \frac{\partial(kx)}{\partial x}\frac{dx}{dt} = 0 \quad\text{or}\quad \omega - k\frac{dx}{dt} = 0$$

$$(5.4.3)$$

where x is the position of the point with constant phase, so that the forward speed (called the *phase speed* for obvious reasons) $c = dx/dt$, from Eq. (5.4.3), is

$$c = \frac{dx}{dt} = \frac{\omega}{k} \qquad\qquad (5.4.4)$$

which, of course, is identical to the well-known expression for the propagation speed of harmonic waves in general, $c = L/T$.

5.4.2 Kinematics

No dynamic aspects of the wave were considered in the derivation of the velocity potential function ϕ of Eq. (5.4.2). It is solely based on the Laplace equation and the kinematic boundary conditions. This is quite remarkable, since it implies that all kinematic aspects, i.e., velocities, accelerations etc., can be derived (as will be shown next) without considering any dynamic aspects, i.e., without the Bernoulli equation or the dynamic surface boundary condition. The expression for the velocity potential function ϕ applies therefore both to free and to forced waves, as long as the surface wave is the harmonic wave of Eq. (5.4.1).

Particle velocity

The particle velocities can readily be obtained from the velocity potential ϕ, just by using the definition of ϕ: the spatial derivatives of ϕ are the velocity components $\partial\phi/\partial x = u_x$ and $\partial\phi/\partial z = u_z$, so that, from Eq. (5.4.2), the particle velocities are given by

$$u_x = \omega a \frac{\cosh[k(d+z)]}{\sinh(kd)} \sin(\omega t - kx) \qquad (5.4.5)$$

$$u_z = \omega a \frac{\sinh[k(d+z)]}{\sinh(kd)} \cos(\omega t - kx) \qquad (5.4.6)$$

or

$$\boxed{u_x = \hat{u}_x \sin(\omega t - kx) \qquad \text{with} \qquad \hat{u}_x = \omega a \frac{\cosh[k(d+z)]}{\sinh(kd)}} \qquad (5.4.7)$$

$$\boxed{u_z = \hat{u}_z \cos(\omega t - kx) \qquad \text{with} \qquad \hat{u}_z = \omega a \frac{\sinh[k(d+z)]}{\sinh(kd)}} \qquad (5.4.8)$$

The velocity in the y-direction is zero since the long-crested, harmonic wave is propagating in the (positive) x-direction. These velocities are called '*orbital velocities*' because they correspond to motion of the particles in closed, circular or elliptical orbits as shown for deep water in Fig. 5.4. Note that the velocities in the crest of the wave ($\eta > 0$) are always oriented in the down-wave direction of wave propagation and that the velocities in the trough of the wave ($\eta < 0$) are always oriented in the up-wave direction. This is very noticeable at sea: in the trough of the wave you are always pulled towards the crest that is approaching you and you are thrown back by the crest.

In *deep* water, i.e., when $kd \to \infty$, the expressions for the amplitudes of the velocity components $\hat{u}_x$ and $\hat{u}_z$ (Eqs. 5.4.7 and 5.4.8) reduce to

$$\hat{u}_x = \omega a e^{kz} \qquad \text{and} \qquad \hat{u}_z = \omega a e^{kz} \qquad \text{deep water} \qquad (5.4.9)$$

and the total velocity, i.e., the magnitude u, is independent of time (because $\hat{u}_x = \hat{u}_z$ and always $\sin^2(\omega t - kx) + \cos^2(\omega t - kx) = 1$ in Eqs. 5.4.7 and 5.4.8):

$$u = \sqrt{u_x^2 + u_z^2} = \omega a e^{kz} \qquad \text{deep water} \qquad (5.4.10)$$

These expressions show that, for deep water, the wave-induced velocities decrease exponentially with the distance to the surface ($z < 0$ below the still-water surface). At the surface, where $z = 0$, the total orbital velocity is

$$u = \omega a \qquad \text{deep water, at the surface} \qquad (5.4.11)$$

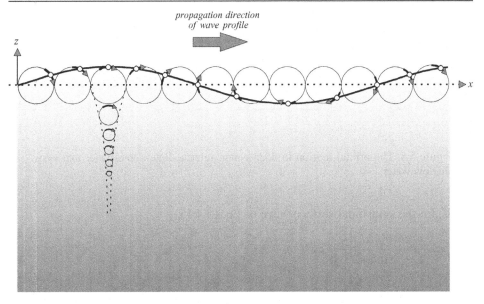

Figure 5.4 The orbital motion of the water particles under a harmonic wave that moves from left to right in deep water.

This is a rather natural result, because, if the particles move in circles (see below), the radius of the circle at the water surface is equal to the amplitude a of the wave. The particles travel along the circumference of that circle (of lenght $2\pi a$) in the period T of the wave, so that the (constant) velocity along that circle must be $u = 2\pi a/T = \omega a$.

In *very shallow* water, i.e., when $kd \to 0$, the expressions for the amplitudes of the velocities reduce to

$$\hat{u}_x = \frac{\omega a}{kd} \qquad \text{and} \qquad \hat{u}_z = \omega a\left(1 + \frac{z}{d}\right) \qquad \text{very shallow water} \qquad (5.4.12)$$

The first of these two expressions shows that, in very shallow water, the amplitude of the horizontal velocity is constant over the vertical, whereas the second expression shows that the amplitude of the vertical velocity varies linearly along the vertical.

Particle path

In general, the path of a particle is obtained by integrating the velocity of the particle in time. A convenient approximation here is to consider a particle located near an arbitrarily chosen position (which we will indicate with $\bar{x}$, $\bar{z}$), and take the velocity at this location. With local co-ordinates x' and z' (centred on $\bar{x}$, $\bar{z}$), the integration

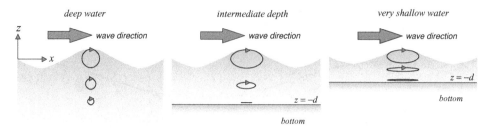

Figure 5.5 The orbital motion in deep water, intermediate-depth water and very shallow water.

yields, for the sinusoidal surface wave of Eq. (5.4.1),

$$x' = -a \frac{\cosh[k(d + \bar{z})]}{\sinh(kd)} \cos(\omega t - k\bar{x})$$

$$(5.4.13)$$

$$z' = a \frac{\sinh[k(d + \bar{z})]}{\sinh(kd)} \sin(\omega t - k\bar{x})$$

Since the horizontal position x' varies as a cosine and the vertical position z' as a sine, each particle goes through an ellipse:

$$\frac{x'^2}{A^2} + \frac{z'^2}{B^2} = 1 \qquad\qquad (5.4.14)$$

with horizontal and vertical semi-main axes

$$A = a \frac{\cosh[k(d + \bar{z})]}{\sinh(kd)} \qquad \text{(horizontal semi-main axis)}$$

$$(5.4.15)$$

$$B = a \frac{\sinh[k(d + \bar{z})]}{\sinh(kd)} \qquad \text{(vertical semi-main axes)}$$

In *deep* water (when $kd \to \infty$), the lengths of the two axes are equal, $A = B$, so that the particles move through circles with the radius decreasing exponentially with the distance to the surface (see Fig. 5.4; $z < 0$ below the still-water surface):

$$r = ae^{kz} \qquad\qquad \text{deep water} \qquad\qquad (5.4.16)$$

In *very shallow* water (when $kd \to 0$), the- lengths of the axes are $A = a/(kd)$ and $B = a(1 + z/d)$, and the particles move in ellipses growing flatter towards the bottom, $B \to 0$ as $z \to -d$, with constant horizontal axis $A = a/(kd)$. At the bottom, the ellipse degenerates into a straight, horizontal line (see Fig. 5.5).

A detailed analysis (not given here) of the motion of the particles at the surface of a finite-amplitude wave shows that the particles crowd together somewhat (horizontally) near the crest, whereas in the trough they separate. In other words, the surface 'shrinks' at the crest and 'expands' in the trough. This is evident in the position of the surface particles in Fig. 5.4 (which is obviously drawn for a wave with finite amplitude). Short waves riding on top of a longer wave will therefore shorten (and thus grow steeper) at the crest of the long wave, and lengthen in the trough. Such modulation of capillary waves is the feature that is observed with imaging radar (e.g., the satellite-borne SAR; see Section 2.4.1).

5.4.3 Dynamics

The dispersion relationship

The above results for the kinematic aspects apply to any harmonic wave propagating at the water surface, be it a free wave (subject only to gravitation; travelling at a precise speed, see below), or a forced wave (subject to additional external forces; it may travel at *any* speed, depending on the forcing). For the harmonic wave to be a free wave, we need to invoke the free-wave condition: the atmospheric pressure at the water surface, $p = $ constant $= 0$. Substituting the harmonic surface profile (Eq. 5.4.1) and the corresponding velocity potential function (Eq. 5.4.2) into the expression for this boundary condition of zero atmospheric pressure (Eq. 5.3.29) gives a relationship between radian frequency ω and wave number k (see Fig. 5.6):

$$\boxed{\omega^2 = gk\tanh(kd)} \qquad \text{or} \qquad L = \frac{gT^2}{2\pi}\tanh\left(\frac{2\pi d}{L}\right) \qquad \text{arbitrary depth}$$

$$(5.4.17)$$

This is called the *dispersion relationship* for reasons to be given in Section 6.4.2.

For *deep* water ($\tanh(kd) \to 1$ for $kd \to \infty$) the dispersion relationship approaches[9]

$$\omega = \sqrt{gk_0} \qquad \text{or } L_0 = gT^2/(2\pi) \qquad \text{(or } L_0 \approx 1.56T^2 \text{ [m, s])} \quad \text{deep water}$$

$$(5.4.18)$$

where k_0 and L_0 are the deep-water wave number and wave length, respectively. The dispersion relation for arbitrary depth can consequently also be written as

$$k_0 = k\tanh(kd) \qquad \text{or} \qquad L = L_0\tanh(2\pi d/L) \qquad \text{arbitrary depth}$$

$$(5.4.19)$$

with k_0 and L_0 related to the radian frequency or period as in Eq. (5.4.18).

---

[9] This deep-water relationship was derived much earlier by Gerstner (1802) in his trochoidal wave theory (see Section 5.6.2).

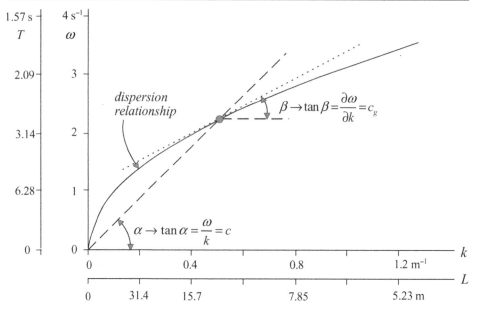

Figure 5.6 The dispersion relationship, the phase velocity c and the group velocity c_g (depth 100 m).

For *very shallow* water ($\tanh(kd) \to kd$ for $kd \to 0$), the dispersion relationship approaches

$$\omega = k\sqrt{gd} \quad \text{or} \quad k = \omega/\sqrt{gd} \quad \text{or} \quad L = T\sqrt{gd} \qquad \text{very shallow water}$$

(5.4.20)

The dispersion relationship of Eq. (5.4.17) is an implicit expression in terms of wave number, which requires an iteration procedure to calculate the wave number for a given frequency and depth. An alternative is to use a look-up table or to use an explicit expression that approximates the solution closely. A good example is given by Eckart (1952):

$$kd \approx \alpha(\tanh\alpha)^{-1/2} \qquad \text{with} \qquad \alpha = k_0 d = \omega^2 d/g \qquad (5.4.21)$$

This expression is exact for the limits of deep and shallow water ($kd \to \infty$ and $kd \to 0$). For all other situations, the error in k is less than 5%. A refinement of this approximation is given by Fenton (1988):

$$kd \approx \frac{\alpha + \beta^2(\cosh\beta)^{-2}}{\tanh\beta + \beta(\cosh\beta)^{-2}} \qquad \text{with} \qquad \beta = \alpha(\tanh\alpha)^{-1/2} \qquad (5.4.22)$$

which is also exact in the deep-water and shallow-water limits and its error in k is less than 0.05% in all other situations.

Literature:
Fenton and McKee (1990), Goda (2000), Hunt (1979), Wiegel (1964).

Phase velocity and group velocity

The propagation speed of the surface wave profile, i.e., the phase speed (see Section 5.4.1), is readily obtained from the dispersion relationship Eq. (5.4.17) with $c = L/T = \omega/k$:

$$\boxed{c = \frac{g}{\omega}\tanh(kd) = \sqrt{\frac{g}{k}\tanh(kd)}} \qquad\qquad \text{arbitrary depth}$$

$$(5.4.23)$$

This expression shows that, in general, the phase speed depends on wave number and therefore on frequency (see Fig. 5.6): long waves travel faster than short waves. Such waves, the propagation speed of which depends on wave length or frequency, are called *dispersive* waves (for reasons to be given in Section 6.4.2). In *deep* water ($\tanh(kd) \to 1$ for $kd \to \infty$), this expression reduces to

$$c_0 = \sqrt{\frac{g}{k_0}} \quad \text{or} \quad c_0 = \frac{g}{\omega} \quad \text{or} \quad c_0 = \frac{g}{2\pi}T \qquad (\text{or } c_0 \approx 1.56T \text{ [m, s]})$$

$$\text{deep water} \qquad (5.4.24)$$

where k_0 is the deep-water wave number. In *very shallow* water ($\tanh(kd) \to kd$ for $kd \to 0$), it reduces to

$$c_{shallow} = \sqrt{gd} \qquad\qquad \text{very shallow water}$$

$$(5.4.25)$$

Equation (5.4.25) shows that, in very shallow water, the phase speed does not depend on wave length or frequency. Under these conditions, the waves are said to be *non-dispersive*.

If we add *two* harmonic waves (η_1 and η_2, see Fig. 5.7), with slightly different frequencies, travelling in the same direction, then these two waves will reinforce each other at one moment (when they are *in* phase; i.e., when the crests of the two component waves coincide) but cancel each other at another moment (when they are 180° *out of* phase, i.e., when the crest of one wave coincides with the trough of the other). This will repeat itself over and over again, in other words, we create a series of wave groups. Taking the amplitudes of the two waves to be equal, the

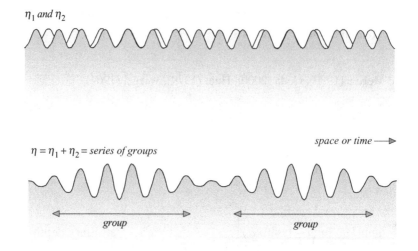

Figure 5.7 Two harmonic waves with slightly different frequencies (or wave numbers) add up to a series of wave groups.

resulting surface elevation is

$$\eta = \eta_1 + \eta_2 = a\sin(\omega_1 t - k_1 x) + a\sin(\omega_2 t - k_2 x) \tag{5.4.26}$$

The group has its maximum surface elevation where η_1 and η_2 are in phase. The propagation speed of this point is by definition the group velocity (it is the phase speed of the *envelope* of the surface elevations). It can be determined by first writing the sum of the two component waves (using standard trigonometric relationships) as

$$\eta = 2a \, \cos\left(\frac{\omega_1 t - k_1 x - \omega_2 t + k_2 x}{2}\right)\sin\left(\frac{\omega_1 t - k_1 x + \omega_2 t - k_2 x}{2}\right)$$

$$\tag{5.4.27}$$

or, by re-arranging the various terms,

$$\eta = 2a \, \cos\left(\frac{\omega_1 - \omega_2}{2}t - \frac{k_1 - k_2}{2}x\right)\sin\left(\frac{\omega_1 + \omega_2}{2}t - \frac{k_1 + k_2}{2}x\right)$$

$$\underbrace{}_{\text{envelope}}\quad\underbrace{}_{\text{carrier wave}} \tag{5.4.28}$$

$$\underbrace{}_{\text{modulating amplitude}}$$

where the sine wave is the carrier wave (with a frequency and wave number equal to the averages of the original values) and the cosine wave is the envelope of the waves (with the difference frequency and difference wave number), which modulates the amplitude of the carrier wave.

The phase speed of the carrier wave is (from the frequency and the wave number of the sine wave in Eq. 5.4.28)

$$c_{carrier} = \frac{(\omega_1 + \omega_2)/2}{(k_1 + k_2)/2} \approx \frac{\omega_1}{k_1}$$ (5.4.29)

and the phase speed of the envelope, called the group velocity for obvious reasons, is (from the frequency and the wave number of the cosine wave in Eq. 5.4.28)

$$c_{envelope} = c_{group} = \frac{(\omega_1 - \omega_2)/2}{(k_1 - k_2)/2} = \frac{\Delta\omega}{\Delta k}$$ (5.4.30)

It follows that, if the difference between the frequencies (and therefore also between the wave numbers) is infinitely small, the group velocity is (see Fig. 5.6)

$$\boxed{c_{group} = c_g = \frac{\partial\omega}{\partial k} = nc}$$ (5.4.31)

where c is the phase speed of the wave and n is (from the dispersion relationship, Eq. 5.4.17)

$$n = \frac{1}{2}\left(1 + \frac{2kd}{\sinh(2kd)}\right)$$ (5.4.32)

Since $0 \leq kd \leq \infty$ and therefore $0 \leq 2kd/\sinh(2kd) \leq 1$, this expression for n shows that n varies between $n = \frac{1}{2}$ (deep water) and $n = 1$ (very shallow water). This implies that the speed of an individual wave (the phase speed) is always larger than or equal to the speed of the group: $c \geq c_g$. One consequence of this is that each wave travels forwards through the group, until it reaches the front of the group, *where it disappears*. This is quite remarkable, but that is not all. The group is kept alive by new *waves that are continuously formed* at the *tail* of the group! All this seems odd, but even a casual look at actual waves in a wave group at sea or in a channel, or even in a pond or a wave flume, shows that this theoretical result is correct. Of course, it is the propagating wave energy that keeps the group alive. Since waves propagate across the ocean as groups, travel times of ocean waves should be calculated with the group velocity, not with the phase speed. In deep water, $n = \frac{1}{2}$, so the group velocity is half the phase speed, $c_g = \frac{1}{2}c$. In very shallow water, $n = 1$ and the group velocity is equal to the phase speed, $c_g = c$, so the individual waves travel as fast as the group. This means that, in very shallow water, each wave maintains its position in the group (no waves disappear at the front, and no waves are generated in the tail).

One important effect of the dependence of the group velocity on frequency is that a field of waves with various frequencies, as is normal for ocean waves, disintegrates slowly into a sequence of wave fields with the longer waves travelling ahead of the shorter waves: the wave energy disperses across the ocean. This phenomenon is therefore called frequency-dispersion (and hence the name 'dispersion' relationship for Eq. 5.4.17 and the term 'dispersive' waves). When storm-generated waves travel across the ocean, it is this frequency-dispersion that transforms the irregular storm waves into regular swell. This phenomenon will be treated further in Section 6.4.2.

Wave-induced pressure

The above motions of the water particles in circles or ellipses imply accelerations that can be caused only by forces acting on these particles. These forces are provided in this case by gradients in the (wave-induced) pressure in the water. The analytical expression for this pressure is readily derived by substituting the solution for the velocity potential (Eq. 5.4.2) into the Bernoulli equation (Eq. 5.3.28), with the result that the total pressure is

$$p = -\rho g z + \rho g a \frac{\cosh[k(d+z)]}{\cosh(kd)} \sin(\omega t - kx) \qquad (z < 0 \text{ below still-water level})$$

$$(5.4.33)$$

(Remember that, in the linear wave theory, the only external force is gravitation. This is evident here by virtue of the presence of g in this expression.) The first term on the right-hand side is the hydrostatic pressure. It is obviously independent of the presence of the wave (at least in the linear approximation; a second-order refinement is given in Section 7.4.2). The second term is due to the wave and therefore represents the wave-induced pressure, denoted as p_{wave}:

$$p_{wave} = \hat{p}_{wave} \sin(\omega t - kx) \qquad \text{with} \qquad \hat{p}_{wave} = \rho g a \frac{\cosh[k(d+z)]}{\cosh(kd)}$$

$$(5.4.34)$$

This is a propagating pressure wave in the water body, *in* phase with the surface elevation and with vertically decreasing amplitude. The expression is valid in the small-amplitude approximation of the linear wave theory, but actual waves will always have finite amplitude, for which the theory breaks down near the surface. Above the still-water line, the pressure is then sometimes, crudely, approximated as hydrostatic (see Fig. 5.8).

This pressure distribution does indeed provide *vertical* accelerations beneath the crest and in the trough of the wave and *horizontal* accelerations beneath the zero-crossings of the wave surface (corresponding exactly to the orbital motion of the water particles). Note the similarity with the expressions for the orbital velocities.

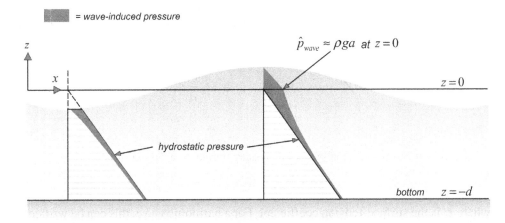

Figure 5.8 The wave-induced pressure superimposed on the hydrostatic pressure along the vertical, beneath a wave crest and beneath a wave trough (amplitude exaggerated for illustrative purposes) with a crude approximation above the still-water line (e.g., the kink in the pressure at $z = 0$ is not realistic).

In *deep water* the amplitude of the wave-induced pressure is ($z < 0$ below the still-water line)

$$\hat{p}_{wave} = \rho g a e^{kz} \qquad \text{deep water} \tag{5.4.35}$$

which represents the same exponential reduction with the distance to the surface as for the orbital velocities and the radius of the particle path (Eqs. 5.4.9 and 5.4.16). In *very shallow water* the wave-induced pressure amplitude is constant along the vertical:

$$\hat{p}_{wave} = \rho g a \qquad \text{very shallow water} \tag{5.4.36}$$

5.4.4 Capillary waves

So far we have assumed that the pressure at the surface is constant (zero) because we wanted the wave to be a free wave, i.e., free of imposed forces such as that exerted by wind. However, for very small waves (centimetre wave length), the water surface itself imposes a force that acts normal to the surface, due to surface *tension*. It is essentially a force that acts *in* the water surface, but it has a component normal to the surface if that surface is curved (see Fig. 5.9):

$$p_{surface\ tension} = -\tau_s \frac{\partial^2 \eta}{\partial x^2} \tag{5.4.37}$$

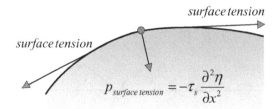

Figure 5.9 Surface tension induces a pressure normal to a curved surface.

where τ_s is the surface-tension coefficient. For a harmonic wave, this can be written as

$$p_{surface\ tension} = +\tau_s k^2 \eta \qquad\qquad (5.4.38)$$

It can be included in the dynamic surface boundary condition (see Eq. 5.3.29), with the result that

$$\frac{\partial \phi}{\partial t} + (g + \tau_s k^2 / \rho)\eta = 0 \qquad \text{at } z = 0 \qquad\qquad (5.4.39)$$

This dynamic surface boundary condition is identical to the dynamic surface boundary condition for a free wave, with g replaced with $g + \tau_s k^2 / \rho$. The other (kinematic) boundary conditions (at the surface and at the bottom) are the same as for a free wave, so all expressions for the kinematic and dynamic aspects of the wave are the same as for a free wave with gravitational acceleration g replaced by $g + \tau_s k^2 / \rho$. For instance, the dispersion relationship becomes

$$\omega^2 = (g + \tau_s k^2 / \rho)k \tanh(kd) \qquad\qquad (5.4.40)$$

and the phase velocity becomes

$$c = \sqrt{\frac{g + \tau_s k^2 / \rho}{k} \tanh(kd)} \qquad\qquad (5.4.41)$$

For a wave length of 0.017 m, the term representing the effect of surface tension is approximately equal to the gravitational acceleration $\tau_s k^2 / \rho \approx g$ (for clean, fresh water at 20 °C, so that $\tau_s \approx 0.073$ N/m). For shorter waves (larger wave numbers) the effect of surface tension will increase. For instance, the phase speed, instead of continuing to decrease for shorter wave lengths, will increase. Waves that are dominated by surface tension are called *capillary waves* or 'ripples'. They can be

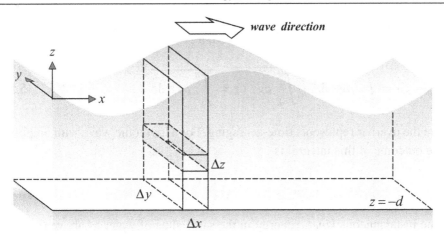

Figure 5.10 The column in the harmonic wave that is used in deriving the expressions for wave energy.

important for remote sensing, since some radar applications operate in the centimetre range. It is the modulation of these capillary waves by the longer waves that creates the image of wind waves on radar (e.g., the satellite-borne SAR; see Sections 2.4.1 and 5.4.2). For waves that engineers are generally interested in (wave lengths longer than 1 m, say) the effect of surface tension is negligible because the value of $\tau_s k^2/\rho$ for these waves is very much less than g (for wave lengths longer than 1 m, $\tau_s k^2/\rho < 0.0003g$).

5.5 Wave energy (transport)

5.5.1 Wave energy

The presence of a wave at the water surface implies that water particles were moved from their position at rest to some other position. This *change of position* requires work done against gravitation and this represents *potential* energy. In addition, the wave particles *move*, which represents kinetic energy. To estimate the *potential* energy, consider a slice of water with thickness Δz in a column with horizontal surface area $\Delta x \, \Delta y$ (see Fig. 5.10). The instantaneous potential energy (i.e., mass $\times$ elevation, at a given moment in time) of this slice of water, relative to $z = 0$, is then $\rho g z \, \Delta x \, \Delta y \, \Delta z$. The corresponding wave-induced potential energy in the entire column, from bottom to surface, is equal to the potential energy in the presence of the wave minus the potential energy in the absence of the wave. Per unit horizontal surface area (divide by the horizontal surface area of the column $\Delta x \, \Delta y$) and time-averaged over one period, it is

$$E_{potential} = \overline{\int_{-d}^{\eta} \rho gz \, dz} - \overline{\int_{-d}^{0} \rho gz \, dz} = \overline{\int_{0}^{\eta} \rho gz \, dz} \qquad (5.5.1)$$

where the overbar represents time-averaging. For a harmonic wave with amplitude a, the outcome of this integral is

$$E_{potential} = \overline{\tfrac{1}{2}\rho g \eta^2} = \tfrac{1}{4}\rho g a^2 \qquad (5.5.2)$$

The instantaneous *kinetic* energy in the same slice of water as above (i.e., $\tfrac{1}{2} \times$ mass $\times$ velocity squared, at a given moment in time) is $\tfrac{1}{2}\rho \, \Delta x \, \Delta y \, \Delta z \, u^2$ (where $u^2 = u_x^2 + u_z^2$). The corresponding time-averaged (over one period) kinetic energy in the entire column, from bottom to surface, is then, per unit surface area,

$$E_{kinetic} = \overline{\int_{-d}^{\eta} \tfrac{1}{2}\rho u^2 dz} \qquad (5.5.3)$$

The result of this integral, for a harmonic wave with amplitude a, using the expressions for u_x and u_z from the linear theory (accurate to second order, see Note 5B) is

$$E_{kinetic} = \tfrac{1}{4}\rho g a^2 \qquad (5.5.4)$$

so that, within the approximations of the linear wave theory, $E_{potential} = E_{kinetic}$. The total time-averaged wave-induced energy density $E = E_{potential} + E_{kinetic}$ is then

$$\boxed{E = \tfrac{1}{2}\rho g a^2}\qquad \text{time-averaged, wave-induced energy (potential plus kinetic)}$$
per unit horizontal area $\qquad (5.5.5)$

Note that energy is proportional to the square of the amplitude; it is therefore a *second-order* property of the wave, estimated with results of the linear wave theory.

5.5.2 Energy transport

As the waves travel across the ocean surface, they carry their potential and kinetic energy with them. To estimate this energy transport (also called energy flux), consider the left-hand vertical side of the same slice of water in the column as above (a window with cross-section $\Delta z \, \Delta y$). The bodily transport of the *potential* energy $\rho g z$, through that window, in the x-direction (with the water particles and therefore with velocity u_x), in a time interval Δt, is $(\rho g z) u_x \, \Delta z \, \Delta y \, \Delta t$. Over the entire depth, from bottom to surface, this transport is

$$f_1 = \left(\int_{-d}^{\eta} (\rho g z) u_x \, dz \right) \Delta y \, \Delta t \tag{5.5.6}$$

The bodily transport of the *kinetic* energy $\frac{1}{2}\rho u^2$, integrated over the entire depth, is similarly

$$f_2 = \left(\int_{-d}^{\eta} (\tfrac{1}{2}\rho u^2) u_x \, dz \right) \Delta y \, \Delta t \tag{5.5.7}$$

NOTE 5B   Integration to second-order accuracy

Determining the order of accuracy of an integral over the vertical beneath a harmonic wave in the linear wave theory is relatively easy with an essentially geometric rationale.

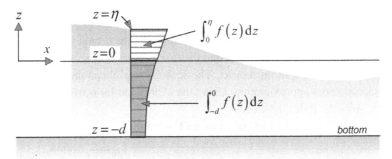

The integral from bottom to surface of a function $f(z)$ beneath a wave, divided into two integrals (amplitude of wave greatly exaggerated): one below the mean surface $(z = 0)$ and one above.

Consider a function that varies over the vertical $f(z)$, which depends on the wave amplitude to a certain power (the order of the function). Interpret the integral of this function from the bottom to the instantaneous surface $\int_{-d}^{\eta} f(z)dz$ as the surface area of the function. This integral can be divided into an integral from the bottom to the *mean* surface and an integral from that mean surface to the instantaneous surface (see the illustration in this note):

$$\int_{-d}^{\eta} f(z)dz = \int_{-d}^{0} f(z)dz + \int_{0}^{\eta} f(z)dz$$

The scale of each of these two surface areas is equal to the vertical scale of the integration interval, times the scale of the function $f(z)$ itself. For the first integral $\int_{-d}^{0} f(z)\,dz$, the vertical scale is the average depth, which is independent of the wave amplitude. The corresponding surface area is therefore proportional to the wave amplitude to the same degree (i.e., the same power) as is the function $f(z)$ itself. In other words: the order of the integral $\int_{-d}^{0} f(z)\,dz$, in terms of the amplitude, is equal to the order of the function $f(z)$. If, for instance, $f(z)$ is of second order in amplitude, then this integral too is of second order. For the second integral, the integration interval is the distance between the mean surface and the moving free surface, and, in contrast to the first integral, this scale is proportional to the amplitude. The scale of this surface area is therefore not only proportional to the wave amplitude to the same power as the function $f(z)$ but also proportional to the amplitude of the wave itself. In other words: the order of the second integral $\int_{0}^{\eta} f(z)\,dz$ is equal to the order of the function $f(z)$ plus one. For instance, if $f(z)$ is of second order in amplitude, then this second integral is of third order. In the integrals in the text of Section 5.5.2, the following functions appear:

(a) $f(z) = \frac{1}{2}\rho u^2$, so the integral (averaged over time) can be written as

$$\overline{\int_{-d}^{\eta} f(z)dz} = \overline{\int_{-d}^{\eta} \frac{1}{2}\rho u^2\,dz} = \overline{\int_{-d}^{0} \frac{1}{2}\rho u^2 dz} + \overline{\int_{0}^{\eta} \frac{1}{2}\rho u^2 dz}$$

$$\approx \overline{\int_{-d}^{0} \frac{1}{2}\rho u^2 dz} = \frac{1}{4}\rho g a^2$$

Since $f(z) = \frac{1}{2}\rho u^2$ is of second order in amplitude (the orbital velocity is proportional to the amplitude), it follows that the first integral is of second order and that the second integral is of third order. The second integral can therefore be ignored in a second-order approximation. Using the expression for u from the linear wave theory gives the result indicated.

(b) $f(z) = (\frac{1}{2}\rho u^2)u_x$, so the integral (averaged over time) can be written as

$$\overline{\int_{-d}^{\eta} f(z)dz} = \overline{\int_{-d}^{\eta} (\tfrac{1}{2}\rho u^2)u_x dz} = \overline{\int_{-d}^{0} (\tfrac{1}{2}\rho u^2)u_x dz} + \overline{\int_{0}^{\eta} (\tfrac{1}{2}\rho u^2)u_x dz} \approx 0$$

Since $f(z) = (\frac{1}{2}\rho u^2)u_x$ is of third order in amplitude (the orbital velocity is proportional to the amplitude), it follows that the first integral is of third order and that the second integral is of fourth order. Both integrals can therefore be ignored in a second-order approximation.

(c) $f(z) = (p_{wave})\,u_x$, so the integral (averaged over time) can be written as

$$\overline{\int_{-d}^{\eta} f(z)dz} = \overline{\int_{-d}^{\eta} (p_{wave})u_x dz} = \overline{\int_{-d}^{0} (p_{wave})u_x dz} + \overline{\int_{0}^{\eta} (p_{wave})u_x dz}$$

$$\approx \overline{\int_{-d}^{0} (p_{wave})u_x dz} = (\tfrac{1}{2}\rho g a^2)\frac{1}{2}\left(1 + \frac{2kd}{\sinh(2kd)}\right)\frac{\omega}{k}$$

Since $f(z) = (p_{wave})u_x$ is of second order in amplitude (both the wave-induced pressure and the orbital velocity are proportional to the amplitude), it follows that the first integral is of second order and that the second integral is of third order. The second integral can therefore be ignored in a second-order approximation. Using the expressions for p_{wave} and u_x from the linear wave theory gives the result indicated.

In addition to this bodily transport of potential and kinetic energy (i.e., with the orbital motion of the water particles), energy is transferred horizontally by work done by the pressure in the direction of wave propagation. This horizontal transfer through a vertical plane (i.e., the left-hand side of the column) in a time interval Δt is equal to the pressure p_{wave} times the distance moved in that interval (in the x-direction $= u_x \Delta t$). Integrated from the bottom to the surface, this is

$$f_3 = \left(\int_{-d}^{\eta} (p \, u_x) dz \right) \Delta y \Delta t \tag{5.5.8}$$

With $p = -\rho g z + p_{wave}$ (from Eqs. 5.4.33 and 5.4.34), we find

$$f_3 = \left(\int_{-d}^{\eta} (-\rho g z + p_{wave}) u_x dz \right) \Delta y \Delta t \tag{5.5.9}$$

Per unit crest length and per unit time (i.e., divided by $\Delta y \Delta t$) and time-averaged, the total energy transport P_{energy} is then the sum of these three contributions:

$$P_{energy} = \overline{f_1} + \overline{f_2} + \overline{f_3}$$

$$= \overline{\int_{-d}^{\eta} (\rho g z) u_x dz} + \overline{\int_{-d}^{\eta} (\tfrac{1}{2} \rho u^2) u_x dz} + \overline{\int_{-d}^{\eta} (-\rho g z + p_{wave}) u_x dz}$$

$$= \overline{\int_{-d}^{\eta} (\tfrac{1}{2} \rho u^2) u_x dz} + \overline{\int_{-d}^{\eta} (p_{wave}) u_x dz} \tag{5.5.10}$$

The first integral on the right-hand side of the last expression ($\overline{f_2}$) is of third order in amplitude (see Note 5B) and may therefore be ignored in a second-order approximation. The second integral is of second order and is therefore the only integral retained in a second-order approximation. This shows that, in such an approximation, all wave energy is transported only by the work done by the wave-induced pressure p_{wave}. This wave-induced pressure is in phase with the horizontal orbital motion and with the surface elevation (see Fig. 5.11): if the water particles move *in* the wave direction, the surface elevation is higher than when the water particles move *against* the wave direction. The net time-averaged effect is therefore a transport of

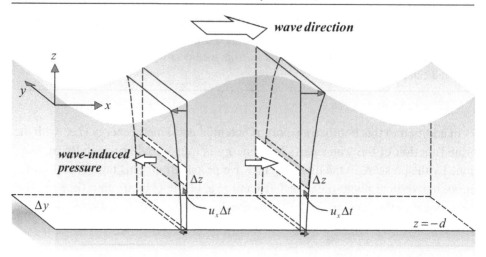

Figure 5.11 The asymmetric, *instantaneous* transport of wave energy by the wave-induced pressure (forward transport over larger water depth than backward transport), results in a net *time-averaged* energy transport in the wave direction.

energy in the wave direction. The value of the corresponding integral is, accurate to second order in amplitude (see Note 5B),

$$\overline{P_{energy}} \approx \int_{-d}^{0} (\overline{p_{wave}\,u_x})\mathrm{d}z = (\tfrac{1}{2}\rho g a^2)\frac{1}{2}\left(1 + \frac{2kd}{\sinh(2kd)}\right)\frac{\omega}{k} \tag{5.5.11}$$

or, since $E = \tfrac{1}{2}\rho g a^2$ and $c = \omega/k$,

$$\boxed{P_{energy} = Enc \quad \text{with} \quad n = \frac{1}{2}\left(1 + \frac{2kd}{\sinh(2kd)}\right)} \tag{5.5.12}$$

energy transport per unit time per unit crest length

Note that the propagation speed nc in this expression is exactly equal to the group velocity (see Eq. 5.4.31). The term 'group velocity' is often used indiscriminately for either the group velocity proper or the transport velocity of the energy. It is therefore widely accepted that one writes

$$\boxed{P_{energy} = Ec_g} \quad \text{in the wave direction} \tag{5.5.13}$$

The direction of energy transport is normal to the wave crest because the water particles move in that direction. This seems trivial, but in the presence of an ambient current this is generally not the case (see Section 7.3.5). Another important effect of an ambient current on the energy transport is readily demonstrated by replacing the particle velocity u_x in the above expressions with $u_x + U_x$ (where U_x is the horizontal component of the ambient current in the x-direction). This adds extra terms to the results of the integrals, which represent (a) the transfer of energy between wave and current and (b) the modified transports of the energies both of the wave and of the current.[10]

5.6 Nonlinear, permanent waves

5.6.1 Introduction

The linear theory of surface gravity waves, as presented in the previous sections, is a theory that matches the spectral description of ocean waves perfectly, because the spectral description is based on the assumption that the wave components are harmonic and independent, in other words, they behave as linear harmonic waves. However, this perfect match also limits the application of the spectral description to the conditions of the linear theory. When the waves are too steep or the water is too shallow, the linear wave theory is no longer valid and the spectrum no longer provides a complete statistical and physical description of the waves. Usually there is no easy alternative. When nonlinear effects are weak, or strong but occur only intermittently, then the waves can be treated on a large scale as linear waves, with relatively small, nonlinear corrections at these scales (hundreds of wave lengths or more). For instance, the process of white-capping (wave breaking in deep water) is locally highly nonlinear, but the related energy dissipation on a larger scale may be treated as a process that is weak in the mean. When nonlinear effects are to be considered on a small scale (a few wave lengths or less, for instance to compute wave forces on a marine structure), then the waves need to be considered locally with a nonlinear theory. A conventional approach is then to treat each wave individually and independently: the wave characteristics are computed on a wave-by-wave basis with a nonlinear theory and the computational results for a large number of such individual waves are analysed statistically to arrive at average characteristics.

In classical nonlinear wave theories, each wave is assumed to be one wave in a train of periodic (but not harmonic) waves, with a constant shape, amplitude and length (permanent waves, i.e., waves that do not evolve as they propagate). These theories are essentially analytical in nature. Here, we introduce three such classical nonlinear theories: the theories of Stokes (1847) and Dean (1965) for steep waves and the cnoidal theory for waves in shallow water (Korteweg and de Vries, 1895). More recent nonlinear theories are based on partial differential equations (rather

---

[10] The transfer of energy between the wave and the current is evident in the expressions by virtue of the appearance of the cross product $(\rho u_x^2 + p_{wave})U_x$ in the integrals, which can be interpreted as work done by the current U_x against a stress with magnitude $\rho u_x^2 + p_{wave}$. Integrated over depth and averaged over time, this work (i.e., energy transfer) is $S_{xx}U_x$, where $S_{xx} = \overline{\int_{-d}^{\eta} (\rho u_x^2 + p_{wave})dz}$ (also known as radiation stress; see Section 7.4.2).

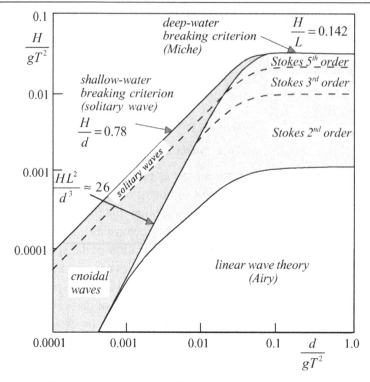

Figure 5.12 The ranges of applicability of the various wave theories (after LeMéhauté, 1976, Kamphuis, 2000, and SPM, 1973; see also Note 5C).

than analytical expressions) that allow the waves to evolve, e.g., as they propagate from deep into shallow water. These theories of evolving waves are introduced in Chapter 7.

The degree of nonlinearity of waves is often quantified with the Ursell number N_{Ursell}, which combines wave steepness and relative water depth (see e.g., Dingemans, 1997a, 1997b and Note 5C):

$$N_{Ursell} = steepness/(relative\,depth)^3 = (H/L)/(d/L)^3 = HL^2/d^3 \qquad (5.6.1)$$

where H is wave height and L is wave length. The cnoidal theory is applicable for $N_{Ursell} > 26$, while the theory of Stokes is applicable for $N_{Ursell} < 10$. Both apply equally well for $10 < N_{Ursell} < 26$. However, this division that is often used by engineers ignores the emergence of breaking when waves grow too steep, in deep or in shallow water. A more detailed division between the applicabilities of Airy, Stokes and cnoidal theories is given in Fig. 5.12.

Literature:
Fenton (1990, 1999).

NOTE 5C   The Ursell number

The Ursell number is essentially the ratio of the amplitude of a harmonic wave and the amplitude of its second-order Stokes correction (see Section 5.6.2). This ratio is approximately equal to

$$N_{Ursell} = HL^2/d^3$$

which can be re-written in different ways, for instance as the ratio of wave steepness over relative depth to the third power, as in Eq. (5.6.1). It can also be written as (with $L = cT$, where the shallow-water phase speed is $c = \sqrt{gd}$)

$$N^*_{Ursell} = \frac{gHT^2}{d^2} = \frac{H/(gT^2)}{(d/gT^2)^2}$$

The numerator and the (square root of the) denominator of this ratio have been used as the variables to characterise the region of applicability of the various wave theories in Fig. 5.12. Another definition, with essentially the same variables, is

$$N^{**}_{Ursell} = \frac{a/d}{(kd)^2} = \frac{H/L}{8\pi^2 (d/L)^3}$$

where $a = H/2$ is amplitude and $k = 2\pi/L$ is wave number. A similar definition is used in Section 9.3.4. Obviously, many variations of the basic definition can thus be found.

5.6.2 Stokes' theory and Dean's stream-function theory

In the linear wave theory, a wave with a harmonic surface profile that conforms to the *linearised* basic equations and boundary conditions is found. It only approximates the *nonlinear* equations and boundary conditions. A better approximation can be found by adding corrections to the harmonic wave profile. This is done in the theories of Stokes (1847) and Dean (1965) by adding extra harmonic waves to the basic harmonic. The differences between these two theories are as follows. First, that the corrections in the theory of Stokes are successive (every higher-order correction is obtained on the basis of the previously obtained lower-order corrections), whereas in the theory of Dean they are obtained simultaneously (and satisfy the dynamic boundary condition exactly). Secondly, the theory of Stokes is formulated in terms of the velocity potential, whereas Dean uses another, but closely related function (the stream function). Neither the theory of Stokes (1847) nor the theory of Dean (1965) performs well in very shallow water (water depth of the order of the wave height or less). For such conditions, the cnoidal theory should be used (see Fig. 5.12 and Section 5.6.3).

In the theory of Stokes (1847), the basic harmonic is written with the wave *steepness* $\varepsilon = ak$ explicitly represented (the cosine notation for the harmonic wave is more convenient here than the sine notation due to the occurrence of higher harmonics):

$$\eta(x, t) = a\cos(\omega t - kx) = \varepsilon\eta_1(x, t) \tag{5.6.2}$$

where $\eta_1(x, t) = k^{-1}\cos(\omega t - kx)$. The first correction in the Stokes theory is an 'extra' harmonic wave, written with the wave steepness raised to the second power (it is therefore a *second-order* correction, i.e. second order in wave steepness ε):

$$\eta(x, t) = \varepsilon\eta_1(x, t) + \varepsilon^2\eta_2(x, t) \tag{5.6.3}$$

where $\varepsilon^2\eta_2$ represents the extra harmonic wave. Using the solution of the linear theory, the nonlinear basic equations are now solved for this extra wave, with the nonlinear boundary conditions (see Appendix B for these equations and boundary conditions). The result is

$$\eta(x, t) = a\cos(\omega t - kx) + ka^2\frac{\cosh(kd)}{4\sinh^3(kd)}[2 + \cosh(2kd)]\cos[2(\omega t - kx)]$$

$$\tag{5.6.4}$$

where the first term on the right-hand side is the Airy wave of the linear wave theory and the second term is the second-order Stokes correction.

The wave represented by Eq. (5.6.4) is called a *second-order Stokes wave*. The phase speed of the extra harmonic (it is called the second harmonic) is equal to the phase speed of the linear wave. It is therefore a *bound* second harmonic (it travels at the speed of the Airy wave, which is called the *primary* wave). The amplitude being constant, and the phase speeds being equal, implies that the surface profile does not evolve in time or space; it is constant. The wave is horizontally symmetrical (around the wave crest) and vertically asymmetrical (around the mean sea level): the wave crest is a little sharper and the wave trough is a little flatter than in a harmonic wave (see Fig. 5.13). In addition, the crests are located at more than half the wave height above the mean water level. This asymmetry is an important deviation from the Gaussian model that is used in the spectral description of ocean waves and it should be considered when determining the maximum crest height in a random sea state (see also Sections 4.2.2 and 4.2.4).

The Stokes expansion can be continued by adding a *third harmonic*:

$$\eta(x, t) = \varepsilon\eta_1(x, t) + \varepsilon^2\eta_2(x, t) + \varepsilon^3\eta_3(x, t) \tag{5.6.5}$$

and the technique is repeated: use the solution of the linear theory and the above second harmonic to solve the nonlinear basic equations for this third harmonic with

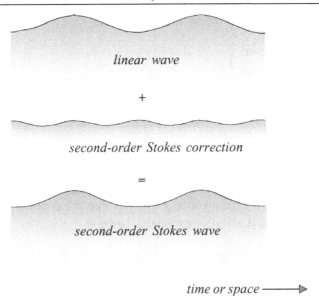

linear wave

+

second-order Stokes correction

=

second-order Stokes wave

time or space ⟶

Figure 5.13 The surface profile of a second-order Stokes wave.

the nonlinear boundary conditions. This gives the *third-order Stokes correction*, which is also a bound harmonic and its wave length and period are a third of those of the basic harmonic wave.[11] The approximation can be expanded indefinitely, so the Stokes theory can be developed to any degree of expansion to get (omitting the dependence on x and t in the notation)

$$\eta = \varepsilon\eta_1 + \varepsilon^2\eta_2 + \varepsilon^3\eta_3 + \varepsilon^4\eta_4 + \varepsilon^5\eta_5 + \cdots \qquad (5.6.6)$$

In practice the expressions become very complicated very rapidly.

In the stream-function theory of Dean (1965), as in the theory of Stokes (1847), the velocity components and the surface profile are written in terms of a series of harmonics (the number of harmonics determining the desired order of approximation) but the nonlinear basic equations are not solved with the velocity potential but with another, closely related function: the stream function ψ. This function is defined in a similar manner to the velocity potential function (but it exists only for

---

[11] The surface profile of this third-order Stokes wave is almost identical to the surface profile of the trochoidal wave in the theory of Gerstner (1802), which is easily constructed graphically as a trochoidal curve turned upside-down (the motion of a point on the side of a wheel rolling over a horizontal surface). However, in the trochoidal wave, the rotation of the water particles is opposite to what it should be; this is perhaps the reason why this theory has not been generally accepted (see Lamb, 1932).

two-dimensional flow, e.g., in the vertical x, z-plane):

$$\frac{\partial \psi}{\partial z} = u_x \qquad \left(= -\frac{\partial \phi}{\partial x} \right)$$

$$-\frac{\partial \psi}{\partial x} = u_z \qquad \left(= -\frac{\partial \phi}{\partial z} \right) \tag{5.6.7}$$

If the stream function is visualised as a hill above a horizontal x, z-plane, then the particle velocities are oriented along the contour lines of the hill, and their magnitudes are equal to the slope of the hill in the direction normal to these contour lines. The existence of a stream function implies that the continuity of the water mass in two dimensions is always guaranteed, because

$$\frac{\partial u_x}{\partial x} + \frac{\partial u_z}{\partial z} = \frac{\partial^2 \psi}{\partial z \partial x} - \frac{\partial^2 \psi}{\partial x \partial z} = 0 \tag{5.6.8}$$

(compare this with the Laplace equation, Eq. 5.3.22). In addition, because the water surface is a streamline (i.e., a line along which $\psi = $ constant), the kinematic surface boundary condition is always satisfied. The mathematics for the nonlinear cases is a little easier with this stream function than with the velocity potential but, like the velocity potential, it is only an auxiliary function without any physical meaning in itself. Similarly to the technique of Stokes, Dean develops the stream function as a sum of harmonics, but, unlike Stokes, Dean determines all coefficients simultaneously. All wave characteristics then follow from this stream function.

Literature:
Cokelet (1977), Dean (1974), Fenton (1985), LeMéhauté (1976), Sakai and Battjes (1980), Skjelbreia and Hendrickson (1960).

5.6.3 Cnoidal and solitary waves

In the above wave theories, the velocity potential function or the stream function is expanded in terms of the wave steepness $\varepsilon = ak$. However, if the depth is small, these theories do not apply. Therefore, in addition to, or instead of, considering nonlinear corrections due to wave *steepness*, corrections need to be applied to account for finite-*depth* effects. In the theory of cnoidal waves (also referred to as the KdV theory after Korteweg and de Vries, 1895), this is done in a manner very similar to that of the Stokes theory and the stream-function theory of Dean: the

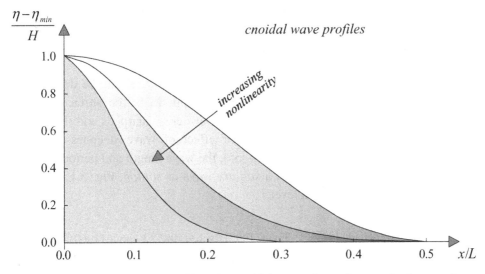

Figure 5.14 The surface profiles of a cnoidal wave, depending on the degree of nonlinearity (after Wiegel, 1960).

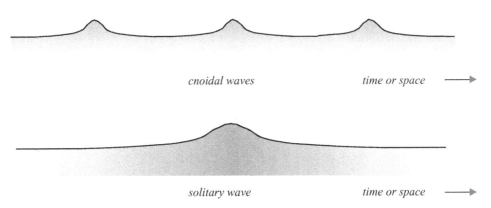

Figure 5.15 The surface profile of a cnoidal wave train and of a solitary wave.

velocity potential is developed in terms of a small parameter (the ratio of amplitude over depth $\beta = a/d$ in this case). Omitting the dependence on x and t in the notation, the surface elevation is then written as

$$\eta = \beta\eta_1 + \beta^2\eta_2 + \beta^3\eta_3 + \beta^4\eta_4 + \beta^5\eta_5 + \cdots \tag{5.6.9}$$

where, however, the basic wave η_1 and the extra waves $\eta_2, \eta_3, \eta_4, \eta_5, \ldots$ are not harmonic waves but *cnoidal* waves. Cnoidal waves are expressed with standard mathematical functions in terms of Jacobian elliptic functions[12] (these are standard functions, just like the sine and cosine functions, i.e., mathematically well-defined functions that can be computed to any degree of accuracy and can be found in standard tables).

The velocity potential, and hence all wave characteristics, can be derived for any given wave amplitude, length and water depth (e.g., the surface profile, Fig. 5.14). As the water depth decreases, the wave crest sharpens and the trough flattens, which is similar to the nonlinear effect of wave steepness. As the depth approaches zero (actually, $L/d \rightarrow \infty$), the wave length and period become infinitely long. This wave is called a *solitary wave* or *soliton*, Fig. 5.15; it rides completely above the mean sea level.

Literature:
CEM (2002), Chappelear (1962), Herman (1992), Isobe (1985), Laitone (1960), Mase and Kirby (1992), Munk (1949a), SPM (1973, 1984), Wiegel (1960, 1964), Yamaguchi (1992).

---

[12] The notation that is used in Jacobian elliptic functions includes the notations *cn*, *sn* and *dn*, hence the name *cn*oidal wave, in analogy with *sinus*oidal wave, which is written with the notation *sin*.

6

Waves in oceanic waters

6.1 Key concepts

- In this book, *oceanic* waters are deep waters (such that the waves are unaffected by the seabed) with straight or gently curving coastlines, without currents or obstacles such as islands, headlands and breakwaters.
- Under certain *idealised* conditions (constant wind blowing perpendicularly off a long and straight coastline over deep water), the significant wave height is determined by the wind, the distance to the upwind coastline (*fetch*) and the time since the wind started to blow (*duration*). So are the significant wave period and the energy density spectrum.
- Under these idealised conditions, the one-dimensional frequency *spectrum* has a universal shape: the JONSWAP spectrum for young sea states or the Pierson–Moskowitz spectrum for fully developed sea states. The (one-sided) directional width of the corresponding two-dimensional spectrum is typically 30°.
- To model waves under more realistic, *arbitrary* oceanic water conditions, the concepts of fetch and duration cannot be used. Instead, the *spectral energy balance* of the waves is used. It represents the time evolution of the wave spectrum, based on the propagation, generation, wave–wave interactions and dissipation of all individual wave components at the ocean surface.
- Conceptually, a Lagrangian approach (based on wave rays) or an Eulerian approach (based on a grid that is projected onto the ocean) can be used to formulate this energy balance. Owing to the interaction amongst the various wave components, the *Eulerian* approach is better suited for computations than the Lagrangian approach.
- Waves are *generated* by air-pressure fluctuations at the sea surface (not by wind friction), which are almost entirely due to wave-induced variations in the airflow (wind) just above the waves.
- In deep water, wave energy is *dissipated* only by breaking (white-capping).
- In deep water, quadruplets of wave components *interact* by resonance.
  - These quadruplet wave–wave interactions redistribute the wave energy over the spectrum. They do not add or remove energy from the spectrum as a whole (the redistribution is said to be conservative).
  - For sufficiently steep waves (waves being generated by wind, i.e., wind sea), the quadruplet wave–wave interactions cause a downshifting of the peak frequency of the spectrum and stabilise the spectral shape (the JONSWAP spectrum).
  - The shape-stabilising capacity of the quadruplet wave–wave interactions is the reason why the JONSWAP spectrum is characteristic for wind sea in oceanic waters. This spectrum is therefore widely accepted as the design spectrum in the engineering community.
- Outside their generation area, the waves degenerate into *swell*, due to frequency- and direction-dispersion. Swell is not steep enough for the quadruplet wave–wave interactions to be effective. The shape of a swell spectrum is therefore not a JONSWAP spectrum. A swell spectrum is narrow in both frequency and direction, but it depends otherwise entirely on the characteristics of the generating storm and the distance to that storm.

6.2 Introduction

Wave observations that are required to estimate past, present or future wave conditions are not always available, either because no instrument was operating at the required time and location (e.g., to investigate an accident) or because the required conditions have not (yet) occurred (e.g., extreme design conditions). The only alternative in the absence of observations is to simulate the wave conditions, using *wind* information. For the short term, waves can be *forecast* with forecast winds (over a period of several days, after which the wind forecasts are no longer reliable), for instance to plan offshore activities such as salvage operations, coastal activities such as dredging, or recreational activities such as surfing. Wave conditions can also be *hindcast*, i.e., computed with archived wind field or wind fields that have been reconstructed by meteorologists in hindsight. With this hindcasting technique, wave conditions in the past can be simulated to generate long-term wave information (see Section 4.3) or to reconstruct special conditions (e.g. an accident at sea). The choice of a wave model to carry out such forecasts or hindcasts depends on the complexity of the case. For a first impression in relatively simple situations, one can use a model that is essentially a generalisation of observations made under idealised conditions. It is simple in the sense that only a few variables are involved and it can be applied without the use of computers. Usually though, the wind, the seabed topography or the tides vary in space or time and hence numerical models that take these variations into account need to be used. These models require a computer and trained personnel.

The wind, on which all forecasts and hindcasts depend, is a complex phenomenon to describe because it is a three-dimensional vector that varies randomly in three space dimensions and time. However, for the purpose of wave modelling, this vector is described by only one *horizontal* component, *averaged* over some time interval (typically 10 min) at a *fixed elevation* above the mean sea surface. The elevation is usually 10 m and the wind is accordingly denoted as $\vec{U}_{10}$. Such a fixed elevation is rather impractical: the anemometer on board a ship can often not be located at that elevation due to structural limitations on the ship and in heavy seas a ship may move vertically over a considerable distance, rendering the notion of a fixed elevation above mean sea level most improbable. Atmospheric models that are used by meteorologists, too, do not always provide the wind at elevation 10 m but instead give it at a level that is consistent with the discretisation of their numerical models, e.g., at levels of constant atmospheric pressure. In actual practice therefore, the wind at elevation 10 m is often *estimated* from the wind at other levels (e.g., by extrapolation with an assumed vertical wind profile). The (horizontal) wind at 10-m elevation is only a function of the two horizontal space dimensions x and y and time t, so $\vec{U}_{10} = \vec{U}_{10}(x, y, t)$.

Sometimes the wind is characterised by an alternative, purely fictitious wind speed (in the sense that it cannot be measured), which is directly related to the shear stress τ of the wind on the ocean surface.[1] It is called the friction velocity and it is denoted by u_*. The relationships are $\tau = \rho_{air} u_*^2 = \rho_{air} C_d U_{10}^2$, where ρ_{air} is the density of air and C_d is called the drag coefficient. An expression relating C_d to U_{10} and a model to compute u_* from the dynamic interaction between wind and waves are given in Section 9.3.2. In this book, we will come across this wind friction velocity only occasionally. Sometimes the use of yet another alternative is advocated: the wind speed U_λ at a dynamically chosen elevation that is a constant fraction of the wave lengths of the ocean waves or U_∞ at the top of the atmospheric boundary layer (the lower part of the atmosphere where the presence of the Earth's surface is noticeable with its upper limit typically at altitude 150–500 m), but that is rarely done.

Literature:
Bidlot *et al.* (1996), CEM (2002), Charnock (1955), Janssen (2004), Kinsman (1965), Phillips (1977), Resio *et al.* (1999), Komen *et al.* (1994), WMO (1998).

6.3 Wave modelling for idealised cases (oceanic waters)

The wave model that is considered first is very simple. It is essentially a generalisation of wave observations under conditions that approximate the following idealisation (see Fig. 6.1): a constant wind (constant in space and time) is blowing over deep water, perpendicularly off a straight and infinitely long coastline ('deep' is taken to mean so deep that the waves are not affected by the bottom, i.e., typically more than half the wave length). The waves are described with only a characteristic wave height (e.g., the significant wave height) and a characteristic period (e.g., the significant wave period or the peak period, i.e., the inverse of the peak frequency of the spectrum) or with a universal one- or two-dimensional spectrum. In this approach of idealised conditions, the waves depend only on the wind speed and the distance to the upwind coastlines (fetch) or the time elapsed since the wind started to blow (the duration; there is no wind before this time). Usually the duration is taken to be infinitely long (in practical applications, sufficiently long that the precise duration is irrelevant), so that wind speed and fetch are the only determining factors.

An entirely different idealised wind field relates to hurricanes.[2] The defining parameters are usually the position, direction and velocity of the centre of the

---

[1] The concept of surface shear stress (friction) in the case of wind-generated waves is not trivial. Actually, the shear stress represents the vertical transport of horizontal momentum from the atmosphere to the ocean in which the waves play a crucial role (see Section 6.4.3).
[2] See Section 1.3 for the alternative terms typhoon and cyclone.

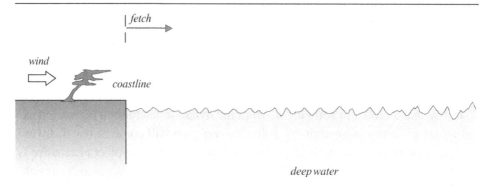

Figure 6.1 The ideal situation of a constant wind blowing over deep water, perpendicularly off a straight and infinitely long coastline.

hurricane (the eye), the atmospheric pressure at the sea surface in the eye relative to
the ambient pressure and the radius to maximum wind (i.e., the horizontal dimension
of the hurricane eye). The wave field is then only a function of these parameters.
This approach to hurricanes will not be considered further.

Literature:
Bretschneider (1959), CEM (2002), Ochi (2003), SPM (1973, 1984).

6.3.1 Idealised wind

In the idealised situation of wave generation described above, the wind is assumed
to be constant. However, even a constant wind blowing off a coast develops an
internal atmospheric boundary layer starting at the coastline (mostly because the
sea surface is much smoother than the land, thus creating an inherent variation in
the wind speed as a function of the distance offshore). These and other deviations
from the ideal case are usually ignored. The only parameters that are assumed
to affect the waves, in addition to the wind, are then fetch (F), duration (t, no
wind or waves for $t < 0$) and gravitational acceleration (g). Other parameters that
may be relevant, such as the viscosity of the water, turbulence in the airflow,
gustiness and atmospheric stability, are usually ignored, thus introducing errors
in estimating the significant wave height of up to 20%, even in such idealised
situations.

In practical applications, the four parameters F, t, U_{10} and g are often reduced
to three by expressing the duration t in terms of an equivalent fetch F_{eq} as follows.
Consider a forecast location P at some distance F from the coast (see Fig. 6.2).
A wave component from an arbitrary direction θ, which arrives at that location P
at time t, has travelled a distance $c_g t$ from that direction since the wind started

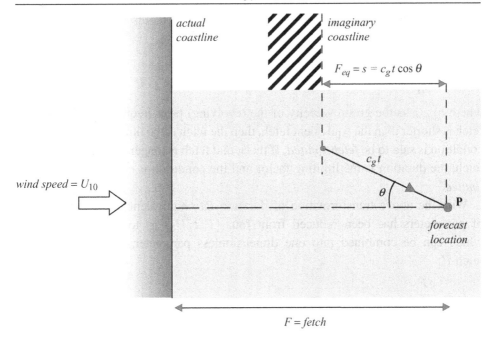

Figure 6.2 The concept of equivalent fetch F_{eq} under the idealised conditions of wind-wave generation.

to blow (c_g is the group velocity of the wave component being considered). If such an individual wave component were to develop independently of the rest of the spectrum, the situation for that component would be equivalent to having the wind blow forever (infinite duration), with the coast at a distance of $s = c_g t \cos\theta$ (see Fig. 6.2). The distance s to this imaginary coast can therefore be seen as an equivalent fetch F_{eq} for that component:

$$F_{eq} = s = c_g t \cos\theta \qquad (6.3.1)$$

The essence here of the equivalence between fetch and duration is that the wind has had the same *time* to transfer energy to the wave component considered. Such equivalence exists for each wave component individually. It is different for each component, because the direction and frequency, and therefore group velocity, are different for each component. For the spectrum *as a whole*, therefore, there is no such unique relationship between fetch and duration. However, one could take the wave component at the peak of the spectrum as the carrier component of the total wave energy and assign an equivalence to this component. This approach is justified in the sense that the energy of young sea states is concentrated around a fairly sharp peak (see Section 6.3.3). The direction of the peak component is equal to the wind direction and therefore constant, but its frequency f_{peak} is not; it evolves over the

duration and the simple expression of Eq. (6.3.1) should therefore be replaced with
an integral:

$$F_{eq} = \int_0^t c_{g,peak}(t)\,dt \tag{6.3.2}$$

where $c_{g,peak}$ is the group velocity of the (evolving) peak frequency. If the actual
fetch is shorter than the equivalent fetch, then the fetch is the limiting factor and the
condition is said to be *fetch-limited*. If the actual fetch is longer than the equivalent
fetch, the duration is the limiting factor and the condition is said to be *duration-limited*.

With the above transformation of duration into equivalent fetch, the number
of parameters has been reduced from four (F, t, U_{10}, g) to three (F, U_{10}, g),
which can be combined into one dimensionless parameter, the dimensionless
fetch $\tilde{F}$:

$$\tilde{F} = \frac{gF}{U_{10}^2} \tag{6.3.3}$$

Literature:
Cavaleri (1994), CEM (2002), Donelan *et al.* (1985), Hurdle and Stive (1989), Kahma and
Calkoen (1992), SPM (1973, 1984), Taylor and Lee (1984), Voorrips *et al.* (1994), Wilson
(1965), Young and Verhagen (1996a), Young (1998).

6.3.2 The significant wave

The significant wave height and the significant wave period have been observed
under conditions that only approximate the idealised situation. Obviously, actual
field conditions are never ideal in the sense that the coastline is never infinitely long
and straight and the wind is never constant and perpendicular to the coastline. The
observations therefore will always have some scatter. However, the observations do
reveal some universal relationships, indicating that the chosen parameters of fetch,
wind speed and gravitational acceleration are indeed the dominating parameters.
These observations have been generalised, not only by using the dimensionless
fetch $\tilde{F}$, but also by using a dimensionless significant wave height,

$$\tilde{H}_{1/3} = \frac{gH_{1/3}}{U_{10}^2} \qquad \text{or} \qquad \tilde{H}_{m_0} = \frac{gH_{m_0}}{U_{10}^2} \tag{6.3.4}$$

and dimensionless significant wave period or peak period,

$$\tilde{T}_{1/3} = \frac{gT_{1/3}}{U_{10}} \qquad \text{or} \qquad \tilde{T}_{peak} = \frac{gT_{peak}}{U_{10}} \tag{6.3.5}$$

(In addition to this notation, I will also use $\tilde{H}$ and $\tilde{T}$, with the context indicating whether these are based on a zero-crossing analysis, $\tilde{H}_{1/3}$ and $\tilde{T}_{1/3}$ or on a spectrum, $\tilde{H}_{m_0}$ and $\tilde{T}_{peak}$, or leaving this indeterminate.) The use of these dimensionless parameters *generalises* the observations since it renders the observations independent of the scale of the observation: the dimensionless conditions in a severe storm at sea would be identical to those in a breeze over a lake, or even in a gentle wind in a laboratory flume, as long as the ratio of fetch over the wind speed squared is the same. This idea has been used extensively in developing this approach: observations are made under relatively mild conditions; they are then made dimensionless and the results can be applied to conditions of a totally different scale.

At short fetches, the waves grow fairly rapidly (*young sea states*), but gradually the growth slows down until it eventually stops (the wave speed of the longest waves approaches the wind speed, and wave breaking balances the energy transfer from wind to waves). In this final stage, the waves are said to be *fully developed*.[3] Pierson and Moskowitz (1964) analysed observations of such fully developed waves in the North Atlantic Ocean. Since fetch is not relevant in such cases (it is sufficiently large that the exact value is irrelevant; it is nominally infinite: $\tilde{F} = \infty$) and the ocean is so deep that depth does not affect the waves, the significant wave height and period depend only on the local wind speed. This implies that, under these fully developed conditions, the *dimensionless* significant wave height and period are universal *constants* (the wind speed is included in the definitions). Pierson and Moskowitz (1964) used the wind speed at the anemometer elevation of the weather ships at which the observations were taken (19.5 m) and they found $g H_{m_0,\infty}/U_{19.5}^2 = 0.21$ and $g T_{peak,\infty}/U_{19.5} = 7.14$ (the subscript ∞ indicates the fully developed state; see also Note 6A). The generally accepted corresponding values in terms of U_{10} are (assuming that $U_{19.5} \approx 1.075 U_{10}$, which is very reasonable)

$$\tilde{H}_\infty = 0.24 \text{ and } \tilde{T}_\infty = 7.69 \text{ deep-water, fully developed sea state,}$$

$$\text{based on } H_{m_0} \text{ and } T_{peak} \qquad (6.3.6)$$

NOTE 6A   Fully developed sea state in deep water

It seems reasonable to assume that the waves stop growing when the phase speed of the waves approaches the wind speed (the relative wind speed is then zero). This implies that $c_{peak} \rightarrow U_{10}$ for fully developed waves (c_{peak} is the phase speed at the peak frequency of the wave spectrum). In deep water, the phase speed can be estimated from linear

---

[3] There is some doubt as to the existence of the fully developed state, although observations seem to be fairly conclusive (see Ewing and Laing, 1987, and Walsh *et al.*, 1989). The uncertainty is of a theoretical nature: wave–wave interactions (see Section 6.4.4) may continue to transfer energy from higher frequencies to ever lower wave frequencies, even if the wave height no longer evolves.

wave theory as $c_{peak} = g T_{peak}/(2\pi)$, so

$$\frac{g T_{peak}}{2\pi} \to U_{10} \qquad \text{from which} \qquad \frac{g T_{peak}}{U_{10}} \to 2\pi \qquad \text{or} \qquad \tilde{T}_\infty \to 2\pi$$

deep water, fully developed

which is in qualitative agreement with the value $\tilde{T}_\infty = 7.69$ observed by Pierson and Moskowitz (1964).

The first systematic observations of the significant wave height and period under *fetch-limited* conditions (on lakes and reservoirs) were made by Sverdrup and Munk (1946, 1947) and, somewhat later, by Bretschneider (1952). Their results have been used widely and, in honour of their contribution, the corresponding parameterisations (analytical functions approximating such data) are called SMB (Sverdrup–Munk–Bretschneider) growth curves. For short fetches, simple power laws are commonly used:

$$\tilde{H} = a_1 \tilde{F}^{b_1}$$

deep water, short fetches
(i.e., young sea states)   (6.3.7)

$$\tilde{T} = a_2 \tilde{F}^{b_2}$$

Later, such observations were carried out by many others. An excellent compilation and re-analysis of a number of such studies has been given by Kahma and Calkoen (1992), who found from the composite data set thus obtained that $a_1 = 2.88 \times 10^{-3}$, $a_2 = 0.459$, $b_1 = 0.45$ and $b_2 = 0.27$ (see Fig. 6.3).[4] To represent these young sea states and the fully developed sea state, and also the *transition* between them, a tanh function is often used because this function has the property that, for small arguments, it approaches its argument: $\tanh x \to x$ for $x \ll 1$; and for large values of its argument, it approaches unity: $\tanh x \to 1$ for $x \gg 1$. Most investigators in this field have therefore fitted the following functions through their observations:

$$\boxed{\tilde{H} = \tilde{H}_\infty \tanh(k_1 \tilde{F}^{m_1})}$$

deep water, all sea states   (6.3.8)

$$\boxed{\tilde{T} = \tilde{T}_\infty \tanh(k_2 \tilde{F}^{m_2})}$$

---

[4] Toba (1972, 1973, 1997) used such relationships to propose the following universal relationship between the dimensionless significant wave height and period (for wind sea): $\tilde{H}_* = \beta (\tilde{T}_*)^{3/2}$, where $\tilde{H}_* = g H_{1/3}/u_*^2$, $\tilde{T}_* = g T_{1/3}/u_*$ and β is a universal constant. These findings are supported by many field observations. With the results of Kahma and Calkoen (1992), the power in this expression of Toba (expressed in terms of U_{10} rather than u_* would be $b_1/b_2 = 1.7$ instead of $3/2$.

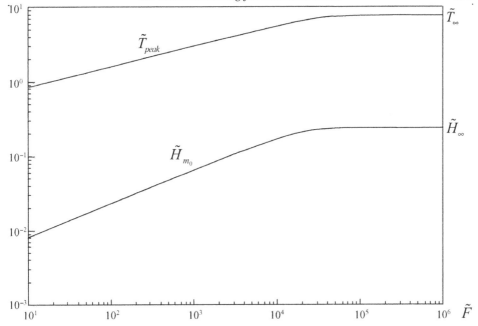

Figure 6.3 The dimensionless significant wave height and period (see Eqs. 6.3.4 and 6.3.5) as a function of dimensionless fetch of Kahma and Calkoen (1992), Pierson and Moskowitz (1964) and Young and Verhagen (1996a; as modified by Breugem and Holthuijsen, 2006).

For long fetches, these expressions reduce to $\tilde{H} = \tilde{H}_\infty$ and $\tilde{T} = \tilde{T}_\infty$ respectively (compare with Eq. 6.3.6); and for short fetches, to $\tilde{H} = \tilde{H}_\infty k_1 \tilde{F}^{m_1}$ and $\tilde{T} = \tilde{T}_\infty k_2 \tilde{F}^{m_2}$, respectively (compare with Eq. 6.3.7). Some recent results are given in Fig. 6.3 and Note 6B.

NOTE 6B   Growth curves of significant wave height and peak period (deep water)

Young and Verhagen (1996a), added two parameters, p and q, to the tanh expressions of Eq. (6.3.8) to control the transition from young sea states to the fully developed sea state:

$$\tilde{H} = \tilde{H}_\infty [\tanh(k_1 \tilde{F}^{m_1})]^p \qquad \text{and} \qquad \tilde{T} = \tilde{T}_\infty [\tanh(k_2 \tilde{F}^{m_2})]^q$$

They estimated the values of these transition parameters from their observations in Lake George (Australia) using the mean wind over the upwind fetch, instead of the local wind speed. Breugem and Holthuijsen (2006) corrected their results by removing observations that seemed to have been affected by the shadowing effect of the lateral coastlines of Lake George. These modified growth curves of Young and Verhagen (1996a) reduce to those of Kahma and Calkoen (1992) for short fetches and to the limit values of Pierson and Moskowitz (1964) for long fetches. The corresponding values of the coefficients[5] are summarised in the following table. The growth curves are given in Fig. 6.3.

The coefficients representing idealised wind-wave growth in deep water

	Pierson and Moskowitz (1964) *fully developed sea state, Eqs. (6.3.8) and this note*	Kahma and Calkoen (1992) *young sea states, Eqs. (6.3.7)*	Young and Verhagen (1996a) modified by Breugem and Holthuijsen (2006) *all sea states, this note*
$\tilde{H} = \tilde{H}_{m_0}$	$\tilde{H}_\infty = 0.24$	$a_1 = 2.88 \times 10^{-3}$ $b_1 = 0.45$	$\tilde{H}_\infty = 0.24$ $k_1 = 4.41 \times 10^{-4}$ $m_1 = 0.79$ $p = 0.572$
$\tilde{T} = \tilde{T}_{peak}$	$\tilde{T}_\infty = 7.69$	$a_2 = 0.459$ $b_2 = 0.27$	$\tilde{T}_\infty = 7.69$ $k_2 = 2.77 \times 10^{-7}$ $m_2 = 1.45$ $q = 0.187$

An intriguing aspect of the idealised case is the applicability of the growth curves to conditions under which the upwind coastline is *not* straight and perpendicular to the wind. A special case occurs when the coastline is straight but slants across the wind direction. Observations show that the mean direction of the waves at the *high* frequencies is then aligned with the wind direction (these frequencies are fully developed and are affected only by the wind) but the mean wave direction at the *lower* frequencies tends to be oriented parallel to the coast (these frequencies are affected by the coastline, which is asymmetrical with respect to the wind direction). A closely related phenomenon is the slower growth of the waves if the lateral coastline limits the *width* of the fetch, for instance in a narrow bay or fjord with the wind blowing along the axis of the bay or fjord (see the comments on Lake George in Note 6B).

It may be noted that fully developed wave conditions are almost unrealisable because very long fetches would be needed. For instance, the dimensionless fetch for fully developed conditions is approximately 2×10^4 (see Fig. 6.3). This corresponds to a fetch of 1800 km for a wind speed of 30 m/s, 460 km for a wind speed of 15 m/s, or 115 km for a wind speed of 7.5 m/s. These fetches are unrealistically large for these (constant) wind speeds. However, fully developed conditions may

---

[5] These coefficients relate to conditions with neutral atmospheric stability. Young (1998) gives corrections for non-neutral atmospheric conditions.

occur when the wind speed suddenly drops at some point along the fetch. The value of the *dimensionless* significant wave height (the significant wave height made dimensionless with the lower wind speed) at that point then increases, possibly to a value higher than the nominal fully developed value of about 0.24 (the waves are said to be over-developed).

Literature:
CEM (2002), Donelan *et al.* (1985), Holthuijsen (1983b), Kahma (1981), Kahma and Pettersson (1994), Kawai *et al.* (1977), Seymour (1977), SPM (1973, 1984), Wyatt (1995), Young and Verhagen (1996a).

6.3.3 *The one-dimensional wave spectrum*

In many of the above studies, not only were $\tilde{H}$ and $\tilde{T}$ obtained as a function of $\tilde{F}$, but so was also the wave spectrum $E(f)$. The most important contribution in this context is from the JOint North Sea WAve Project (JONSWAP; Hasselmann *et al.*, 1973). An example observation obtained during JONSWAP is shown in Fig. 6.4. The upper panel of Fig. 6.4 shows that, under idealised conditions of wave generation, the spectrum evolves from the high frequencies to lower frequencies. A remarkable feature of this evolution is that the spectrum retains its shape along the fetch. At first glance, the spectral shape seems to sharpen with increasing fetch, but normalising the observed spectra reveals the evolution of the *shape* of the spectrum. This is shown in the lower panel of Fig. 6.4, where the same spectra as in the upper panel are normalised with the maximum variance density $E(f_{peak})$ and peak frequency f_{peak}.

The high-frequency *tails* of these fetch-limited spectra have the same shape as the *tails* of the fully developed spectra observed by Pierson and Moskowitz (1964). This shape was suggested as early as 1958 by Phillips (1958), who derived this supposedly universal characteristic with a dimensional analysis based on an assumed behaviour of the high frequencies. His hypothesis was that wave *breaking* limits the spectral level of the high frequencies. Since breaking is dominated by gravitational acceleration, the variance density would then depend only on the frequency (f) and the gravitational acceleration (g):

$$E(f) \sim g^a f^b \qquad \text{high-frequency tail, deep water} \qquad (6.3.9)$$

Since the dimension of the variance density $E(f)$ is $[\mathrm{m^2/Hz}] = [\mathrm{m^2\,s}]$, it follows from Eq. (6.3.9) that

$$[\mathrm{m^2 s}] = [\mathrm{ms^{-2}}]^a\,[\mathrm{s^{-1}}]^b \qquad (6.3.10)$$

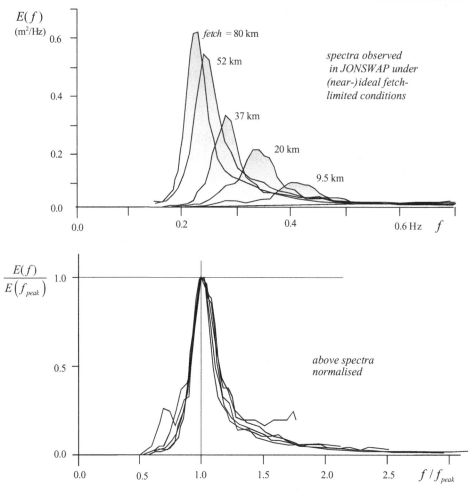

Figure 6.4 Spectra observed during the Joint North Sea Wave Project (JON-SWAP) under idealised, deep-water conditions: upper panel, observed spectra (after Hasselmann *et al.*, 1973); lower panel, same spectra but normalised.

Since the powers of the units m and s should be equal on both sides of the $=$ sign, it follows that $a = 2$ and $b = -5$. The expression for a breaking-dominated spectral tail would therefore be[6]

$$\boxed{E(f) \sim g^2 f^{-5}}$$ high-frequency tail, deep water   (6.3.11)

It must be stressed that these arguments are based on a plausible but unsubstantiated physical argument, namely that the shape of the tail of the spectrum is dominated

___

[6] An alternative reasoning, with the same result, is given by Thornton (1977), who argued that breaking occurs when the forward speed of the water particles at the surface exceeds the propagation speed of the wave itself. This leads to a shape of the spectral tail $E(f) \sim c^2 f^{-3}$, which in deep water, with $c^2 = g^2 (2\pi f)^{-2}$, is $E(f) \sim g^2 f^{-5}$.

by wave breaking. We will see that the truth is more complicated than that (other processes are involved too; see Section 6.4). The result of the analysis of Phillips (1958) may well therefore be invalid; in fact, Phillips has repeatedly expressed his surprise that his arguments are still considered valid after all these years (e.g., Phillips, 1985). However, the analysis of Phillips is a nice illustration of a dimensional analysis and, in spite of its flawed basis, the conclusion that the shape of the tail is f^{-5} has been confirmed in many studies, including Pierson and Moskowitz (1964) and JONSWAP (Hasselmann *et al.*, 1973). However, others have argued that the high-frequency tail depends not only on g but also on the wind speed. For instance, Toba (1973) included the friction velocity u_* and found

$$E(f) \sim g u_* f^{-4} \qquad \text{high-frequency tail, deep water} \qquad (6.3.12)$$

The question of whether the shape of the spectral tail under these idealised deep-water conditions is f^{-5} or f^{-4} is still being discussed (see Note 6C), but the answer to this question does not seem to be important for engineering applications because the effect is often barely noticeable. The engineering community has always used the f^{-5}-tail and will presumably do so for the foreseeable future.

NOTE 6C   The f^{-4}-shape of the spectral tail

The theoretical grounds for the f^{-4}-shape of the spectral tail are based on the physical processes of generation, wave–wave interactions and dissipation involved (e.g., Kitaigorodskii, 1983; Phillips, 1985; Resio, 1987; Young and van Vledder, 1993; Komatsu and Masuda, 1996; Perrie and Zakharov, 1999; Resio *et al.*, 2004). There is also an abundance of observations that supports the f^{-4}-shape (e.g., Kawai *et al.*, 1977; Mitsuyasu, 1977; Mitsuyasu *et al.*, 1980; Kahma, 1981; Forristall, 1981; Donelan *et al.*, 1985; Hwang *et al.*, 2000a; Resio *et al.* 2004). Even the original observations of JONSWAP and Pierson and Moskowitz agree better with the f^{-4}-shape than with the f^{-5}-shape (see Battjes *et al.*, 1987; Alves *et al.*, 2003). Others have found values of the power of f between 2 and 10 (e.g., Mitsuyasu, 1977; Forristall, 1981; Huang *et al.*, 1981; Hansen *et al.*, 1990). Banner (1990a) finds the f^{-4}-tail for the frequency spectrum but an f^{-5}-tail for those wave components in the *two-dimensional* spectrum that travel downwind.

   If the spectral tail were indeed proportional to f^{-4}, then the proportionality coefficient α_{Toba} in $E(\omega) = \alpha_{Toba} g u_* \omega^{-4}$ (as originally formulated by Toba, 1973) would be a universal constant. This seems to be supported by the review of Battjes *et al.* (1987), who found for the spectra observed in JONSWAP and six other field studies $\alpha_{Toba} = 0.096$ (average of their Table 2 for field observations) and no dependence on the dimensionless peak frequency (see also Phillips, 1985 and Janssen, 2004; but Donelan *et al.*, 1985 did find a dependence).

In view of the support for the f^{-4}-tail, Donelan *et al.* (1985) suggested modifying the JONSWAP spectrum (see Eq. 6.3.15) accordingly, simply by replacing f^{-5} with $f^{-4}f_{peak}$ (and replacing the coefficient 5/4 with unity to retain f_{peak} as the peak frequency):

$$E_{Donelan}(f) = \alpha_{Donelan}g^2(2\pi)^{-4}f^{-4}f_{peak}^{-1}\exp\left[-\left(\frac{f}{f_{peak}}\right)^{-4}\right]$$

$$\times\gamma_{Donelan}^{\exp\left[-\frac{1}{2}\left(\frac{f/f_{peak}-1}{\sigma_{Donelan}}\right)^2\right]}$$

Comparing this expression with the above expression of Toba (1973) (remember to include the Jacobian when transforming the expression of Toba from the ω-domain to the f-domain) shows that $\alpha_{Donelan} = \alpha_{Toba}u_*\omega_{peak}/g$. Other alternatives to the JONSWAP spectrum have also been proposed, e.g., the Mitsuyasu spectrum and the Wallops spectrum (see Huang *et al.*, 1990a). However, the JONSWAP spectrum is still by far the most widely used spectrum.

The variance density in the above expression for the spectral tail has no upper limit: it approaches infinity for low frequencies (see Fig. 6.5). To approximate observed spectra therefore, over the *entire* frequency range, the f^{-5}-expression needs to be cut off at some low frequency. Pierson and Moskowitz (1964) used a smooth cut-off function, which is zero at low frequencies and unity at high frequencies (thus retaining the f^{-5}-shape for these high frequencies). The fully developed spectrum, called the Pierson–Moskowitz spectrum, thus obtained (see Fig. 6.5) is[7]

$$E_{PM}(f) = \alpha_{PM}g^2(2\pi)^{-4}f^{-5}\exp\left[-\frac{5}{4}\left(\frac{f}{f_{PM}}\right)^{-4}\right] \qquad (6.3.13)$$

fully developed spectrum in deep water

where α_{PM} and f_{PM} are the energy scale and the peak frequency, respectively.[8]

Since the Pierson–Moskowitz spectrum (or PM spectrum) is assumed to represent fully developed conditions in deep water, the peak frequency can depend only on

---

[7] The original expression was formulated in terms of the *radian frequency* ω. If re-written in terms of frequency f, the transformation factor is $(2\pi)^{-5}$ multiplied by a Jacobian of 2π, giving the factor $(2\pi)^{-4}$ in Eq. (6.3.13).

[8] This expression can be generalised, using the peak frequency to retain the proper dimensions: $E(f) = \alpha g^2(2\pi)^{1-m}f_{peak}^{m-5}f^{-m}\exp[-(m/n)(f/f_{peak})^{-n}]$ with the zeroth-order moment $m_0 = \Gamma[(m-1)/n]$ $\times[\alpha g^2 f_{peak}^{-4}(2\pi)^{1-m}(n/m)^{(m-1)/n}/n]$, where $\Gamma[.]$ is the gamma function, which is available in any mathematical reference book, e.g., Abramowitz and Stegun (1965). Alternatively, the expression can also be generalised by using the wind speed: $E(f) = \alpha g^{m-3}U^{5-m}(2\pi)^{1-m}f^{-m}\exp[-(m/n)(f/f_{peak})^{-n}]$, where U is a wind speed (it may be U_{10} or u_*). The zeroth-order moment then is $m_0 = \Gamma[(m-1)/n]\left[\alpha g^{m-3}U^{5-m}f_{peak}^{1-m}(2\pi)^{1-m}(n/m)^{(m-1)/n}/n\right]$. The coefficient α is not necessarily identical in these two generalisations. For a Pierson–Moskowitz-type spectrum ($m = 5$, $n = 4$), both expression for m_0 reduce to $m_0 = (1/5)\alpha g^2(2\pi)^{-4}f_{peak}^{-4}$.

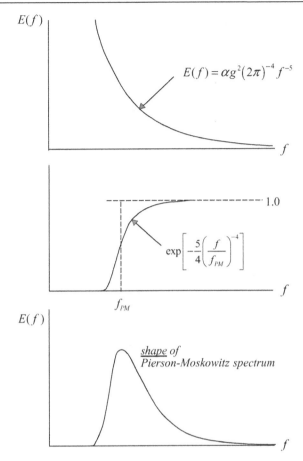

Figure 6.5 The high-frequency tail for the wind-sea spectrum suggested by Phillips (1958) and the smooth low-frequency cut-off suggested by Pierson and Moskowitz (1964), creating the Pierson–Moskowitz spectrum.

the wind speed. By fitting Eq. (6.3.13) to their observations, Pierson and Moskowitz (1964) found $\alpha_{PM} = 0.0081$ and the dimensionless peak frequency $f_{PM}U_{19.5}/g = 0.14$. For $U_{19.5} \approx 1.075\,U_{10}$ (see Section 6.3.2), it follows that $\tilde{f}_{PM} = f_{PM}U_{10}/g = 0.13$. The corresponding values for the dimensionless significant wave height and period are $\tilde{H}_\infty = 0.24$ and $\tilde{T}_\infty = 1/\tilde{f}_{PM} = 7.69$ (see Section 6.3.2). Alves *et al.* (2003) carefully re-analysed the observations of Pierson and Moskowitz (1964) and confirmed these values *in these observations*. Ewing and Laing (1987) found from *other* observations that these values overestimate fully developed wave conditions for wind speeds lower than 16 m/s.

The spectra observed during JONSWAP appear to have a sharper peak than the Pierson–Moskowitz spectrum. To account for this in a parameterisation of the observations, the scientists of JONSWAP chose to take the *shape* (!) of the

Pierson–Moskowitz spectrum (not its energy scale or frequency scale, of course) and to enhance its peak with a peak-enhancement function $G(f)$:

$$G(f) = \gamma^{\exp\left[-\frac{1}{2}\left(\frac{f/f_{peak}-1}{\sigma}\right)^2\right]}$$ (6.3.14)

in which γ is a peak-enhancement factor and σ is a peak-width parameter ($\sigma = \sigma_a$ for $f \leq f_{peak}$ and $\sigma = \sigma_b$ for $f > f_{peak}$ to account for the slightly different widths on the two sides of the spectral peak; see Fig. 6.6). This sharpens the spectral peak, but has no effect on other parts of the spectrum. This idealised spectrum is called the JONSWAP spectrum. Its complete expression is[9]

$$E_{JONSWAP}(f) = \alpha g^2 (2\pi)^{-4} f^{-5} \exp\left[-\frac{5}{4}\left(\frac{f}{f_{peak}}\right)^{-4}\right] \gamma^{\exp\left[-\frac{1}{2}\left(\frac{f/f_{peak}-1}{\sigma}\right)^2\right]}$$

$$\longleftarrow\quad\text{Pierson–Moskowitz }shape\quad\longrightarrow$$

$$\longleftarrow\qquad\qquad\text{JONSWAP}\qquad\qquad\longrightarrow$$

fetch-limited in deep water   (6.3.15)

Since the fetch during JONSWAP was rather limited, no transition of the spectrum to the fully developed sea state was observed. However, results from many studies have confirmed this result of JONSWAP over fetches that are most relevant to the engineer. In addition, the JONSWAP spectrum has been shown to be rather universal, not only for idealised fetch-limited conditions but also for arbitrary wind conditions in deep water, including storms and hurricanes. The reason for this is that, for sufficiently steep waves, the quadruplet wave–wave interactions (see Section 6.4.4) tend to stabilise the shape of the spectrum into the JONSWAP shape.[10] Since design conditions are often storm conditions (i.e., with relatively steep waves), the JONSWAP spectrum is the design spectrum for many engineers. The JONSWAP spectrum does not apply to swell[11] because the steepness of swell is low and the shape-stabilising capacity of the quadruplet wave–wave interactions is therefore weaker or practically absent.

The values of the energy scale parameter α, the frequency scale parameter f_{peak} and the shape parameters γ, σ_a and σ_b develop as the spectrum develops.

---

[9] To the best of my knowledge, this expression cannot be integrated analytically. However, good approximations for zero-, first- and second-order moments, m_{-2}, m_{-1}, m_0, m_1 and m_2, are given by Yamaguchi (1984) for $\sigma_a = 0.07$, $\sigma_b = 0.09$ and $\gamma = 1-10$, e.g., $m_0 = \alpha g^2 (2\pi)^{-4} f_{peak}^{-4}(0.065\,33\gamma^{0.8015} + 0.134\,67)$, which is accurate to within 0.25% for $1 \leq \gamma \leq 10$.

[10] Actually, the Donelan spectrum with an f^{-4}-tail (see Note 6C), but for engineering purposes this spectrum closely resembles the JONSWAP spectrum.

[11] See Section 6.4.2.

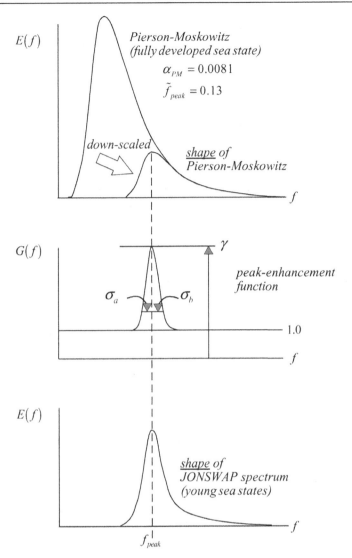

Figure 6.6 The Pierson–Moskowitz spectrum, the shape of the Pierson–Moskowitz spectrum and the shape of the JONSWAP spectrum.

Kahma and Calkoen (1992), on the basis of their compilation (see Section 6.3.2), suggested the following dependence of $\tilde{f}_{peak} = f_{peak}U_{10}/g$ on dimensionless fetch $\tilde{F}$:

$$\tilde{f}_{peak} = 2.18\tilde{F}^{-0.27} \tag{6.3.16}$$

which does not account for the transition to the fully developed situation, in which $\tilde{f}_{peak} = 0.13$. Instead, Eq. (6.3.8) with $\tilde{f}_{peak} = 1/\tilde{T}_{peak}$ may be used for this. The energy scale parameter α is equally a function of dimensionless fetch $\tilde{F}$ but it can also be expressed in terms of the dimensionless peak frequency, e.g., from Lewis and Allos (1990):

$$\alpha = 0.0317\, \tilde{f}_{peak}^{0.67} \tag{6.3.17}$$

In JONSWAP, the scatter in the values of the shape parameters γ, σ_a and σ_b was so large that no dependence on the dimensionless fetch could be discerned. The average values were $\gamma = 3.3$, $\sigma_a = 0.07$ and $\sigma_b = 0.09$. Others have repeated the JONSWAP study at different times and locations with essentially the same results. The transition to fully developed sea states is apparently poorly defined but, if required, it can be obtained with (Lewis and Allos, 1990; see also Eqs. 8.3.10 and 8.3.11)

$$\gamma = 5.870\, \tilde{f}_{peak}^{0.86}$$

$$\sigma_a = 0.0547\, \tilde{f}_{peak}^{0.32} \tag{6.3.18}$$

$$\sigma_b = 0.0783\, \tilde{f}_{peak}^{0.16}$$

These relationships are consistent with the JONSWAP observations but they have been forced to be equal to the values of $\alpha = 0.0081$ and $\gamma = 1.0$ for the Pierson and Moskowitz spectrum at $\tilde{f}_{peak} = 0.13$ (the values of σ_a and σ_b are irrelevant when $\gamma = 1.0$).

Literature:
Alves *et al.* (2003), Ewing and Laing (1987), Huang *et al.* (1981, 1990a), Mitsuyasu (1968, 1969), Mitsuyasu *et al.* (1980), Phillips (1985), Resio *et al.* (1999), Toba (1973, 1997).

6.3.4 *The two-dimensional wave spectrum*

The two-dimensional frequency–direction spectrum is difficult to observe, as noted in Chapter 2. Usually, only some *overall* directional characteristics are observed, notably the mean direction and the directional *width* of the spectrum (representing the degree of short-crestedness of the waves). This concept of directional width requires the introduction of the directional distribution $D(\theta; f)$. It is essentially the cross-section through the two-dimensional spectrum *at a given frequency*, normalised such that its integral over the directions is unity. In other words, it is a normalised, circular transect through the two-dimensional spectrum (see Fig. 6.7):

$$D(\theta; f) = \frac{E(f, \theta)}{E(f)} \tag{6.3.19}$$

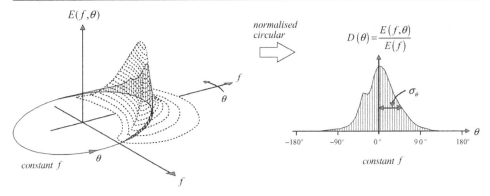

Figure 6.7 The directional energy distribution at a given frequency under arbitrary conditions and its (one-sided) directional width σ_θ.

That the integral over directions of this function is unity is readily shown as follows:

$$\int_0^{2\pi} D(\theta; f)\,d\theta = \int_0^{2\pi} \frac{E(f, \theta)}{E(f)}\,d\theta = \frac{\int_0^{2\pi} E(f, \theta)\,d\theta}{E(f)} = \frac{E(f)}{E(f)} = 1 \qquad (6.3.20)$$

Note that the directional distribution $D(\theta; f)$ gives the *normalised* distribution of the wave energy density over directions at *one* frequency, whereas the two-dimensional spectrum $E(f, \theta)$ gives the *non-normalised* distribution over both frequency and direction. Obviously $D(\theta; f)$ may vary with frequency. Very often such frequency-dependence is ignored in the notation, so that $D(\theta; f)$ is often written as $D(\theta) = D(\theta; f)$. Strictly speaking $D(\theta; f)$ is dimensionless, but it can be considered to have a *unit* of [1/angle], i.e., [1/rad] or [1/degree].

The directional spreading of the waves can be defined as the (one-sided) directional width of $D(\theta)$, denoted as σ_θ (see Fig. 6.7), in analogy with the conventional definition of standard deviation: $\sigma_\theta^2 = \int_{-\pi}^{+\pi} \theta^2 D(\theta)d\theta$ (where θ is taken relative to the mean wave direction). However, for various reasons, it is better to replace θ with $\sin \theta$, or better still, by $2\sin\left(\frac{1}{2}\theta\right)$, so that

$$\sigma_\theta^2 = \int_{-\pi}^{+\pi} \left[2\sin\left(\tfrac{1}{2}\theta\right)\right]^2 D(\theta)\,d\theta \qquad (6.3.21)$$

Young *et al.* (1996) and Ewans (1998) have published a large number of observations of σ_θ, which are summarised in Fig. 6.8, showing that σ_θ varies from approximately 30° at the peak frequency f_{peak} to about 60° at $3 f_{peak}$ (but the scatter in the observations is rather large). They report finding little or no dependence of

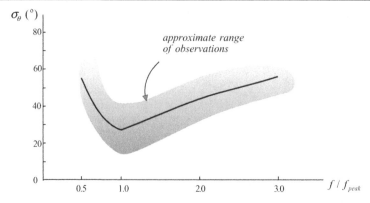

Figure 6.8 The directional width (one-sided) of the directional energy distribution $D(\theta; f)$ as a function of normalised frequency and the expression of Eq. 6.3.22. Observations of Young *et al.* (1996) and Ewans (1998).

σ_θ on wind speed. A reasonable fit to their observations is

$$\sigma_\theta = \begin{cases} 26.9(f/f_{peak})^{-1.05} & \text{in degrees, for } f < f_{peak} \\ 26.9(f/f_{peak})^{0.68} & \text{in degrees, for } f \geq f_{peak} \end{cases} \tag{6.3.22}$$

The *shape* of the distribution $D(\theta)$ is not well known, not even in the idealised situation that we consider here. It is usually speculated that this distribution has a maximum in the wind direction (most of the wave energy travelling downwind) and that it falls off gradually to the offwind directions (see Fig. 6.9, but see Note 6D). Several expressions with this character have been suggested to describe $D(\theta)$. The best-known and probably most widely used is the $\cos^2\theta$ model (e.g., Pierson *et al.*, 1952):

$$D(\theta) = \begin{cases} \dfrac{2}{\pi} \cos^2\theta & \text{for } |\theta| \leq 90° \\ 0 & \text{for } |\theta| > 90° \end{cases} \tag{6.3.23}$$

where the direction θ is taken relative to the mean wave direction. Its directional width $\sigma_\theta \approx 30°$. As Eqs. (6.3.22) show, this value agrees well with observations near the peak frequency. Moreover, it is a constant, i.e., independent of wind and frequency, which is convenient for many engineering applications. To obtain more flexibility, this model has been generalised to the $\cos^m\theta$ model:

$$D(\theta) = \begin{cases} A_1 \cos^m\theta & \text{for } |\theta| \leq 90° \\ 0 & \text{for } |\theta| > 90° \end{cases} \tag{6.3.24}$$

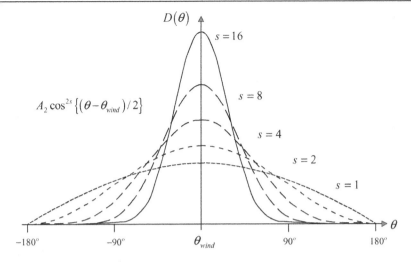

Figure 6.9 The $\cos^{2s}(\tfrac{1}{2}\theta)$ model for the directional energy distribution under idealised conditions.

where $A_1 = \Gamma(\tfrac{1}{2}m + 1)/[\Gamma(\tfrac{1}{2}m + \tfrac{1}{2})\sqrt{\pi}]$ is a normalisation coefficient (in which $\Gamma(.)$ is the gamma function, see footnote in Section 6.3.3) so as to have $\int_0^{2\pi} D(\theta)d\theta = 1$. The power m in this model controls the width of the distribution. A similar model for the directional distribution is (e.g., Longuet-Higgins et al., 1963; Mitsuyasu et al., 1975)

$$D(\theta) = A_2 \cos^{2s}\left(\tfrac{1}{2}\theta\right) \qquad \text{for } -180° < \theta \le 180° \qquad (6.3.25)$$

where $A_2 = \Gamma(s + 1)/[\Gamma(s + \tfrac{1}{2})2\sqrt{\pi}]$ and the power s controls the width of the distribution (see Fig. 6.9).

The relationship between the directional width σ_θ and the width parameter s is

$$\sigma_\theta = \sqrt{\frac{2}{s + 1}} \qquad \text{in radians} \qquad (6.3.26)$$

The variation of s over frequencies is readily determined from Eqs. (6.3.22) and (6.3.26). The most remarkable difference between the $\cos^m \theta$ model and the $\cos^{2s}(\tfrac{1}{2}\theta)$ model is that the former limits wave propagation to a sector of 90° on either side of the mean wave direction, whereas the latter allows waves to propagate against the wind (i.e., θ larger than 90°). Which of the two directional models is better in this respect is simply not known, because observations and theory are not clear on this issue. Other models for the directional distribution have also been suggested, e.g., the wrapped-normal distribution, which has the shape of the Gaussian distribution (see Note 6D).

Literature:
Apel (1994), Banner and Young (1994), Benoit *et al.* (1997), Cote *et al.* (1960), Donelan *et al.* (1985), Elfouhaily *et al.* (1997), Ewans (1998), Forristall and Ewans (1998), Goda (1997), Hasselmann *et al.* (1980), Holthuijsen (1983b), Hwang *et al.* (2000b), Kuik *et al.* (1988), Mardia (1972), Mitsuyasu *et al.* (1975), Pierson *et al.* (1955), Tucker and Pitt (2001), Young (1994), Young *et al.* (1995).

NOTE 6D   The bimodal directional distribution of wave energy

Observations show that the directional distribution at the peak frequency in the two-dimensional wave spectrum is relatively sharp and unimodal, but that it flattens towards higher frequencies and becomes *bimodal* at frequencies higher than approximately twice the peak frequency, i.e., the directional distribution at these frequencies has maxima in two symmetrical directions slightly off the wind direction (see illustration below; e.g., Holthuijsen, 1983b; Jackson *et al.*, 1985; Banner and Young, 1994; Young *et al.*, 1995; Ewans, 1998; Hwang *et al.*, 2000b; Wang and Hwang, 2001a, 2001b). This corresponds well to the *visual* impression of waves that are being generated by a local wind. Under these conditions, the waves seem to create a diamond pattern of wave crests and troughs (I spent many hours flying over the sea surface, watching waves under such conditions and found this confirmed in stereo-photographs taken during these flights; e.g. Holthuijsen, 1983b). Sverdrup and Munk (1946) already alluded to this phenomenon when they observed that '. . . the waves travel in different directions . . . and the consequent crisscrossing leads to a checkerboard pattern of crests and troughs . . .' This bimodality seems to be generated by the quadruplet wave–wave interactions (see Section 6.4.4, e.g., Longuet-Higgins, 1976; Young and van Vledder, 1993; Banner and Young, 1994).

Ewans (1998) proposed to parameterise the directional distribution correspondingly as the sum of two identical wrapped-around Gaussian distributions, each symmetrically shifted relative to the mean wave direction. The expression for this model is (slightly simplified, for didactic reasons only; see the illustration below)

$$D(\theta) = \frac{1}{\sqrt{8\pi}\sigma_\theta^*} \left\{ \exp\left[-\frac{1}{2}\left(\frac{\theta - \theta_{peak}^*}{\sigma_\theta^*} \right)^2 \right] + \exp\left[-\frac{1}{2}\left(\frac{\theta + \theta_{peak}^*}{\sigma_\theta^*} \right)^2 \right] \right\}$$

$$\text{for} -\infty < \theta < \infty$$

where θ_{peak}^* is the direction of each of the two directional peaks, relative to the mean wave direction, and σ_θ^* is the directional width of each of the two Gaussian distributions. From his observations, Ewans (1998) deduced (see also Wang and Hwang, 2001a, 2001b and Hwang and Wang, 2001), in degrees,

for $f < f_{peak}$: $\theta^*_{peak} = 7.5$

$$\sigma^*_\theta = 11.4 + 5.4 \left(\frac{f}{f_{peak}} \right)^{-7.9}$$

for $f \geq f_{peak}$: $\theta^*_{peak} = 0.5 \exp\left[5.45 - 2.75 \left(\frac{f}{f_{peak}} \right)^{-1} \right]$

$$\sigma^*_\theta = 32.1 - 15.4 \left(\frac{f}{f_{peak}} \right)^{-2}$$

Such a function, composed of two Gaussian distributions, is relatively sharp and uni-modal if the peak directions are relatively close together ($\theta^*_{peak} < \sigma^*_\theta$), rather flat for $\theta^*_{peak} \approx (1\text{--}2)\sigma^*_\theta$ and bimodal for $\theta^*_{peak} > 2\sigma^*_\theta$:

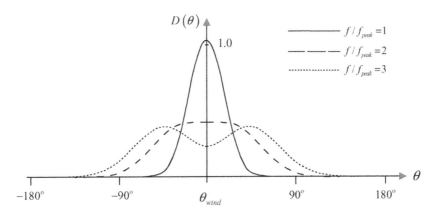

The directional energy distribution of ocean waves deduced by Ewans (1998).

6.4 Wave modelling for arbitrary cases (oceanic waters)

Wave predictions are hardly ever required in the above idealised situation. Such a situation is normally considered only in order to get a first estimate of wave conditions in simplified situations or to calibrate and verify theoretical or numerical wave models. This may be reasonable for small scales or in trade-wind or monsoon areas where the wind is more or less constant but, in the most energetic ocean regions of the world, the wind varies rather rapidly in both time and space. This simple approach is then totally inadequate. This is certainly true for the mid-latitude (extra-tropical) storms and obviously also for hurricanes.

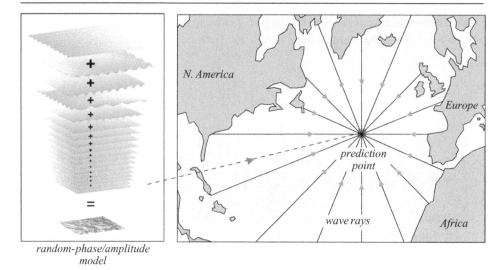

random-phase/amplitude
model

Figure 6.10 Following all wave components, as they travel across the ocean, along wave rays, to a prediction point, while accounting for all processes of generation, wave–wave interactions and dissipation, gives the spectrum at that location (on an oceanic scale, the wave rays are great-circles, which appear as straight lines on a great-circle map).

To introduce wave modelling for arbitrary cases, we will assume the random-phase/amplitude model (the sea-surface elevation is the summation of a large number of independent wave components; see Section 3.5.2 and Fig. 6.10). A wave prediction is then based on predicting each of these independent wave components individually: the spectral density $E = E(f, \theta)$ of each wave component is considered as it varies in time (t) and horizontal position[12] (x, y): $E(f, \theta) = E(f, \theta; x, y, t)$.

If we wish to predict the spectrum at a certain location in the ocean (see Fig. 6.10), we need only follow each and every wave component across the ocean from its point of inception (at a coastline) to the prediction point and account for all effects of generation, wave–wave interaction and dissipation that it encounters. More properly formulated: we need to integrate the evolution equation of the wave energy, while travelling along the wave ray at the group velocity, from the coast to the prediction point.

Literature:
Donelan and Hui (1990), Komen *et al.* (1994), Sobey (1986).

---

[12] On a large scale we should interpret the horizontal co-ordinates as spherical co-ordinates; on a smaller scale we may use Cartesian co-ordinates. For convenience's sake, I will use Cartesian co-ordinates and show the slightly more complicated spherical alternative only when required.

6.4.1 The energy balance equation

The evolution of the energy density of each wave component (f, θ) can be obtained by integrating an energy evolution equation while propagating with the group velocity along a wave ray:

$$\frac{\mathrm{d}E(f, \theta; x, y, t)}{\mathrm{d}t} = S(f, \theta; x, y, t) \tag{6.4.1}$$

where the term on the left-hand side is the rate of change of the energy density, and $\mathrm{d}x/\mathrm{d}t = c_{g,x}$ and $\mathrm{d}y/\mathrm{d}t = c_{g,y}$ (where $c_{g,x}$ and $c_{g,y}$ are the x- and y-components of the group velocity of the wave component under consideration), and frequency and direction are constant (in deep water). The term on the right-hand side (called the source term) represents all effects of generation, wave–wave interactions and dissipation. Conceptually this (Lagrangian) approach is very straightforward because in deep water the wave rays are straight lines or great-circles[13] and Eq. (6.4.1) needs only to be integrated along these lines. For one prediction point, the set of all relevant wave rays (all directions and frequencies) is a fan of straight lines or great-circles centred at that prediction point (see Fig. 6.10).

The integration of the source term along each of these rays would not be difficult if that term were known along the rays. That, unfortunately, is not the case: it will be shown later that, at each point along the ray, the source term depends not only on the component that is being followed, but also on the entire, two-dimensional spectrum, at that point, i.e., on wave components that cross the wave ray (on their way across the ocean). The energy densities of these other components are not known (they travel along other wave rays), so the Lagrangian approach cannot be used for *computations*. It is conceptually attractive, but we need to use another approach for computations. Two alternatives are available: (a) use another formulation that avoids the problem or (b) simplify the source term so that it is *in*dependent of the other wave components. The first alternative is provided by the Eulerian approach, in which the spectrum is computed not only at a single prediction point but rather at a large number of locations in the ocean simultaneously with a *local* energy balance at each of these locations. This approach is fundamentally correct and it is used in advanced wave modelling (second- and third-generation wave models; see Section 6.4.7). The second alternative is to formulate the source term such that it depends only on the wave component that is being followed along the ray and on external parameters such as the wind but not on other wave components (first-generation wave models; see Section 6.4.7). This is a rather simple and economical

---

[13] On large (oceanic) scales, the straight lines should be interpreted as great-circles (the cross-section of the globe with a flat plane through the centre of the globe; the shortest distance between two points on the globe is measured along a great-circle). The wave direction, in the corresponding system of spherical co-ordinates, then slowly varies along a great-circle as the wave travels across the ocean (due to the convergence of the meridians towards the poles); thus, for spherical propagation, the wave direction cannot be said to be constant.

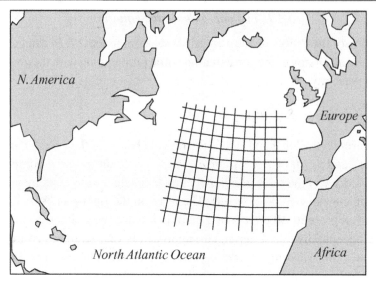

Figure 6.11 A regular (longitude, latitude) grid for the Eulerian approach of wave modelling in oceanic waters. The energy balance is considered for *each* individual wave component in *each* individual geographic cell.

approach and it does provide reasonable results, but with present-day computers it is no longer needed and it will not be treated here.[14]

The Eulerian formulation treats the energy balance of the waves on a regular geographic grid, either a Cartesian x, y-grid (for small areas) or a longitude-latitude λ, φ-grid (for larger areas; see Fig. 6.11). To derive the local energy balance for this approach, consider one cell of the geographic grid (size Δx in the x-direction and Δy in the y-direction); see Fig. 6.12. The energy balance for this cell (and all others in the grid) is essentially the bookkeeping of the energy of an arbitrary wave component (f, θ) travelling through this cell, i.e., balancing the change of energy in the cell over time interval Δt against the net import and the local generation of energy:

change of energy in cell = net import of energy

+ local generation of energy   (6.4.2)

The term on the left-hand side of this balance is equal to the energy in the cell at the end of the interval, minus the energy in the cell at the start of the interval.

---

[14] Another alternative, which is not used very often, is a hybrid approach, which combines the Lagrangian approach (for propagating the waves) with an Eulerian approach (to determine the source term).

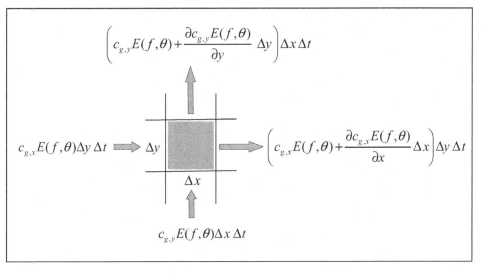

Figure 6.12 The energy propagation through one cell of the regular grid projected onto the ocean in the Eulerian approach.

Ignoring the dependence on x, y and t in the notation, this can be written as (see Fig. 6.12)

$$change\ of\ energy\ in\ cell = \left(E(f, \theta)\Delta x\, \Delta y + \frac{\partial E(f, \theta)}{\partial t} \Delta x \Delta y \Delta t \right)$$
$$- E(f, \theta)\Delta x \Delta y$$
$$= \frac{\partial E(f, \theta)}{\partial t} \Delta x \Delta y \Delta t \qquad (6.4.3)$$

The first term on the right-hand side of the energy balance of Eq. (6.4.2) is the *net* import of energy into the cell during interval Δt. For the x-direction it is equal to the energy import through the left-hand side of the cell (with propagation speed $c_{g,x} = c_g \cos\theta$; the cell width is Δy) minus the energy export through the right-hand side of the cell (with an energy transport that has evolved over the distance Δx; see Fig. 6.12):

$$net\ import\ of\ energy\ in\ the\ x\text{-}direction = c_{g,x} E(f, \theta)\Delta y \Delta t$$
$$- \left(c_{g,x} E(f, \theta) + \frac{\partial c_{g,x}\, E(f, \theta)}{\partial x} \Delta x \right)\Delta y \Delta t$$
$$= -\frac{\partial c_{g,x}\, E(f, \theta)}{\partial x} \Delta x \Delta y \Delta t \qquad (6.4.4)$$

Similarly, the net import of energy in the y-direction during the interval Δt is

$$net\ import\ of\ energy\ in\ the\ y\text{-}direction = -\frac{\partial c_{g,y}\,E(f,\theta)}{\partial y}\Delta x\Delta y\Delta t \qquad (6.4.5)$$

The second term on the right-hand side of the energy balance of Eq. (6.4.2) represents the locally generated energy in the cell, during the time interval Δt:

$$locally\ generated\ energy = S(f,\theta)\Delta x\Delta y\Delta t \qquad (6.4.6)$$

where $S(f,\theta)$ is the source term, representing all effects of generation, wave–wave interactions and dissipation per unit time per unit surface area. So, in total, the energy balance for the cell $\Delta x\Delta y$ over the time interval Δt is (substitute Eqs. 6.4.3–6.4.6 into Eq. 6.4.2)

$$\frac{\partial}{\partial t}E(f,\theta)\Delta x\Delta y\Delta t = -\frac{\partial c_{g,x}\,E(f,\theta)}{\partial x}\Delta x\Delta y\Delta t$$

$$-\frac{\partial c_{g,y}\,E(f,\theta)}{\partial y}\Delta x\Delta y\Delta t + S(f,\theta)\Delta x\Delta y\Delta t \qquad (6.4.7)$$

where $c_{g,x} = c_g\cos\theta$ and $c_{g,y} = c_g\sin\theta$ and c_g is the propagation speed of wave energy (the group velocity, see Eqs. 5.4.31 and 5.5.13). Dividing all terms by $\Delta x\,\Delta y\,\Delta t$ and moving the transport terms to the left-hand side gives the Eulerian spectral energy balance equation for each wave component, each cell, at each moment in time. Adding the dependence on time and horizontal space again in the notation gives

$$\boxed{\;\frac{\partial E(f,\theta;x,y,t)}{\partial t} + \frac{\partial c_{g,x}\,E(f,\theta;x,y,t)}{\partial x} + \frac{\partial\,c_{g,y}\,E(f,\theta;x,y,t)}{\partial y} \\ = S(f,\theta;x,y,t) \qquad \text{deep water}\;} \qquad (6.4.8)$$

> **This spectral energy balance is the third most important concept in this book.**

In *deep* water (not in shallow water), the propagation speeds $c_{g,x}$ and $c_{g,y}$ are independent of x and y, and they can be taken out of the derivatives (in Eq. 6.4.8):

$$\frac{\partial E(f,\theta)}{\partial t} + c_{g,x}\frac{\partial E(f,\theta)}{\partial x} + c_{g,y}\frac{\partial E(f,\theta)}{\partial y} = S(f,\theta) \qquad (6.4.9)$$

This equation is mathematically identical to the energy evolution equation along a wave ray in *deep* water (see Eq. 6.4.1): $dE(f,\theta;x,y,t)/dt = S(f,\theta;x,y,t)$. In coastal waters, where the wave length and direction are affected by the bottom topography, see Chapters 7 and 8, such an identity does not exist.

The source term $S = S(f, \theta; x, y, t)$ is often written as

$$S = S_{in} + S_{nl4} + S_{wc} \qquad (6.4.10)$$

with S_{in}, S_{nl4} and S_{wc} representing, separately, the processes of wave generation by wind, quadruplet wave–wave interaction and dissipation by white-capping (see Note 6E). These processes will be treated in the following but it must be noted that our understanding of these processes is far from complete. The quadruplet wave–wave interactions in deep water are well understood; the generation by wind is only reasonably well understood; and dissipation by white-capping is barely understood. The formulations that represent the last two processes in operational wave models are therefore to a large extent empirical (i.e., based on observations, intuition, speculation and calibrations).

NOTE 6E   Nonlinear processes and the random-phase/amplitude model (oceanic waters)

The source term $S = S_{in} + S_{nl4} + S_{wc}$ represents two types of processes:
(a) *Wind–wave interaction*, i.e., the process of wave generation by wind (S_{in}), which is usually represented by the feedback theory of Miles (1957; see Section 6.4.3). In this theory, the wave components are treated as independent, in compliance with the random-phase/amplitude model of the waves (see Section 3.5.2). The extension of this theory by Janssen (1991a) involves some degree of wave–wave interaction (the energy transfer from wind to one wave component depends on other wave components). Strictly speaking, this violates the basis of the random-phase/amplitude model.
(b) *Wave–wave interactions* are processes that, for fairly steep waves, only, (1) transfer energy from one wave component to another and (2) may couple the phases of the wave components involved. In deep water, these processes are mostly quadruplet wave–wave interactions and white-capping ($S_{nl4} + S_{wc}$). Such energy transfer and phase-coupling is a violation of the random-phase/amplitude model, in which the components are assumed to be independent. An obvious example is provided by the Stokes wave (see Section 5.6.2), in which the phases of the primary harmonic and its higher harmonics are equal. Such phase-coupling is also evident in the shape of breaking waves (sharp crests) and in the deviation from the Gaussian model for the instantaneous surface elevation (see Section 4.3).
In advanced wave-prediction models (the third generation, see Section 6.4.7), energy transfer is accounted for with a separate source term in the spectral energy balance equation, but phase-coupling is not. Ignoring the phase-coupling may be *locally* important for describing the shape of individual waves and for the short-term statistics of the waves. However, considering the very reasonable results of these oceanic wave models, it does not seem to be important on a larger scale, in deep water.

For larger areas and certainly for global scales, where longitude–latitude co-ordinates are required, the formulation of the energy balance of Eq. (6.4.8) needs to be modified to account for the effects of propagation on a sphere (i.e., great-circle propagation). The energy balance equation is then formulated, in spherical co-ordinates, as

$$\frac{\partial E(f,\theta;\lambda,\varphi,t)}{\partial t} + \frac{\partial c_{g,\lambda} E(f,\theta;\lambda,\varphi,t)}{\partial \lambda} + (\cos\varphi)^{-1}\frac{\partial c_{g,\varphi}\cos\varphi\, E(f,\theta;\lambda,\varphi,t)}{\partial \varphi}$$
$$+ \frac{\partial c_{\theta} E(f,\theta;\lambda,\varphi,t)}{\partial \theta} = S(f,\theta;\lambda,\varphi,t) \qquad (6.4.11)$$

where λ and φ are longitude and latitude, respectively, and $c_{g,\lambda}$ and $c_{g,\varphi}$ are the group velocity components in longitude and latitude directions respectively, $c_{g,\lambda} = (c_g \sin\theta)/(R\cos\varphi)$ and $c_{g,\varphi} = (c_g \cos\theta)/R$, and c_{θ} is the turning rate of the wave direction due to the change in (nautical) direction as the wave travels along a great-circle: $c_{\theta} = c_g \sin\theta \tan\varphi/R$, where R is the Earth's radius (the oblateness of the Earth is normally ignored in wave models).

The above Eulerian approach of modelling waves in the ocean is represented by only one equation: the energy balance equation Eq. (6.4.8) or Eq. (6.4.11), but the integration of this equation over space and time involves a very large number of points in geographic space and time and a large number of wave components. For each combination of these points and components this equation must be computed. *This* number of equations is very large: it is equal to the number of frequencies in the spectrum ($\approx$30, say), times the number of directions in the spectrum ($\approx$36, say), times the number of grid points in the geographic grid ($\approx$10 000, say). The total is therefore easily 10 000 000 equations, which need to be computed at every time step of (typically) 15 min for the integration in time (i.e., about 500 times for a five-day forecast)! This illustrates the considerable computing power that is needed for an Eulerian oceanic wave model. The problem of the wave components being *inter*dependent, which forced the move from a Lagrangian approach to this Eulerian approach, is now properly solved, but at a considerable price.

Literature:
Cavaleri and Malanotte-Rizzoli (1981), Gelci *et al.* (1956), Komen *et al.* (1994), Lavrenov (2003), Snodgrass *et al.* (1966), SWAMP (1985).

6.4.2 Wave propagation and swell

The linear theory of surface gravity waves shows that in deep water the propagation speed of the wave energy depends on the frequency of the wave component considered. The energy of low-frequency waves therefore travels faster than the energy of

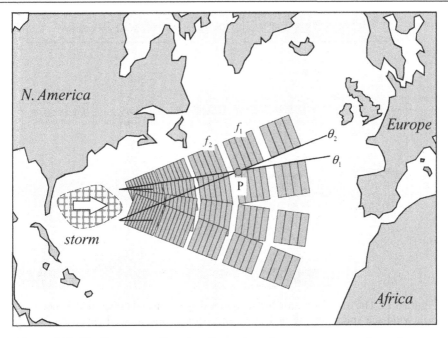

Figure 6.13 The frequency–direction dispersion of ocean waves transforms irregular, short-crested wind-sea waves in a storm into regular, long-crested swell outside the storm.

high-frequency waves. The initially random wave field, as generated in a storm, will therefore disintegrate when it moves out of the storm. It will disintegrate in fields of more regular waves in the direction of propagation, with the low frequencies in the lead and the high frequencies in the trailing edge (see Fig. 6.13). This process is called frequency-dispersion. In addition to growing more and more regular, the waves will also change from short-crested to long-crested because the waves in the storm travel in a range of directions and the initial wave field will disintegrate in these directions (see Fig. 6.13). This is called direction-dispersion. Waves that have thus dispersed across the ocean are called 'swell'.

To illustrate these dispersion phenomena with a simple, qualitative model, consider a storm off the coast of Florida with a dominant westerly wind direction and no wind outside that storm (see Fig. 6.13). Each wave component that is generated in the storm leaves the storm more or less as a rectangular field of wave energy with a width that is roughly equal to the width of the storm and a length that is determined by the length and duration of the storm. Each such wave field leaving the storm travels at the group velocity of the corresponding wave component and in the direction of that component. The fastest travelling waves (i.e., the lowest frequencies) will lead the forward propagation, while the directional spreading in

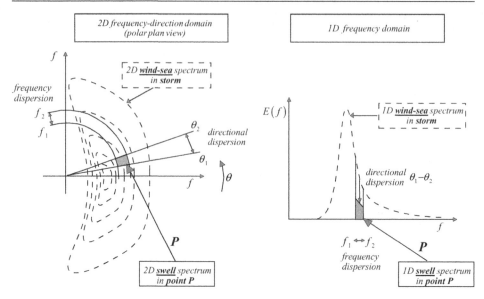

Figure 6.14 The transformation by frequency dispersion and directional dispersion of a wind-sea spectrum (dashed lines; in a storm) into a swell spectrum (solid blocks; at geographic location P in Fig. 6.13).

the storm spreads the wave fields laterally. The initially random, short-crested wave field therefore disintegrates into a large number of individual regular, long-crested wave fields due to these differences in propagation speed and direction.[15]

At some moment in time, the leading low-frequency wave components have arrived at an arbitrarily chosen point P in the ocean (see Fig. 6.13). After some time, these components have passed that point, to be followed by other components with slightly higher frequencies. These too will pass. After a while, therefore, all components with frequencies lower than f_1, say, have passed point P, whereas wave components with higher frequencies (higher than f_2, say) have not yet arrived. Only components with frequencies between f_1 and f_2 are present at point P. This implies that, at point P, wave energy is present only in the frequency range $f_1 - f_2$ (see Fig. 6.14). This causes the waves at point P to be much more regular than those in the storm. Since frequency-dispersion becomes more pronounced as the waves travel further across the ocean, this regular character increases with distance from the storm. A beautifully regular swell on a sunny tropical beach (to make it a little poetic) is the result of this frequency-dispersion of waves that have been generated in a far-away storm.

---

[15] This would be the case if the spectrum consisted of discrete frequencies and directions, as in numerical wave models. This effect is called the 'garden-sprinkler effect' (SWAMP, 1985; Booij and Holthuijsen, 1987). In the real ocean, the spectrum is continuous and the disintegration is evident only as a continuous spreading of the wave field.

The direction-dispersion has a similar filtering effect. The wave fields that spread out across the ocean are generated in the storm in many different directions (centred on some mean wind direction). Depending on their direction of propagation, some of these wave fields will pass *south* of point P (all wave fields with directions south of θ_1; see Figs. 6.13 and 6.14), while others will pass *north* of point P (all wave fields with directions north of θ_2). This implies that, at point P, wave energy is present only between directions θ_1 and θ_2, i.e., only in the angle of view from point P to the storm. This limited directional sector $\theta_1 - \theta_2$ causes the waves to be more long-crested at point P than they were in the storm. Since this directional dispersion is more pronounced as the waves travel further across the ocean, this long-crested character increases with distance from the storm. The beautifully regular swell on the sunny tropical beach is therefore not only regular but also long-crested.

Frequency- and direction-dispersion implies that, at any one time, the spectrum at point P contains energy only within a narrow frequency band and a narrow directional sector. The energy in this narrow spectrum is therefore only a fraction of the energy in the initial, broader spectrum in the storm. The waves at point P are therefore (much) lower than those in the storm, *solely due to dispersion; no dissipation is involved.*[16] This relatively simple model of frequency–direction-dispersion without dissipation has been well verified by tracking swell across the ocean, sometimes over more than half the Earth's circumference, e.g., from the southern Indian Ocean (near Antarctica) to Alaska along great-circles that pass through the gap between Australia and New Zealand.

Literature:
Barber and Ursell (1948), Barber (1969), Booij and Holthuijsen (1987), Gelci *et al.* (1964, 1970), Munk *et al.* (1963), Snodgrass *et al.* (1966).

6.4.3 Generation by wind

One candidate model for the initial generation of waves is the instability of the water surface layer in which the wind generates a current. Two fluids with different speeds, i.e., water and air in this case, will generate instabilities at their interface if the densities and current speeds differ enough (e.g., Lamb, 1932). This is obvious when the wind starts to blow over still water: the first waves to appear are small and very short, slowly getting longer and higher. Another candidate is a mechanism suggested by Phillips (1957), in which waves are generated by resonance between

---

[16] Ocean waves barely lose energy outside storms, because the waves are not steep enough to break and the effects of the viscosity of water are negligible. Early attempts at ocean-wave forecasting (in the 1950s) often ignored the dispersion effect and (erroneously) simulated the reduction in wave height with dissipation.

propagating wind-induced pressure waves (i.e., air pressure) at the water surface and freely propagating water waves. The actual situation is not well understood, but such understanding is fortunately not crucial for wave predictions since these are not very sensitive to the initial conditions (as long as *some* initial waves are present). When initial waves have thus been generated, Miles (1957) finds that these waves modify the airflow and hence the wind-induced pressure at the water surface in such a way that they enhance their own growth. Waves are therefore generated by wind-induced surface *pressure*, not by wind-induced surface *friction*.[17] Janssen (1991a, 1991b) extended the theory of Miles (1957) to include the effect of the waves on the entire lower atmosphere, which in turn affects the waves again. Advanced wave-prediction models are based on the Miles theory, sometimes supplemented with this extension of Janssen. Of these models, some also include Phillips' theory, but only to initiate wave growth.

Phillips' theory can be briefly summarised as follows. Initially, i.e., when the water surface is flat, the wind, by its very nature, induces a turbulent pressure on the water surface, propagating as a nearly frozen (random) field (see Fig. 6.15). In analogy with the random-phase/amplitude model of the water waves, this pressure field can be seen as the superposition of many harmonic *air-pressure* waves, all oriented in many different directions but all propagating in the wind direction (in contrast to the water waves, each of which travels in its own direction, normal to its crest). In the moving but (nearly) frozen pattern of wind pressure, some pressure components have the same speed, wave length and direction (i.e., normal to the pressure crest) as freely propagating water-wave components. These matching pressure waves transfer energy to their counterpart water waves by resonance. For a constant wind, Phillips (1957) estimates this transfer to be constant in time, resulting in a linear growth in time:[18]

$$S_{in,1}(f, \theta) = \alpha \qquad \text{with } \alpha = \alpha(f, \theta; \vec{U}_{wind}) \tag{6.4.12}$$

In most operational wave models, this resonance mechanism is ignored because, as in the real ocean, small waves are always present in these models to trigger wave growth. Alternatively, initial, small waves can simply be imposed in the model, or an empirical expression to generate initial waves can be used, for instance, the

---

[17] This statement should be a little more subtle than that. The atmosphere exerts a certain friction on the ocean surface (generating currents and a tilt of the mean ocean surface) and the ocean surface exerts a corresponding friction on the atmosphere (generating an atmospheric boundary layer). The problem is in the definition of 'friction'. It is essentially (in this case) the transfer of horizontal momentum in the vertical direction across the ocean/atmosphere interface. For instance, in a breaking wave, a mass of water with a horizontal velocity component is injected into the upper layer of the ocean, with a vertical movement. This, and many similar motions on (much) larger and smaller scales, may be seen as a horizontal 'friction' on the ocean surface.

[18] There is no relation between this coefficient α and the energy-scale parameter α of the high-frequency tail of the spectrum of Phillips (see Section 6.3.3).

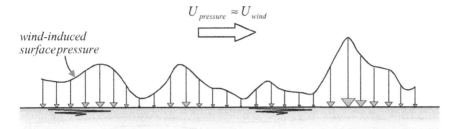

Figure 6.15 The normal wind-induced pressure moving as a (nearly) frozen distribution across the water surface.

expression of Cavaleri and Malanotte-Rizzoli (1981):

$$S_{in,1}(f, \theta) = \begin{cases} C_\alpha [u_* \cos(\theta - \theta_{wind})]^4 & \text{for } |\theta - \theta_{wind}| \leq 90° \text{ and } f \geq f_{PM} \\ 0 & \text{for all other wave components} \end{cases}$$

(6.4.13)

where f_{PM} is the Pierson–Moskowitz peak frequency (see Section 6.3.3), C_α is a (tunable) coefficient (see also Section 9.3.2), u_* is the (friction-) wind velocity (see Section 6.2) and θ_{wind} is the wind direction. The frequencies and directions are limited in range here, to ensure that only waves that are affected by the wind are generated by this mechanism.

In Phillips' theory it is assumed that the wind-induced surface pressure is a natural aspect of the wind itself, without influence of the waves. However, such influence is unavoidable and will increase as the waves evolve. Consider (as in Phillips' theory) only one wave component independently of the others (this makes it a linear approximation of the problem). The average wind profile above the water surface will then be disturbed by this harmonic water wave (see Fig. 6.16), the disturbance being greatest at the water surface itself and rather smaller at higher elevations. In his theoretical model, Miles (1957) finds that the air pressure *at the water surface* attains a maximum on the windward side of the wave crest and a minimum on the leeward side of the wave crest.[19] This implies that the wind effectively pushes the water surface down where the wave surface is moving down (the windward side of the crest) and pulls the water surface up where the surface is moving up (the leeward side of the crest). This out-of-phase coupling between pressure and surface motion transfers energy to the waves. Since this transfer depends on the amplitude of the water wave, it becomes more effective as the wave evolves. In other words, as

---

[19] Jeffreys (1925, 1926) suggested such a mechanism in a much-simplified form (with the wave-induced pressure being proportional to the local surface *slope*, i.e., 180° out of phase with the surface elevation) but it did not compare well with observations and has been abandoned.

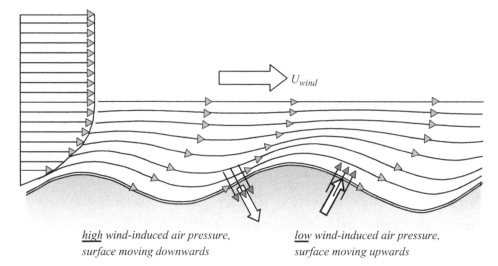

high wind-induced air pressure,   low wind-induced air pressure,
surface moving downwards   surface moving upwards

Figure 6.16 The wave-induced wind-pressure variation over a propagating harmonic wave.

the wave grows by this mechanism, the mechanism becomes more effective, so the wave can therefore grow faster, which in turn makes the mechanism even more effective, etc.

The process enforces itself: it is a positive-feedback mechanism. Miles (1957) formulated it as

$$S_{in,2}(f, \theta) = \beta E(f, \theta) \tag{6.4.14}$$

where the coefficient β depends on the speed and direction of the wind and the wave: $\beta \sim [U \cos(\theta - \theta_{wind})/c]^2$, where U is a reference wind speed (not necessarily U_{10}) and c is the phase speed of the water-wave component. Since this source term depends on the energy density itself, this formulation results in an exponential growth of $E(f, \theta)$ in time (for a constant wind). Measurements have qualitatively confirmed this effect and the coefficient β has been estimated from such measurements, for example as (see Fig. 6.17)

$$\beta = \varepsilon_1 \frac{\rho_{air}}{\rho_{water}} \left[\frac{u_* \cos(\theta - \theta_{wind})}{c} \right]^2 2\pi f \qquad \text{Plant (1982)} \tag{6.4.15a}$$

or

$$\beta = \varepsilon_2 \frac{\rho_{air}}{\rho_{water}} \left[28 \frac{u_*}{c} \cos(\theta - \theta_{wind}) - 1 \right] 2\pi f \qquad \text{Snyder } et\ al.\ (1981)$$

$$\tag{6.4.15b}$$

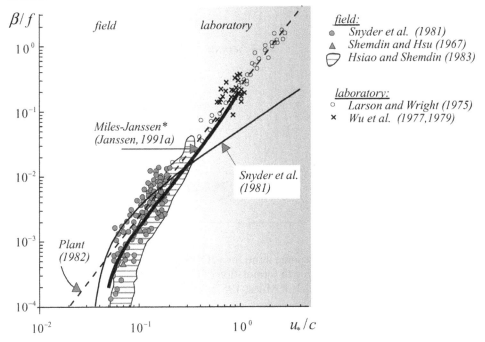

Figure 6.17 The coefficient of wind-induced wave growth (the Miles theory and inferred from measurements; after Hsiao and Shemdin, 1983; Belcher *et al.*, 1994 and Young**, 1999). * The solid line representing the Miles–Janssen theory is the line originally computed by Janssen (the line published in Janssen, 1991a, inadvertently deviates slightly from the above original line; personal communication P. A. E. M. Janssen, 2005). ** *Wind Generated Ocean Waves*, Elsevier Ocean Engineering Book Series, Vol. 2, by I. R. Young, p. 52, Copyright 1999, with permission from Elsevier.

where ρ_{air} and ρ_{water} are the densities of air and water, respectively, and ε is a tunable coefficient. Most wave models cut off the value of β in Eq. (6.4.15b) at $\beta = 0$, so as to avoid negative growth (i.e., energy transfer from the waves to the wind).[20]

In the above theory of Miles (1957), the wind is decoupled from the waves in the sense that the waves do not affect the mean wind (they affect only the wind-pressure fluctuations in the surface layer). Janssen (1991a) has developed a version of this theory in which the wave–atmosphere system is treated as a coupled system and shows that the waves actually have some influence on the entire lower atmosphere

---

[20] A negative value would imply transfer of wave energy to the wind, i.e., waves would generate wind. This may well be realistic, although the transfer would be only a small fraction of the energy transfer from wind to waves. To illustrate this: I felt a light but noticeable *breeze* when standing at the end of the 80 m × 50 m *indoor* Ocean Basin of the Norwegian Marine Technology Research Institute (Trondheim), in which breaking waves of height 0.5–1 m were being generated mechanically (with computer-controlled flaps). Such transfer of energy from waves to wind is predicted by some alternative wave-generation theories, e.g., Chalikov (1986).

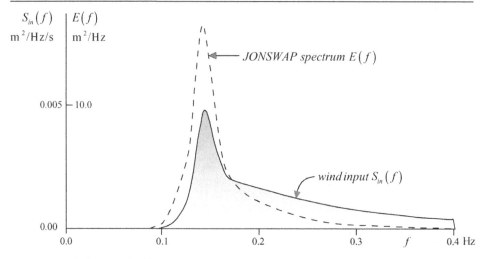

Figure 6.18 The wind input source term, for a JONSWAP spectrum in deep water (computed with the initial-growth formulation of Cavaleri and Malanotte-Rizzoli, 1981, and the feedback model of Miles, 1957; for $H_{m_0} = 3.5$ m, $T_{peak} = 7$ s and $U_{10} = 20$ m/s).

(the wave-induced surface friction for the lower atmosphere is larger for young sea states than for older sea states), thus somewhat affecting the evolution of mid-latitude storms.

In summary, the source term for the generation of waves by wind can be written as

$$\boxed{S_{in}(f, \theta) = \alpha + \beta E(f, \theta)}$$ (6.4.16)

The shape of this source term (integrated over directions) for a JONSWAP spectrum is shown in Fig. 6.18, using the formulations of Cavaleri and Melanotte-Rizzoli (1981) and Snyder *et al.* (1981). Apparently, most of the energy transfer from wind to waves occurs at the spectral peak and on its high-frequency side.

The observed total input of energy to the waves seems to be explained sufficiently well by the above theory of Miles. However, the measurements that were used to validate the theory (measuring the wave-induced variations of the wind-induced air pressure at the wave surface, using such instruments as a wave follower; see Section 2.3.3) are very difficult to carry out and a fairly large uncertainty still exists. For instance, the scatter in the values of β inferred from such measurements is of the order of a factor of ten (see Fig. 6.17), which illustrates the uncertainty in our understanding of the generation of wind waves and the room that is still left for alternative or supplementary theories. For instance, wave breaking induces flow separation of the air motion at the crest of the wave, which locally strongly enhances the transfer of energy from wind to waves. Another example is the result of detailed numerical simulations of the airflow in the lower atmospheric boundary layer (the layer just above the waves with a thickness of about the significant wave height). This work has shown that the turbulence of the airflow (which in turn is determined by the atmospheric stability) is important and that the actual energy transfer to the waves would be only a fraction of the energy transfer given by Eqs. (6.4.15). Such lower transfer would be compensated by less dissipation due to white-capping to achieve the same overall wave growth.

Literature:
Banner and Melville (1976), Banner (1990b), Belcher *et al.* (1994), Benoit *et al.* (1996), Booij *et al.* (1999), Burgers and Makin (1993), Chalikov (1986), Elliott (1972a, 1972b), Hsiao and Shemdin (1983), Janssen and Komen (1985), Janssen and Viterbo (1996), Jones and Toba (2001), Kawai (1982), Komen *et al.* (1984), Miles (1960), Peirson and Belcher (2004), Phillips (1977), Plant (1982), Shemdin and Hsu (1967), Snyder *et al.* (1981), Tolman and Chalikov (1996), Voorrips *et al.* (1994), WAMDI group (1988), Willmarth and Wooldridge (1962), Yan (1987), Young and Sobey (1985, 1988), Young (1999).

6.4.4 Nonlinear wave–wave interactions (quadruplet)

The second mechanism that affects wave growth in deep water is the transfer of energy *amongst* the waves, i.e., from one wave component to another, by resonance. To visualise such nonlinear wave–wave interactions, consider a large wave tank with constant water depth (Fig. 6.19). One machine generates harmonic waves in one corner of the tank. Another machine generates waves in another corner with a different frequency and in a different direction. The resulting waves in the tank create a diamond pattern of crests and troughs, which has its *own* wave length, speed and direction (and hence its own wave number; see Fig. 6.19).

Such a diamond pattern would interact with a third, freely propagating wave component (i.e., one obeying the dispersion relationship of the linear wave theory) if this third wave had the same wave length, speed and direction as the diamond pattern. The original pair of wave components would thus interact with this third wave component if the proper conditions were met (triad wave–wave interaction). Such interaction between freely propagating waves is called resonance. At the end of the tank we would see emerging not only the two mechanically generated waves but also the third wave generated by the resonance interaction. Or, if the third wave already existed, the three waves would have exchanged energy amongst themselves. Each of the components would thus have lost or gained energy, but the total energy (the *sum* of the energy of the three components at each point in the tank) would remain constant. Now, this was only to visualise this type of nonlinear wave–wave interaction. In deep water, the resonance conditions (the

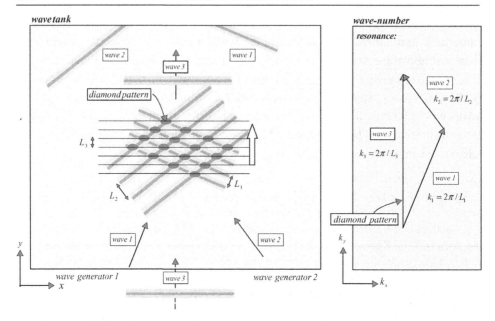

Figure 6.19 *Triad* wave–wave interactions (*not realisable in deep water*). A hypo-thetical wave-tank experiment with one pair of mechanically generated, freely propagating waves, interacting with a third, freely propagating wave. The wave-number vectors of the three wave components and of the diamond pattern are shown in the right-hand panel in wave-number space: the wave-number vector of the third wave is equal to the wave number of the diamond pattern, which is equal to the sum of the wave numbers of the original two wave components: $\vec{k}_3 = \vec{k}_1 + \vec{k}_2$. (For the concept of wave-number vectors, see Section 3.5.8).

matching of wave speed, length and direction) *cannot* be met by three freely prop-agating wave components. Triad wave–wave interactions therefore do not occur in deep water. However, it is possible in deep water to have one *pair* of wave compo-nents interacting with another *pair*, if the wave numbers (and frequencies) of the two corresponding diamond patterns match (see Fig. 6.20). The reason is that, in deep water, two such pairs, i.e., *four* wave components, *can* fulfil the resonance conditions and can thus resonate. This matching of frequencies and wave numbers is expressed with the resonance conditions:

$$f_1 + f_2 = f_3 + f_4$$
$$\vec{k}_1 + \vec{k}_2 = \vec{k}_3 + \vec{k}_4$$

(6.4.17)

(For the concept of wave-number vectors, see Section 3.5.8.)

These resonance conditions state that, *if* the frequency, wave number and direc-tion of one diamond pattern coincide with those of another diamond pattern, then

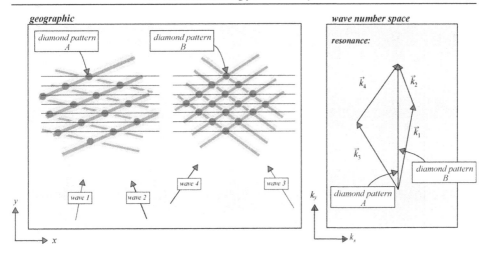

Figure 6.20 *Quadruplet* wave–wave interactions (realisable in deep water). Two pairs of wave components can create two diamond patterns with identical wave lengths and directions and therefore identical wave numbers. When the four waves are superimposed (not shown here), they can thus resonate. The wave-number vectors of the four wave components are shown in the right-hand panel in wave-number space with $\vec{k}_1 + \vec{k}_2 = \vec{k}_3 + \vec{k}_4$.

energy is transferred amongst the four free components involved. Such a set of four wave components is called a quadruplet and the interactions are called *quadruplet wave–wave interactions*. The full expressions for these interactions have been given by Hasselmann (1962). They can be written in the following form (a Boltzmann integral; see Hasselmann, 1960, 1962, 1968):

$$
S_{nl4}(\vec{k}_4) = \iiiint T_1(\vec{k}_1, \vec{k}_2, \vec{k}_1 + \vec{k}_2 - \vec{k}_4) E(\vec{k}_1) E(\vec{k}_2) E(\vec{k}_1 + \vec{k}_2 - \vec{k}_4) d\vec{k}_1 d\vec{k}_2
$$
$$
- E(\vec{k}_4) \iiiint T_2(\vec{k}_1, \vec{k}_2, \vec{k}_4) E(\vec{k}_1) E(\vec{k}_2) d\vec{k}_1 d\vec{k}_2
$$

(6.4.18)

where $\vec{k}_4$ is the vector wave number of the wave component considered in the source term, $\vec{k}_1$, $\vec{k}_2$ and $\vec{k}_3 = \vec{k}_1 + \vec{k}_2 - \vec{k}_4$ are the three other wave components involved, subject to the resonance conditions of Eq. (6.4.17), and T_1 and T_2 are transfer coefficients, which are complicated functions of the wave-number vectors involved. The first integral represents the 'passive' part of the interactions, i.e., it is independent of the energy density $E(\vec{k}_4)$ of wave component $\vec{k}_4$. The second integral corresponds to the 'active' part, i.e., it *does* depend on $E(\vec{k}_4)$. Whether the energy of the wave component $\vec{k}_4$ grows or decays depends on the balance between the active and passive terms.

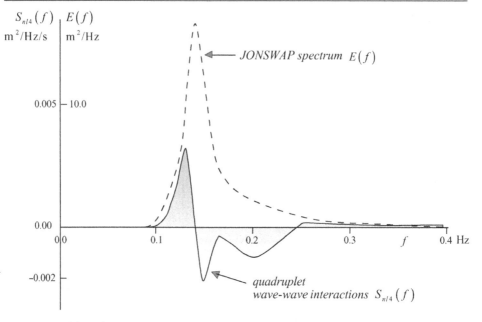

Figure 6.21 The source term for quadruplet wave–wave interactions, for a JON-SWAP spectrum in deep water (computed with the near-exact WRT technique coded by van Vledder, 2006, see also van Vledder and Bottema, 2002; for $H_s = 3.5$ m and $T_{peak} = 7$ s). The pronounced 'deficit' of interactions, in this example around 0.165 Hz, occurs at the transition between the peak enhancement function and the f^{-5}-tail of the spectrum (see Section 6.3.3). It indicates that the quadruplet wave–wave interactions are trying to smooth this transition. The WRT technique is named after the developers, Webb, Resio and Tracy (see Webb, 1978, and Tracy and Resio, 1982; see also Resio and Perrie, 1991).

It should be emphasised that the quadruplet wave–wave interactions only redistribute energy over the spectrum. No energy is added or withdrawn from the spectrum as a whole. The shape of this source term (integrated over directions) for a JONSWAP spectrum is shown in Fig. 6.21.

The $+ / - / +$ character of this source term, at least for a JONSWAP-type spectrum (with zero at the peak frequency; see Fig. 6.21), implies that the quadruplet interactions transfer a significant fraction of the wind input from the mid-range frequencies to lower frequencies and a small fraction to higher frequencies. At the high frequencies, white-capping dissipates this energy. At the low frequencies, the energy is absorbed without appreciable dissipation. The energy at the low frequencies therefore grows, shifting the peak of the spectrum to lower frequencies, and thus dominating the evolution of the spectrum.

A remarkable property of the quadruplet interactions is their capacity to stabilise the shape of the spectrum of steep waves, i.e., waves that are being generated by

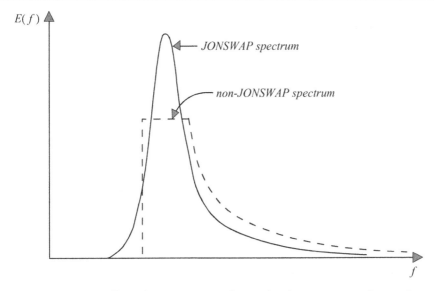

Figure 6.22 For sufficiently steep waves, the quadruplet wave–wave interactions will force a non-JONSWAP spectrum into a JONSWAP spectrum (deep water).

wind.[21] If, for instance, the tail of the spectrum deviates locally from a smooth f^{-4}-shape (e.g., there is a small hump in the tail), the quadruplet interactions will smooth the tail back to approximately the universal f^{-4}-shape. Alternatively, if the shape deviates considerably from the JONSWAP shape (see Fig. 6.22), the quadruplet interactions will force it (back) into a JONSWAP shape (with an f^{-4}-tail, see Note 6C). This is the main reason why the JONSWAP spectrum is often observed in storms, which do not even approximate the idealised situation in which the JONSWAP spectrum was first observed (see Section 6.3.3). Of course, it is the balance amongst the wind input, the quadruplet wave–wave interactions and the white-capping that determines the actual shape of the spectrum. If the wind is strong and highly variable (in speed or direction), the wind input may have a stronger effect on the shape of the spectrum than the quadruplet wave–wave interactions and a non-JONSWAP spectrum may evolve. However, if the wind varies sufficiently slowly, the quadruplet wave–wave interactions will dominate and a JONSWAP spectrum evolve. This is usually the case in a storm and even in large parts of a hurricane (at distances from the hurricane's centre less than about ten times the radius to maximum wind; at larger distances the wave field is a mix of wind sea and swell).

This shape-stabilising effect of the quadruplet wave–wave interactions is the reason why the JONSWAP spectrum is widely accepted as the design spectrum

---

[21] The quadruplet interactions seem to be also responsible for the directional bimodal shape of the wave spectrum under idealised conditions; see Note 6D.

for engineering purposes (see also Section 6.3.3). However, if the wind drops, or the waves leave their generation area, the steepness of the waves reduces sharply (due to frequency-dispersion and direction-dispersion; see Section 6.4.2) and the quadruplet wave–wave interactions decrease accordingly. Under swell conditions, therefore, a JONSWAP spectrum is not be expected. In fact, the spectral shape then depends entirely on the history of the individual wave components, which may be very different for different portions of the spectrum. At sea, this usually results in a spectrum with multiple swell peaks and a locally generated JONSWAP spectrum at higher frequencies.

The computation of the quadruplet wave–wave interactions requires considerable computer resources because of the large number of quadruplets involved (each wave component interacts in a large number of quadruplet combinations). Considerable efforts have therefore been devoted to finding approximations that will reduce such computer requirements but retain the basic characteristics of these interactions. One of these approximations has been adopted by advanced operational wave models, namely the discrete-interaction approximation (DIA) of Hasselmann *et al.* (1985a), which considers the interactions of each wave component in the spectrum in only two quadruplets. For each of these two quadruplets, one computes the interactions between the wave component under consideration and itself (self–self interaction) and two other components (bringing the total formally to four wave components). The two quadruplets differ only in that the directions of two of the four components have reversed sign (i.e., mirror directions with respect to the first quadruplet). Both quadruplet configurations are applied to all wave components in the spectrum. The DIA thus reduces the computational time to manageable proportions and enables operational wave models to incorporate these interactions. Such explicit calculation of the quadruplet wave–wave interactions is the factor discriminating between third-generation wave models on the one hand and first- and second-generation wave models on the other (in which these interactions are absent or severely parameterised; see Section 6.4.7).

Literature:
Hasselmann (1960, 1962, 1963a, 1963b), Phillips (1960), Young and van Vledder (1993), Young (1997, 1999).

6.4.5 Dissipation (white-capping)

Wave breaking in deep water (called white-capping) is a very complicated phenomenon, which so far has defied theoretical understanding. It involves highly nonlinear hydrodynamics on a wide range of scales, from gravity surface waves to capillary waves, down to turbulence. A complicating factor is that there is no generally accepted precise definition of breaking and quantitative observations are very difficult to carry out. Not surprisingly, therefore, breaking is the least understood of

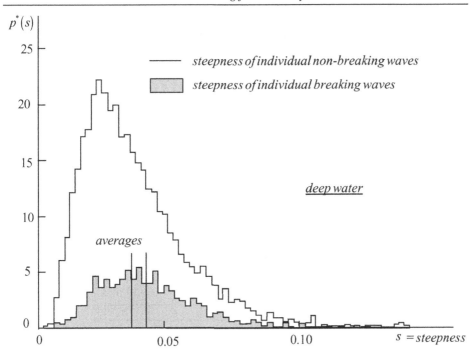

Figure 6.23 The distribution of observed steepness $(s = H/L)$ of individual breaking and non-breaking waves under high-wind conditions in deep water (both probability density functions have been scaled such that $\int p^*(s)ds$ is the probability of breaking or non-breaking; after Holthuijsen and Herbers, 1986). The average steepness of non-breaking waves is 0.036; the average steepness of breaking waves is 0.042; the probability of breaking is 0.12.

all processes affecting waves. Some speculations as to what controls wave breaking have been made and it seems reasonable to assume that it is the wave steepness. For instance, Miche (1944) has shown theoretically that the maximum wave height H_{max} for a fixed-form, periodic wave (i.e. the maximum wave height of an individual wave) is determined by the fact that the particle velocity u_x in the crest cannot be larger than the forward speed of the wave c (i.e., always $u_x \leq c$). This results in

$$H_{max} \approx 0.14L \tanh\left(\frac{2\pi d}{L}\right) \qquad (6.4.19)$$

which for *deep* water gives a maximum steepness of the *individual* wave $s_{max} = H_{max}/L \approx 0.14$ (wave breaking in shallow water is addressed in Section 9.3.4). However, *observations* at sea (deep water) have shown that whether an individual wave is breaking or not is almost independent of the steepness of that wave (but $H/L = 0.14$ seems to be an upper limit; see Fig. 6.23). Perhaps the degree of randomness or short-crestedness is just as important as steepness.

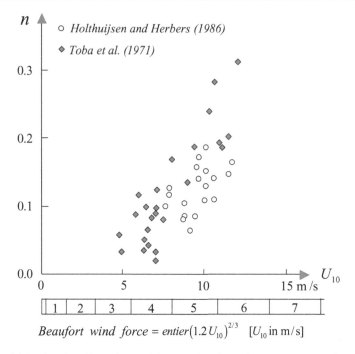

Figure 6.24 The visually estimated fraction n of breaking waves as a function of wind speed (n is the number of observed breaking waves/total number of observed waves, at a fixed horizontal position); after Holthuijsen and Herbers (1986).

The breakers in the open ocean are called '*white-caps*' and sometimes '*white horses*' (in other languages, other poetic names are used, e.g., '*jumping rabbits*' in Japanese). The occurrence of white-caps is essentially a characteristic of the sea state itself, but white-capping is obviously closely related to the wind (see Fig. 6.24). In fact, the Beaufort wind-force scale is based, to a large extent, on the white-cap coverage of the ocean surface. White-caps are not only important for the evolution of ocean waves; they also play a key role in the exchange of gas across the air–sea interface and the production of airborne droplets and aerosols (which are needed as condensation particles for precipitation over the ocean and over land; see Monahan and MacNiocaill, 1986).

The (dissipative) effect of white-capping on the evolution of the waves is *locally* highly nonlinear, but on *average*, i.e., averaged over a large number of waves, it is rather weak. In wave models it is therefore represented as a source term in the energy balance of the waves. In spite of the uncertainty about the relevance of wave steepness, several approaches to deriving such a source term are based on this assumption. The best-known is the theory of Hasselmann (1974) in which each white-cap acts as a pressure pulse on the sea surface, just downwind of the

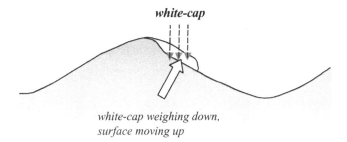

Figure 6.25 The white-cap as a pressure pulse at the lee-wind side of the crest of a breaking wave.

crest (see Fig. 6.25). At that location in the wave, the weight of the white-cap acts against the rising sea surface, thus draining energy from the wave. This is almost the mirror-image of the feedback mechanism for wind-induced growth proposed by Miles (1957): the white-cap drains energy *from* the wave (transporting it to surface currents and turbulence) at roughly the same location as where the wind transfers energy *to* the wave. In other words, the weight of the white-cap counteracts, to some extent, the pulling effect of the deficit in air pressure on the lee side of the wave crest.

The theory of Hasselmann (1974) gives only a general form for the white-capping source term:

$$S_{wc}(f, \theta) = -\mu k E(f, \theta)$$ (6.4.20)

where k is the wave number and μ is a coefficient representing some statistical property of the white-caps. The value of μ is expressed in terms of (unknown) integrals over the entire spectrum:

$$\mu = \mu \left[\iint \cdots E(f, \theta) \mathrm{d}\theta \, \mathrm{d}f \right]$$ (6.4.21)

which makes the source term for white-capping quasi-linear: Eq. (6.4.20) is linear in the spectral density but the coefficient μ depends on the entire spectrum. The value of the white-capping coefficient μ has been estimated with some informed speculation on its character (e.g., that it should depend on the overall wave steepness) and by calibrating a numerical wave model with observed wave conditions (Bouws and Komen, 1983; Komen *et al.*, 1984):

$$\mu = C_{wc} \left(\frac{\tilde{s}}{\tilde{s}_{PM}} \right)^4 \frac{\tilde{f}}{\tilde{k}}$$ (6.4.22)

where C_{wc} is a tunable coefficient, $\tilde{s}$ is the overall wave steepness (involving integrals over the spectrum), $\tilde{s}_{PM}$ is the value of $\tilde{s}$ for the Pierson–Moskowitz spectrum and $\tilde{f}$

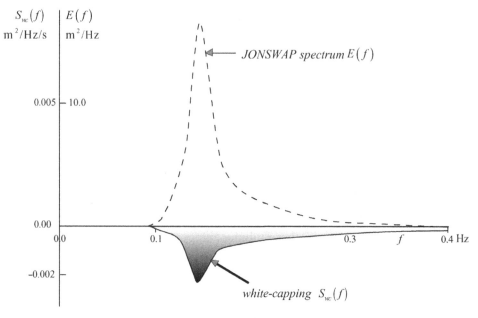

Figure 6.26 The white-capping source term, for a JONSWAP spectrum in deep water (computed with the pulse model of Hasselmann, 1974; for $H_s = 3.5$ m and $T_{peak} = 7$ s).

and $\tilde{k}$ are the mean frequency and wave number, respectively (see details in Section 9.3.4). The corresponding shape of the source term is shown in Fig. 6.26 for a JONSWAP spectrum.

Literature:
Alves and Banner (2003), Babanin *et al.* (2001), Banner and Grimshaw (1991), Banner *et al.* (2000), Booij and Holthuijsen (2001), Bortkovskii (1983), Bouws and Komen (1983), Ding and Farmer (1994), Donelan and Yuan (1984), Douglass (1990), Hasselmann (1974), Holthuijsen and Herbers (1986), Janssen (1991a, 1991b), Katsaros and Atatürk (1991), Komen *et al.* (1984), Longuet-Higgins (1969, 1987), Monahan and MacNoicaill (1986), Nepf *et al.* (1998), Ochi and Tsai (1983), Oh *et al.* (2004), Peregrine (1999), Resch *et al.* (1986), Ross and Cardone (1974), Woolf and Thorpe (1991).

6.4.6 Energy flow in the spectrum

The source terms have been illustrated in the above Sections for a JONSWAP spectrum as functions of frequency at one moment in time. A summary of this is given in Fig. 6.27. The transfer of energy from the wind to the waves occurs mostly near the peak of the spectrum and at the mid-range frequencies but the corresponding energy gain at these frequencies is rapidly removed by white-capping and quadruplet wave–wave interactions (transporting the energy to the higher and

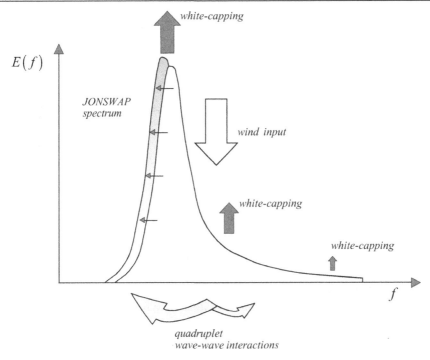

Figure 6.27 The flow of energy through an evolving JONSWAP spectrum (deep water).

lower frequencies). The total effect for the mid-range frequencies is negative: they lose energy. At the higher frequencies, where energy is received from the mid-range frequencies, white-capping immediately dissipates this energy, thus balancing, at these frequencies, the three processes of generation, quadruplet wave–wave interactions and white-capping. The energy level at these frequencies is therefore more or less in equilibrium (it slowly oscillates a little). At the low frequencies (below the peak frequency), the energy that is received from the mid-range frequencies is absorbed. Together with some energy transfer from the wind, this makes the spectral peak migrate towards these frequencies. Surprisingly, therefore, the spectral growth is not due to a *direct* interaction with the wind (that is only a minor contribution). Rather, it is due to an *indirect* interaction: the wind energy is received by the mid-range frequencies and then transferred by the quadruplet wave–wave interactions to the lower frequencies, where the spectrum grows.

It is also possible to consider the evolution of a *single* wave component separately, as a function of time. If one such wave component is followed as it travels through the area of generation (take one fixed frequency f_0 and one fixed direction θ_0), its evolution follows directly from the above (see Fig. 6.28). Initially, the wave component is in the low-frequency range of the spectrum (because the initial

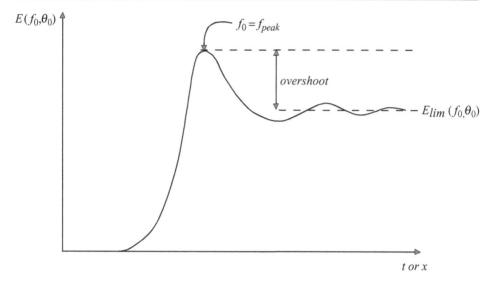

Figure 6.28 An artist's impression of the evolution of the energy density of a wave component under constant-wind conditions, as a function of time or distance (deep water).

spectrum is located at high frequencies). As the spectrum evolves, i.e., migrates to lower frequencies, the frequency under consideration (which remains constant) becomes the peak frequency and then enters the mid-frequency range of the spectrum. Eventually it ends up in the high-frequency range of the evolving spectrum (all relative to the ever-decreasing peak frequency). As a low-frequency component in the spectrum, it grows exponentially due to the feedback wind-generation mechanism and the quadruplet wave–wave interactions which are being fed by energy from higher frequencies, but only up to a certain maximum when the frequency under consideration becomes the peak frequency of the evolving spectrum. The energy then decays to a lower level, due to wave–wave interactions (transporting energy to higher and lower frequencies) and white-capping. Eventually, the frequency under consideration ends up in the high-frequency tail of the spectrum, which is more or less in equilibrium (with small oscillations). The phenomenon of the temporary high level of energy (when the component considered is at or near the peak frequency) is called 'overshoot'.

6.4.7 First-, second- and third-generation wave models

Wave models in which the above theories of generation, quadruplet wave–wave interactions and white-capping are explicitly implemented demand considerable computer capacity, mostly because of, as mentioned earlier, the large number of wave components that have to be considered for the wave–wave interactions (even if the DIA approximation is used for this). Other models try to avoid this problem by exploiting a similarity between the effect of these interactions and the effect of the feedback wind-generation mechanism: both give exponential growth but with different growth rates (as numerical experiments have shown). By adjusting the coefficients for the wind-induced growth (see Eq. 6.4.14), the *measured net growth*, which includes the effect of the wave–wave interactions, can be approximated with this term alone. The combined source term, assuming that white-capping is included for the growing waves, is then

$$S_{in+nl4+wc} = A + BE(f, \theta) \qquad \text{for } E(f, \theta) \leq E_{lim}(f, \theta) \qquad (6.4.23)$$

in which the values of A and B are chosen such that a credible approximation of the observed growth of waves is obtained. As indicated earlier, in many operational oceanic wave models, linear growth is ignored so that $A = 0$. The value of B is typically $B = 5\beta$, with β from Eqs. (6.4.15), which indicates that the effect of the quadruplet wave–wave interactions is about four times as strong as the effect of the direct wind input (if we ignore white-capping for these growing waves). The expression of Eq. (6.4.23) would be valid as long as the wave component grows, i.e., as long as the energy density is less than some upper limit $E_{lim}(f, \theta)$ (see Fig. 6.28). This level is usually assumed *a priori* to be the level of the f^{-5}-tail of the JONSWAP or Pierson–Moskowitz spectrum, with a certain directional spreading, e.g., the $\cos^{2s}(\theta/2)$ distribution, centred on the local wind direction. Once this level has been reached, the growth is terminated:

$$S_{in+nl4+wc} = 0 \qquad \text{for } E(f, \theta) \geq E_{lim}(f, \theta) \qquad (6.4.24)$$

White-capping is thus simulated by imposing a maximum on the energy level of the spectrum. This approach does not allow the phenomenon of overshoot to occur and the universal shape of the spectral tail is imposed rather than computed (thus losing the ability to adjust the spectral shape to the actual balance amongst the three processes of wind generation, quadruplet wave–wave interactions and white-capping). Some of these models permit some dissipation for over-developed waves, i.e., $S_{in+nl4+wc} < 0$ if $E(f, \theta) > E_{lim}(f, \theta)$, which can happen when the wind speed drops or the wind direction changes and the value of $E_{lim}(f, \theta)$ correspondingly drops. Wave models that are based on this approach of enhancing the wind input and imposing a high-frequency tail to simulate the effects of quadruplet wave–wave interactions and white-capping are called first-generation wave models.

Another approach that allows one to avoid computing the quadruplet wave–wave interactions explicitly is to use an approximation that is much simpler than that of the DIA. For instance, by using pre-computed quadruplet wave–wave interactions for a JONSWAP spectrum for non-JONSWAP spectra. This may be supplemented in the wave model by imposing the universal f^{-5}-or f^{-4}-tail, or numerically smoothing the computed tail into this shape. These models are called second-generation models. Alternatively, if for the wind sea (and not for the swell) a JONSWAP spectrum is assumed *a priori*, only the parameters of this spectrum need to be computed, while the rest of the spectrum (swell) only needs to be propagated in the model (wind sea may be conveniently defined as those wave components that are directly affected by the local wind, e.g., those propagating slower than the wind speed, within 90° of the wind direction; swell is then the remaining part of the spectrum). This approach leads to a combination of a model in which, for the wind-sea part, the energy balance equation is replaced with evolution equations for the JONSWAP parameters and another model for swell propagation (the combination is therefore called a hybrid model). In such a hybrid model, transition from wind sea to swell and back occurs when the wind moves the boundaries between wind sea and swell (in frequency and direction). These models are also considered to be a second-generation models.

The most advanced *operational* wave models compute the quadruplet wave–wave interactions explicitly with the DIA of Hasselmann *et al.* (1985a). In such models *the spectrum is free to develop without any shape imposed a priori*. The prototype of these models, which are called third-generation wave models, is the WAM model (WAMDI group, 1988; Komen *et al.*, 1994). Some experimental models use near-exact computations of the quadruplet wave–wave interactions, e.g., the WRT technique in the Xnl code of van Vledder (2006).

Literature:
First-generation models: Cardone *et al.* (1976), Cavaleri and Malanotte-Rizzoli (1981), Chen and Wang (1983), Collins (1972), Gelci *et al.* (1956), Gelci and Devillaz (1970), Pierson *et al.* (1966), SWAMP (1985).

Second-generation wave models: Barnett (1968), Ewing (1971), Ewing and Hague (1993), Golding (1983), Graber and Madsen (1988), Günther *et al.* (1979), Hasselmann *et al.* (1976), Holthuijsen *et al.* (1989), Sanders (1976), Sobey and Young (1986), SWAMP (1985), Yamaguchi *et al.* (1988), Young (1988a, 1988b), Zakharov and Pushkarev (1999).

Third-generation wave models: Bender (1996), Benoit *et al.* (1996), Booij *et al.* (1999), Komen *et al.* (1994), Li and Mao (1992), Sørensen *et al.* (2004), Suzuki *et al.* (1994), SWAMP (1985), Tolman and Chalikov (1996), WAMDI group (1988).

7

Linear wave theory (coastal waters)

7.1 Key concepts

- In this book, coastal waters are waters that are shallow enough to affect the waves, adjacent to a coast, possibly with (small) islands, headlands, tidal flats, reefs, estuaries, harbours or other features, with time-varying water levels and ambient currents (induced by tides, or river discharge).
- Horizontal variations in water *depth* cause shoaling and refraction. Horizontal variations in *amplitude* cause diffraction.
- *Shoaling* is the variation of waves in their direction of propagation due to depth-induced changes of the group velocity in that direction. These changes in group velocity generally increase the wave amplitude as the waves propagate into shallower water (the propagation of wave energy slows down, resulting in 'energy bunching').
- *Refraction* is the turning of waves towards shallower water due to depth- or current-induced changes of the phase speed in the *lateral* direction (i.e., along the wave crest). For harmonic, long-crested waves in situations with parallel depth contours, Snel's law can be used to compute the wave direction. If the depth contours are not parallel, the wave direction should be computed with wave rays.
- *Diffraction* is the turning of waves towards areas with lower amplitudes due to *amplitude* changes along the wave crest. Diffraction is particularly strong along the geometric shadow line of obstacles such as islands, headlands and breakwaters. For long-crested, harmonic waves, propagating over a horizontal bottom, Huygens' principle, or a generalisation thereof, can be used to compute the diffraction pattern.
- A long-crested, harmonic wave that *reflects* off an obstacle, with or without energy dissipation, creates a (partially) standing wave.
- The simultaneous occurrence of shoaling, refraction, diffraction and reflection of long-crested, *harmonic* waves can be computed with the mild-slope equation.
- Waves transport not only energy but also momentum. This momentum transport acts as a horizontal stress in the water (*radiation stress*). Gradients in these stresses act as forces that may generate a *slope* of the mean water surface (set-up or set-down) or a *current*.
- In the surf zone, the combination of wave-induced set-up and wave groupiness generates low-frequency waves that radiate out to sea as infra-gravity waves (*surf beat*).
- When waves enter water that is so shallow that the linear wave theory no longer holds, the non-linear Boussinesq equations are available. These equations implicitly include shoaling, refraction, diffraction and reflections and also nonlinear wave–wave interactions. Even depth-induced breaking can be included.

7.2 Introduction

When ocean waves enter coastal waters, their amplitude and direction will be affected by the limited water depth. The phenomenon of the waves changing in the longitudinal direction (i.e., in the direction of propagation) due to variations in the *group* velocity in that direction is called *shoaling*. Near the coast, it generally results in an increase in wave height. The phenomenon of the wave direction changing due to depth-induced variations in the *phase* speed in the *lateral* direction (i.e., along the wave crest) is called *refraction*. It turns the wave direction towards shallower water and results in either an increase or a decrease in wave height, depending on the actual changes in wave direction. These depth-induced changes in amplitude and direction are usually sufficiently slow (small over the distance of one wave length) that *locally* the linear wave theory for waters with a horizontal bottom can be used. However, occasionally the variations in amplitude are not slow and the linear wave theory, as given in Chapter 5, needs to be expanded. This is particularly true for waves propagating around obstacles such as small islands, reefs and headlands, or breakwaters, where the wave amplitude may vary rapidly across the geometric shadow line of such obstacles. This rapid variation in amplitude causes the waves to turn into the areas with lower amplitude. This phenomenon is called *diffraction*. Such horizontal variations of the wave amplitude, in both forward and lateral directions, also affect the transport of *momentum* of the waves. This in turn may generate currents and a set-up or set-down of the mean water surface, particularly in the surf zone. All these phenomena are due to transportation characteristics of the waves. They can be accommodated by the linear wave theory as long as the waves are not too steep or not in very shallow water. When nonlinear effects have to be accounted for, two alternatives are available. If these effects are strong, Boussinesq models, in which the actual surface elevation is computed with *high* spatial resolution (a small fraction of the wave length), can be used (the phase-resolving approach; see Section 1.3). If these effects are sufficiently weak, it is possible to compute the spectrum of the waves at each point of the area of interest with a spectral energy balance, with nonlinear sources and sinks to represent effects of generation, wave–wave interactions and dissipation (see Chapter 8). In these models, the wave spectrum is computed with *low* spatial resolution (usually several dozen wave lengths; this is the phase-averaged approach; see Section 1.3).

Literature:
Arcilla and Lemos (1990), Barber (1969), Crapper (1984), Dean and Dalrymple (1998), Dingemans (1997a 1997b), Goda (2000), Kamphuis (2000), Khandekar (1989), Kinsman (1965), Lamb (1932), Leblond and Mysak (1978), LeMéhauté (1976), Lighthill (1978), Massel (1996), Mei (1989), Mei *et al.* (2006), Phillips (1977), Sorenson (1993), Stoker (1957), Svendsen (2006), Tucker and Pitt (2001), Whitham (1974), Wiegel (1964), Young (1999).

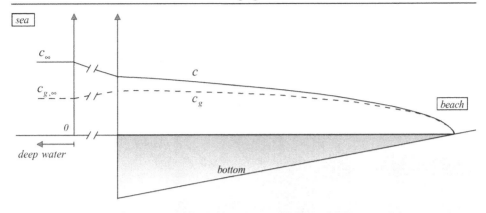

Figure 7.1 The phase speed and group velocity of a harmonic wave as the wave approaches the coast over a flat, gently sloping bottom (c_∞ and $c_{g,\infty}$ are the phase speed and group velocity in deep water). Deep water, where $c_{g,\infty} = \frac{1}{2}c_\infty$, is far to the left of this picture (indicated by the interruption in the horizontal axis).

7.3 Propagation

7.3.1 Shoaling

A harmonic wave, propagating over a fixed seabed topography with gentle slopes and no currents, retains its frequency, but, since the dispersion relationship remains valid (repeated from Chapter 5),

$$\omega^2 = gk \tanh(kd) \tag{7.3.1}$$

its wave length will decrease, if the depth decreases and the phase speed will correspondingly decrease (repeated from Chapter 5; see Fig. 7.1):

$$c = \sqrt{\frac{g}{k} \tanh(kd)} \tag{7.3.2}$$

Initially, the *group velocity* c_g increases slightly, but then it also decreases (repeated from Chapter 5; see Fig. 7.1):

$$c_g = nc \qquad \text{with} \qquad n = \frac{1}{2}\left(1 + \frac{2kd}{\sinh(2kd)}\right) \tag{7.3.3}$$

As the wave propagates into shallower water, the phase speed approaches the group velocity and the wave becomes less and less dispersive (phase speed becomes less dependent on frequency). Both the phase speed and the group velocity approach zero at the waterline. This has serious consequences for the applicability of the linear

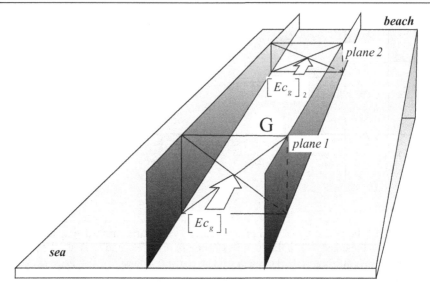

Figure 7.2 A wave approaching a straight coastline at normal incidence under stationary conditions. Under such conditions, in the absence of wave generation or dissipation, the wave energy *leaving* volume G through plane 2 is equal to the wave energy *entering* volume G through plane 1.

wave theory under such conditions because it causes the wave amplitude to go to infinity (see below).

Such variations in the group velocity cause variations in local wave energy and hence in amplitude. To illustrate this, consider a wave propagating through shallow water towards a straight coastline (i.e., parallel bottom contours) at normal incidence (i.e., perpendicular to the coastline; see Fig. 7.2). Since for normal incidence no variations occur along the wave crest, refraction is absent and the wave direction remains perpendicular to the coastline (the bottom need not be flat; gentle depth variations in the wave direction are allowed). In this situation, which we will assume to be stationary, the variation of the wave amplitude can be determined from a simple energy balance. To that end, consider a volume G that is defined by two vertical sides in the wave direction (see Fig. 7.2) plus two vertical planes, one on the seaward side and another on the beachward side, both normal to the wave direction (planes 1 and 2 in Fig. 7.2), and the bottom and the mean water surface. In the absence of any generation or dissipation of wave energy, no energy enters volume G through the water surface or the bottom. In addition, no energy enters or leaves through the lateral sides (the direction of energy transport is in the wave direction). Wave energy can therefore only *enter* the volume through plane 1 and *leave* through plane 2 (at rates $P_1 b$ and $P_2 b$ per unit time, respectively, where $P = Ec_g$ is the energy transport per unit crest length and b is the distance between the two lateral sides).

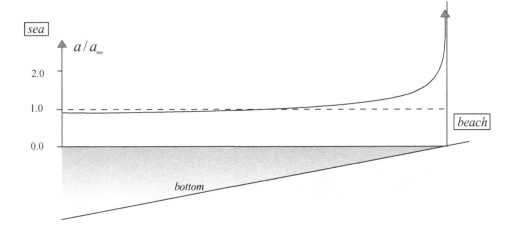

Figure 7.3 The amplitude evolution due to shoaling as a harmonic wave approaches the coast at normal incidence over a flat, gently sloping bottom (a is the amplitude; a_∞ is the amplitude in deep water). Deep water, where $a/a_\infty = 1$, is far to the left of this picture.

Because of the conservation of energy, $P_2 b = P_1 b$ and the amplitude in plane 2 can be readily obtained from the amplitude in plane 1:

$$P_2 b = P_1 b \rightarrow [Ec_g]_2 = [Ec_g]_1 \rightarrow \frac{1}{2}\rho g a_2^2 c_{g,2} = \frac{1}{2}\rho g a_1^2 c_{g,1} \tag{7.3.4}$$

so that

$$\boxed{a_2 = \sqrt{\frac{c_{g,1}}{c_{g,2}}}\, a_1} \tag{7.3.5}$$

If we take the up-wave boundary in deep water, and correspondingly replace the index 1 with ∞, and drop the index 2, then, the coefficient $K_{sh} = \sqrt{c_{g,\infty}/c_g}$ is called the shoaling coefficient.

The effect of shoaling (isolated from all other effects of propagation, generation and dissipation) over a flat, sloping bottom is initially to decrease but then to increase the amplitude (see Fig. 7.3). This effect may be referred to as 'energy bunching' (the horizontal compacting of energy; just like the compacting of a traffic jam when the cars are slowing down). The above energy balance shows that, as the group velocity approaches zero at the waterline, the wave amplitude theoretically goes to infinity. Obviously, the theory breaks down long before that. In addition, other processes such as refraction and wave breaking may well cause a totally different evolution of the waves over an arbitrary seabed topography.

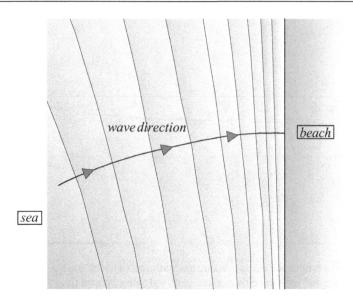

Figure 7.4 A wave always turns to the region with lower propagation speed, i.e., a wave generally turns towards the coast.

7.3.2 Refraction

If a harmonic wave approaches the same straight coast as in the situation above, but now at an angle (oblique incidence), the wave will slowly change direction as it approaches the coast (refraction; see Fig. 7.4). This is due to the *depth* variation along the wave *crest* with a corresponding variation in *phase speed* along that crest (repeated from Chapter 5):

$$c = \sqrt{\frac{g}{k} \tanh(kd)} \qquad (7.3.6)$$

This is readily seen as follows. The crest moves faster in deeper water than it does in shallow water (see Fig. 7.5) so that, in a given time interval, the crest moves over a larger distance in deeper water than it does in shallower water. The effect is that the wave turns towards the region with shallower water, i.e., towards the coast. This is a universal characteristic of waves: a wave always turns towards the region with lower propagation speed. This is true for water waves but also for sound waves, light waves and any other kind of wave. It is also true when the medium in which the waves propagate is moving, e.g., when an ambient current is present (see Section 7.3.5). The corresponding *rate* of directional turning (the speed at which the wave direction changes) can be derived with a physically oriented, geometric

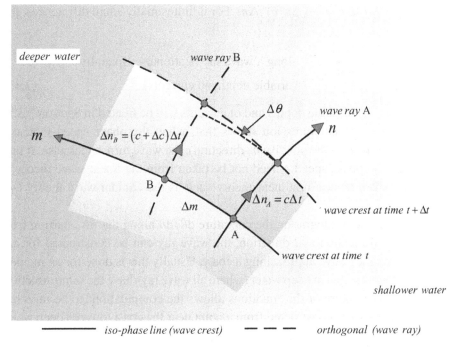

Figure 7.5 The turning of a wave crest towards the region with lower phase speed (i.e., shallower water).

argument. Such a treatment of wave propagation without diffraction is called 'the geometric-optics approximation'.

Consider, in an arbitrary situation, a line of equal phase of the wave (an iso-phase line: a crest is a special iso-phase line; see Fig. 7.5), along which the phase speed varies. For the derivation, we will use a local, left-turning system of orthogonal m, n co-ordinates (counter-clockwise rotations are positive), with m along the iso-phase line (call it a crest) and n along a line oriented normal to the crest (this line is called an orthogonal or *wave ray*). Two points A and B on the crest, separated by a distance Δm, move normal to the crest, i.e., in the wave direction. In a time interval Δt they move over distances $\Delta n_A = c\Delta t$ and $\Delta n_B = (c + \Delta c)\Delta t$, respectively; point A along its orthogonal and point B along its orthogonal. If the phase speed increases along the crest (i.e., Δc positive with increasing m), then the corresponding directional turning of the crest $\Delta\theta$ is clockwise (negative), so $\Delta\theta = -(\Delta n_B - \Delta n_A)/\Delta m = -\Delta c\Delta t/\Delta m$ (see Fig. 7.5). This turn in direction of the crest is obviously equal to the turn of the wave direction (the two directions being normal to each other). The spatial *rate* of turning, i.e., the change in direction per unit forward distance $\Delta\theta/\Delta n$ (which is the curvature of the wave ray) is correspondingly $\Delta\theta/\Delta n = -\Delta c\Delta t/(\Delta m\Delta n)$. Or, since $\Delta n = c\Delta t$, the spatial turning rate is $\Delta\theta/\Delta n = -(\Delta c/c)/\Delta m$. For infinitesimally small differences, this becomes[1]

$$\left(\frac{d\theta}{dn}\right)_{ref} = -\frac{1}{c}\frac{\partial c}{\partial m}$$
along a wave ray (stationary, spatially variable depth; no currents)   (7.3.7)

(the effects of time-variable depths and of currents will be treated in Section 7.3.5). The minus sign in this expression shows that, when the phase speed increases along the wave crest (in the positive m-direction), the wave turns clockwise. It may be noted that the phase speed c need not be taken from the linear wave theory; it may also be taken from a nonlinear theory such as the cnoidal wave theory (see Section 5.6.3).

By computing and integrating the curvature $d\theta/dn$ along the ray, starting from any given initial location and direction, the wave ray can be constructed for any incident (harmonic) wave approaching a coast. Usually this is done for an incident wave that is long-crested in deep water (where all wave rays have the same direction; see Fig. 7.6). The nature of the equations allows the computations to be *reversed*, i.e., the ray can also be constructed from a point near the coast towards deep water. This reverse-refraction technique (also called 'back tracing') is sometimes used as an alternative for computing the refraction effects for a single location on the coast, by computing the set of wave rays fanning out in many different directions from that location to deeper water (see Dorrestein, 1960). Wave rays used to be constructed manually, with pen and ink on paper, but computers are now used for this task (although mostly for illustrative purposes, because the computation of wave refraction is more and more based on an alternative, Eulerian, approach; see below). For a situation with *parallel depth contours*, there is the simple alternative of *Snel's Law*[2] (see Note 7A):

$$\frac{d}{dn}\left(\frac{\sin\theta}{c}\right) = 0$$ (7.3.8)

or

$$\boxed{\frac{\sin\theta}{c} = \text{constant}}$$
along a wave ray for parallel depth contours   (7.3.9)

where the angle of propagation θ is taken between the ray and the normal to the depth contours (see Fig. 7.7). In such a case of parallel depth contours, the wave direction

---

[1] With $c = \omega/k$ and $\omega = $ constant substituted into Eq. (7.3.7), the spatial turning rate can also be written as $(d\theta/dn)_{ref} = (1/k)\partial k/\partial m$. Note that the minus sign is absent from this expression (because c is inversely proportional to k or, to put it differently, k is larger in shallower water, in contrast to c, which is smaller).

[2] Willebrod Snel van Royen (1580–1626) was a Dutch scientist (concerned with mathematics, optics, cartography, astronomy and navigation) who is best known by the Latin version of his name Snellius. The sine law which describes refraction and carries his name is therefore properly referred to as Snel's Law (not Snell's Law).

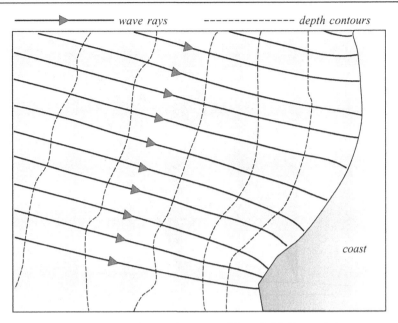

Figure 7.6 Wave rays for a harmonic wave with an initially straight crest over a simple seabed topography.

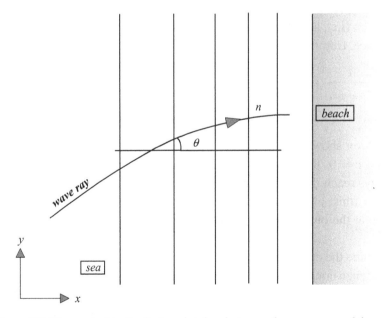

Figure 7.7 The angle θ in Snel's Law is taken between the wave ray and the normal to the straight and parallel depth contours.

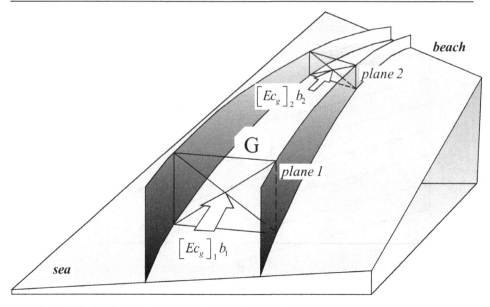

Figure 7.8 Under stationary conditions, in the absence of wave generation or dissipation, the wave energy *leaving* volume G through plane 2 is equal to the wave energy *entering* volume G through plane 1.

at any location (i.e., any depth) can readily be obtained from the deep-water wave direction since, from Eq. (7.3.9),

$$\sin \theta = \frac{c}{c_{deep\ water}} \sin \theta_{deep\ water} \qquad \text{along a wave ray for parallel depth contours}$$

(7.3.10)

This expression shows that, when the wave approaches the waterline, where the phase speed $c = 0$, the wave direction will be $\theta = 0$. In other words, all ocean waves always reach the shore at a right angle, independently of their direction in deeper water (this, of course, is a theoretical result; actually the wave direction, for instance at the outer edge of the surf zone, will not be exactly normal to the shore).

To determine the effect of refraction on the wave *amplitude*, consider again the above one-dimensional situation but now for oblique incidence. The volume *G* is now defined by two curved, vertical sides through *two parallel neighbouring wave rays* and two vertical planes normal to the wave direction (planes 1 and 2 in Fig. 7.8), and the bottom and the mean sea surface.

Again, the direction of energy propagation is in the wave direction, so no energy enters or leaves through the lateral sides. Therefore, in a stationary situation and in the absence of any generation or dissipation of wave energy, the energy that is *leaving* volume G per unit time through plane 2, $P_2 b_2$, is equal to the energy that is *entering* the volume per unit time through plane 1, $P_1 b_1$ (where P is the energy transport per unit crest length, see Section 5.5, and b is the distance between the

---

NOTE 7A   Snel's Law

The rate of change of the wave direction θ along a wave ray can be written as (see Eq. 7.3.7)

$$\frac{d\theta}{dn} = -\frac{1}{c}\frac{\partial c}{\partial m} = -\frac{1}{c}\left(-\frac{\partial c}{\partial x}\sin\theta + \frac{\partial c}{\partial y}\cos\theta\right)$$

Consider now a situation with parallel depth contours, normal to the x-axis (see Fig. 7.7). In this one-dimensional situation, all derivatives in the y-direction are zero, so $d\theta/dn = (1/c)(dc/dx)\sin\theta$. In general we have $dc/dn = (\partial c/\partial x)(dx/dn) + (\partial c/\partial y)(dy/dn)$, which in this case reduces to $dc/dn = (dc/dx)(dx/dn) = dc/dx\cos\theta$, so that $dc/dx = (1/\cos\theta)dc/dn$ and we find by substitution

$$\frac{d\theta}{dn} = \frac{1}{c}\frac{dc}{dn}\frac{\sin\theta}{\cos\theta}$$

Multiplying the left- and right-hand sides by $\cos\theta/c$ gives

$$\frac{1}{c}\cos\theta\frac{d\theta}{dn} = \frac{1}{c^2}\sin\theta\frac{dc}{dn} \qquad \text{so} \qquad \frac{1}{c}\frac{d(\sin\theta)}{dn} + \sin\theta\frac{d(1/c)}{dn} = 0$$

Inversing the rule of chain differentiation gives Snell's Law:

$$\frac{d(\sin\theta/c)}{dn} = 0 \quad \text{or} \quad \sin\theta/c = \text{constant}$$

A shorter derivation can be based on the fact that the wave-number field is rotation-free (see the note in Appendix D):

$$\frac{\partial k_y}{\partial x} - \frac{\partial k_x}{\partial y} = 0$$

Since all derivatives in the y-direction are zero, it follows from this expression that $dk_y/dx = 0$. With a constant frequency ω and $k_y = k\sin\theta$ we find from this that $d(k\sin\theta/\omega)/dx = 0$. The phase velocity is $c = \omega/k$, so $d(\sin\theta/c)/dx = 0$. If $\sin\theta/c$ is constant in the x-direction (in other words, everywhere), as stated by this expression, it is also constant along the wave ray (in this one-dimensional situation), so $d(\sin\theta/c)/dn = 0$.

two wave rays). The amplitude in plane 2 can therefore be readily obtained from the amplitude in plane 1:

$$P_2 b_2 = P_1 b_1 \rightarrow [Ec_g]_2 b_2 = [Ec_g]_1 b_1 \rightarrow \frac{1}{2}\rho g a_2^2 c_{g,2} = \frac{1}{2}\rho g a_1^2 c_{g,1} \frac{b_1}{b_2}$$

(7.3.11)

so

$$a_2 = \sqrt{\frac{c_{g,1}}{c_{g,2}}} \sqrt{\frac{b_1}{b_2}} a_1$$

(7.3.12)

If we take the up-wave boundary in deep water, and correspondingly replace the index 1 with ∞, and drop the index 2 then the coefficient $\sqrt{b_1/b_2} \rightarrow K_{ref} = \sqrt{b_\infty/b}$ is called the refraction coefficient (the shoaling coefficient K_{sh} was defined earlier, see Section 7.3.1). .

The above approach of estimating the wave amplitude from wave rays (called the Lagrangian approach) is simple and effective if the seabed topography is fairly smooth. However, the seabed topography near a coast is often rather complicated. It may contain large-scale features such as sub-marine canyons or shoals, like the Hudson Canyon off New York, the Dogger Bank in the central North Sea and the Grand Banks off the east coast of Canada. Such large-scale features will cause large-scale refraction. The seabed may also contain small-scale features such as shoals and channels in coastal regions. This will result in local, small-scale refraction. Over long distances the effects of such local, small-scale refraction may accumulate and result in a scattering of the rays (K. Hasselmann calls this scintillation, in analogy with the passage of light from a star through the Earth's atmosphere, making the stars scintillate in the night sky). In all these cases, irrespective of whether the bottom features are large, small, well defined or more or less random, many wave rays will generally cross many other wave rays at many different locations (see Figs. 7.9 and 8.8).

The distance between initially adjacent wave rays approaches zero at the intersections of such rays, and the refraction coefficient in Eq. (7.3.12) approaches infinity: $K_{ref} \rightarrow \infty$. In other words, the wave amplitude grows infinitely large at these intersections. Under some conditions, the crossing wave rays create an envelope, called a caustic, where the wave height would theoretically approach infinity along a line (see the lower-right-hand corner in Fig. 7.9). Obviously, the theory breaks down near such points and caustics. Decreasing the initial distance between the rays in the computations does not solve the problem (on the contrary, it usually makes the situation worse).

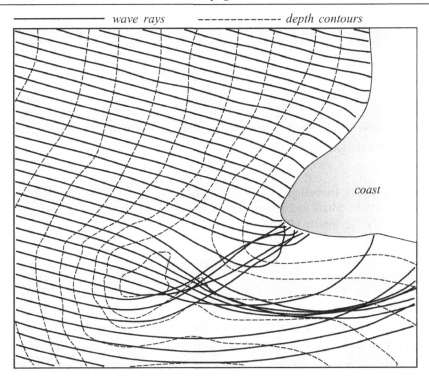

——————————— *wave rays*   – – – – – – – – – – – *depth contours*

coast

Figure 7.9 Wave rays for a harmonic wave over a complicated bottom topography.

We have seen earlier that, for *oceanic* waters, the Lagrangian approach to wave computations is not suited because of the computation of nonlinear source terms in the spectral energy balance equation (see Section 6.4.1). In *coastal* waters this appears to be also unsuited, for the same reason, but also because of the crossing of wave rays. Again, as in oceanic waters, the Eulerian approach offers an alternative by discretising the geographic space in cells. This gives an average wave condition per geographic cell, thus *avoiding* the crossing of wave rays and smoothing the wave conditions in each cell.[3] The mathematical formulation of this Eulerian approach requires the determination of the turning rate of the wave direction per unit *time*, $d\theta/dt$ moving with the wave energy (rather than per unit distance, $d\theta/dn$; see Eq. 7.3.7). The expression for $d\theta/dt$ can be obtained from the expression for $d\theta/dn$, using the propagation speed of the wave energy, i.e., the group velocity c_g. Consider two positions of the crest at a distance Δn in the forward direction (see Fig. 7.5). The directional turn of the crest over this distance is $\Delta\theta = d\theta/dn\,\Delta n$. This distance is travelled by the wave energy in a time interval $\Delta t = \Delta n/c_g$, so $\Delta n = c_g\Delta t$ and, therefore, $\Delta\theta = d\theta/dn\,c_g\Delta t$. Taking infinitesimally small intervals

[3] However, the proper solution is to take diffraction effects into account; see Section 7.3.3.

and substituting in the expression for $d\theta/dn$ of Eq. (7.3.7) gives the turning rate in time in a frame of reference moving with the wave energy $c_{\theta,ref} = d\theta/dt$:[4]

$$c_{\theta,ref} = -\frac{c_g}{c}\frac{\partial c}{\partial m} \qquad \text{stationary, spatially variable depth; no currents}$$

(7.3.13)

This expression will be used to represent refraction in the spectral energy balance for coastal waters (in Chapter 8).

Literature:
Arthur *et al.* (1952), Dorrestein (1960), Holthuijsen and Booij (1994), Liu (1990), Munk and Traylor (1947), Munk and Arthur (1952), O'Reilly and Guza (1991), Skovgaard and Petersen (1977), Yamaguchi (1986).

7.3.3 Diffraction

To introduce the phenomenon of diffraction, consider a harmonic, long-crested wave, travelling in water of *constant* depth, around a headland or breakwater (see Fig. 7.10). In the absence of refraction (since the bottom is horizontal), the waves will travel into the shadow of the obstacle in an almost circular pattern of crests with rapidly diminishing amplitudes. Owing to the shadowing effect of the headland, large variations in amplitude will occur across the geometric shadow line of the headland. If diffraction were *ignored*, the wave would propagate along straight wave rays (since depth is constant), no energy would cross the shadow line and no waves would penetrate the shadow area behind the headland. With diffraction *accounted for*, the wave rays (defined here as orthogonals of the wave crests) curve into the shadow area behind the headland (see Fig. 7.11).

In the above derivation of the refraction expressions, the linear theory was used for a wave with constant amplitude. We now consider a case in which the amplitude varies rapidly in horizontal space and we need to include spatial derivatives of the amplitude in the Laplace equation of the linear wave theory (see Section 5.4). This extension of the Laplace equation gives rise to subsequent extra terms in

---

[4] An alternative expression is found as follows: in the absence of an ambient current, the frequency (which depends on wave number k and depth d) is constant along a wave crest:

$$\frac{d\omega}{dm} = \frac{\partial\omega}{\partial k}\frac{\partial k}{\partial m} + \frac{\partial\omega}{\partial d}\frac{\partial d}{\partial m} = 0 \qquad \text{so} \qquad \frac{\partial\omega}{\partial k}\frac{\partial k}{\partial m} = -\frac{\partial\omega}{\partial d}\frac{\partial d}{\partial m}$$

From Eq. (7.3.13) and the footnote near Eq. (7.3.7), we find:

$$c_{\theta,ref} = \frac{c_g}{k}\frac{\partial k}{\partial m} = \frac{1}{k}\frac{\partial\omega}{\partial k}\frac{\partial k}{\partial m}$$

Substitution then gives

$$c_{\theta,ref} = -\frac{1}{k}\frac{\partial\omega}{\partial d}\frac{\partial d}{\partial m}.$$

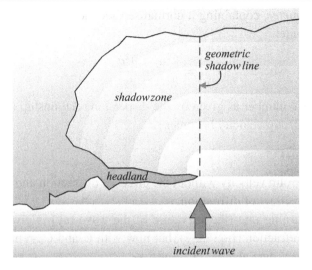

Figure 7.10 Diffraction around a headland with a circular wave pattern in the shadow zone (constant depth and no reflections).

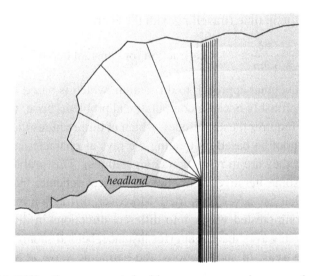

Figure 7.11 Diffraction represented with wave rays curving around a headland (constant depth, no reflections).

the analytical expressions of the linear wave theory. For instance, the phase speed becomes

$$C = c(1 + \delta_a)^{-1/2} \tag{7.3.14}$$

where c is the phase speed of the harmonic wave without the effect of diffraction and C is the phase speed *with* the effect of diffraction. The parameter δ_a is the *diffraction parameter*, containing a normalised second-order spatial derivative of the wave amplitude:

$$\delta_a = \frac{\nabla^2 a}{k^2 a} \qquad \text{where} \qquad \nabla^2 a = \frac{\partial^2 a}{\partial x^2} + \frac{\partial^2 a}{\partial y^2} \qquad (7.3.15)$$

and k is the wave number as given by the dispersion relationship (see Eqs. 5.4.17 and 7.3.1). The group velocity becomes

$$C_g = c_g(1 + \delta_a)^{1/2} \qquad (7.3.16)$$

where c_g is the group velocity without the effect of diffraction and C_g is the group velocity with the effect of diffraction.

The spatial turning rate (the curvature of the wave ray) is given by the same equation as for refraction (Eq. 7.3.7), but now with C and C_g replacing c and c_g, respectively. After some algebra, we find

$$\left(\frac{\mathrm{d}\theta}{\mathrm{d}n}\right)_{dif} = \frac{1}{2(1 + \delta_a)} \frac{\partial \delta_a}{\partial m} \qquad \text{diffraction for constant depth} \qquad (7.3.17)$$

and the turning rate in time (travelling with the energy along the wave ray) is

$$c_{\theta,dif} = \frac{C_g}{2(1 + \delta_a)} \frac{\partial \delta_a}{\partial m} \qquad \text{diffraction for constant depth} \qquad (7.3.18)$$

The above conceptual approach to diffraction, which is based on wave rays, is unconventional, probably because computational problems arise: the computation of the wave rays requires the amplitudes, which in turn requires the wave rays (the computations cannot be based on treating the rays as characteristics of the basic equations, which they are in refraction computations). This problem can perhaps be solved with some iterative or implicit numerical scheme, but to the best of my knowledge that has never been tried. Moreover, diffraction may turn initially unidirectional, long-crested waves into different directions in different areas. In regions where such waves meet again (for instance, behind shoals or islands), this would give cross-seas. In such regions, the harmonic wave at a given location is the sum of two (or more) harmonic waves from different directions, requiring phase information to determine the amplitude. This complicates the computations further. However, an approximation will be given later for the Eulerian approach of the energy balance equation for random short-crested waves (the phase-decoupled refraction–diffraction approximation in Section 8.4.2).

Because of the need for phase information, diffraction is more conveniently computed with phase-*resolving* models (see Section 1.3) based on Huygens'[5] principle.

---

[5] Christiaan Huygens was a Dutch scientist (1629–1695) who spent a great deal of his time on observations with telescopes (he discovered the first moon and the rings of Saturn) and on the development of the clock so as to help maritime navigation (the longitude problem).

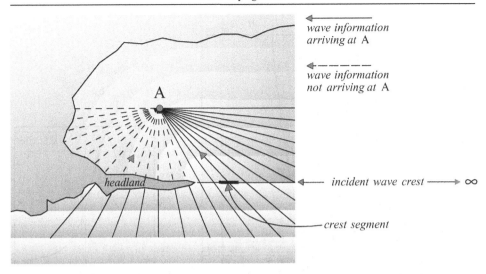

Figure 7.12 The effect of diffraction on the wave field behind a headland estimated with Huygens' principle.

Consider a point of interest A, in the sheltered region behind the headland in the above example, and an incident wave crest up-wave from that point extending from the tip of the headland to infinity (see Fig. 7.12).

Each *segment* of this up-wave crest is considered to be a source, sending information on amplitude and phase along straight lines (water of constant depth) to point A. All fictitious crest segments on the headland (if the crest were to extend over land) obviously do not send that information. In other words, wave information, up-wave from the headland, is blocked from propagating to point A. It is the removal of this information that causes diffraction effects at point A. The wave at point A can be reconstructed by adding the wave information that *does* arrive at point A (amplitudes and phases), radiating from the up-wave crest. Repeating this reconstruction for all points in the area down-wave from the incident crest gives the entire wave field in this area, including the shadow area. Since this approach provides both the amplitude and the phase of the wave at all points in the down-wave area, it also provides the wave direction in the down-wave area (the wave direction being normal to lines of equal phase by definition).

An analytical solution to the diffraction problem is available for the simple case of a straight, semi-infinite breakwater in water of constant depth (with reflections off the breakwater but no other reflections, such as off a nearby coastline; this solution was originally developed for light waves by Sommerfeld in 1896). The solution depends on the wave direction relative to the breakwater, which is modelled as a straight, thin, rigid, reflecting, vertical screen. For each point in the domain of interest, the solution is formulated as an integral that represents the

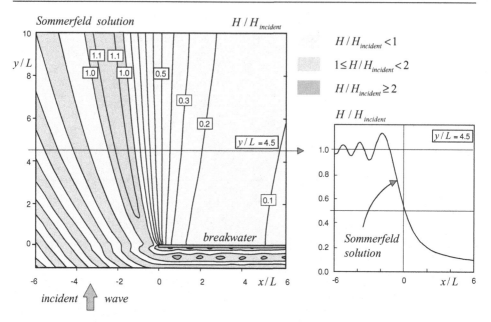

Figure 7.13 Diffraction (normalised wave height) of a normal incident, harmonic wave around a semi-infinite, straight breakwater in water with constant depth (Sommerfeld solution; L is the wave length and H is the wave height).

contributions (amplitude and phase) of the up-wave sources along the wave crest (as in Huygens' principle), including the wave that is reflecting off the breakwater and subsequently diffracting around the tip of the breakwater into the shadow area. In this manner the wave height can be found at any location affected by the breakwater (see Fig. 7.13). Without the reflected wave, the integral can be represented graphically, providing an approximate graphical solution to this simple diffraction problem (the Cornu spiral, e.g., Lacombe, 1951, 1965). It is possible to compute the wave conditions for slightly more complex situations by combining Sommerfeld solutions for several individual breakwaters, e.g., to simulate a gap in a breakwater. Alternatively, diagrams and tables with computed or observed diffraction patterns for situations with simple layouts of breakwaters have been published widely. However, it must be stressed that the diffraction patterns for harmonic, long-crested waves as treated here are rather different from those for random, short-crested waves. The reason is that, in the summation of many different harmonic waves (to simulate random, short-crested waves), the diffraction effects of the *individual* harmonic waves are partially cancelled out. Random, short-crested waves therefore create smoother diffraction patterns of the significant wave height than would an equivalent harmonic, long-crested wave (see Goda, 2000 and Section 8.4.2).

For situations with a more complicated layout of obstacles, a numerical computer model is needed. It is based on the same principle as above, but instead of considering the crest of the incident wave as the only source of amplitude and phase information, it considers all points on a closed boundary encompassing the area of interest as sources of wave amplitude and phase, i.e., all points on the opensea boundary and on the obstacle(s) and the coast. This method is known as the boundary-element method. The model computes the effect of these sources on all points of the boundary and on any arbitrarily chosen point within the enclosed area. In this approach, all effects of reflections off obstacles and the coast can be taken into account. This is particularly important for computing wave penetration into a harbour where vertical walls may reflect 100% of the wave energy, thus creating patterns of standing waves, which are properly accounted for in such a model. The basic equation for this type of model is relatively simple to derive. Instead of the harmonic wave with a straight crest in the y-direction, which was considered in the derivation of the linear wave theory in Chapter 5, with a constant amplitude a and a phase that varies horizontally as $-kx$ (so as to have a wave propagating everywhere in the positive x-direction), consider now a harmonic wave with amplitude a and phase α that both vary arbitrarily horizontally, $a(x, y)$ and $\alpha(x, y)$:

$$\eta(x, y, t) = a(x, y)\sin[\omega t + \alpha(x, y)] \tag{7.3.19}$$

This harmonic wave may propagate in any direction, depending on $\alpha(x, y)$, and need therefore not be cylindrical (i.e., need not have an infinitely long, straight crest). The velocity potential function for this wave (for a horizontal bottom) would be (analogously to Section 5.4.1)

$$\phi(x, y, z, t) = \frac{\omega\, a(x, y)}{k} \frac{\cosh[k(d + z)]}{\sinh(kd)} \cos[\omega t + \alpha(x, y)] \tag{7.3.20}$$

The diffraction computation consists essentially of solving, for any particular layout of the coastline and obstacles and constant depth, the Laplace equation, expressed in terms of this potential function (repeated from Eq. 5.3.22):

$$\frac{\partial^2\phi}{\partial x^2} + \frac{\partial^2\phi}{\partial y^2} + \frac{\partial^2\phi}{\partial z^2} = 0 \tag{7.3.21}$$

For the solution in terms of a propagating harmonic wave, it is convenient to separate the dimensions, i.e., to separate the horizontal x, y-dimension from the time t-dimension (the vertical z-dimension is already separated in Eq. 7.3.20). This is readily achieved by re-writing the expression for the harmonic wave (see Eq. 7.3.19) as

$$\eta(x, y, t) = A(x, y)\cos(\omega t) + B(x, y)\sin(\omega t) \tag{7.3.22}$$

with[6] $a = \sqrt{A^2 + B^2}$, $A = a\sin\alpha$ and $B = a\cos\alpha$, so that $\tan\alpha = A/B$. Writing the velocity potential function correspondingly and substituting it into the Laplace equation gives

$$\left(\frac{\partial^2 A}{\partial x^2} + \frac{\partial^2 A}{\partial y^2} + k^2 A\right)f(z)\cos(\omega t) - \left(\frac{\partial^2 B}{\partial x^2} + \frac{\partial^2 B}{\partial y^2} + k^2 B\right)f(z)\sin(\omega t) = 0$$

$$(7.3.23)$$

where $f(z) = (\omega/k)\cosh[k(d+z)]/\sinh(kd)$. This equation can be true for all values of t only if the expressions between large parentheses in this equation are equal to zero:

$$\frac{\partial^2 A}{\partial x^2} + \frac{\partial^2 A}{\partial y^2} + k^2 A = 0$$

$$\frac{\partial^2 B}{\partial x^2} + \frac{\partial^2 B}{\partial y^2} + k^2 B = 0$$

$$(7.3.24)$$

These equations are called the Helmholtz[7] equations[8] and they can be solved with numerical models with the proper boundary conditions (vertical walls, where the wave may reflect but not penetrate, and free radiation of wave energy back to open sea). As indicated above, the numerical techniques for solving these equations are based on Huygens' principle. A more advanced model, in the sense that the depth need *not* be constant (the bottom need not be horizontal), so that refraction and diffraction need to be combined, is briefly introduced in the next section.

Literature:

Battjes (1968), Booij *et al.* (1997), Briggs *et al.* (1995), CEM (2002), Dean and Dalrymple (1998), Dingemans (1997a), Goda *et al.* (1978), Goda (2000), Holthuijsen *et al.* (2003), Lacombe (1951, 1965), Penney and Price (1952), Rivero *et al.* (1997), SPM (1973, 1984), Wiegel (1964).

---

[6] To find this, consider $x(t) = a\sin(\omega t + \alpha)$, which may also be written (standard trigonometry) as $x(t) = a[\sin(\omega t)\cos\alpha + \cos(\omega t)\sin\alpha]$, so that, if we write $x(t) = A\cos(\omega t) + B\sin(\omega t)$, then $A = a\sin\alpha$ and $B = a\cos\alpha$. From this we find $a = \sqrt{A^2 + B^2}$ and $\tan\alpha = A/B$.

[7] Hermann Ludwig Ferdinand von Helmholtz (1821–1894) was a German scientist who (like Daniel Bernoulli, see Section 5.3.4) started his career in medicine, showing that muscle force was derived from chemical and physical principles rather than from vital (non-physical) forces as it was fashionable to believe at that time. His work then took him more and more to physiology (acoustics, optics) and physics (hydromechanics, electrodynamics) based on mechanical principles and mathematics. He established the principle of the conservation of energy.

[8] The equations are usually written in complex notation resulting in one complex differential equation of the same appearance: $\partial^2 G/\partial x^2 + \partial^2 G/\partial y^2 + k^2 G = 0$, or, in vector notation, $\nabla^2 G + k^2 G = 0$, where $G = A + iB$ is a complex wave function with A and B defined as in the main text, so that $\eta(t) = \text{Re}\{Ge^{-i\omega t}\}$. This differential equation is called 'the' Helmholtz equation (at least in the world of hydromechanics).

7.3.4 Refraction and diffraction

If the waves propagate in shallow water over a non-horizontal bottom, with rapid spatial variations in wave amplitude, both refraction and diffraction need to be accounted for. Conceptually, this is readily achieved by combining the effects of varying water depth and varying amplitude in the computation of the wave rays (see Eq. 7.3.7 and its footnote, and Eq. 7.3.17). The spatial turning rate (the curvature of the wave ray) is then

$$\left(\frac{d\theta}{dn}\right)_{ref+dif} = \frac{1}{k}\frac{\partial k}{\partial m} + \frac{1}{2(1+\delta_a)}\frac{\partial \delta_a}{\partial m} \tag{7.3.25}$$

with a modified diffraction parameter δ_a:

$$\delta_a = \frac{\nabla cc_g \nabla a}{k^2 cc_g a} \quad \text{with } \nabla cc_g \nabla a = \frac{\partial}{\partial x}\left(cc_g \frac{\partial a}{\partial x}\right) + \frac{\partial}{\partial y}\left(cc_g \frac{\partial a}{\partial y}\right) \tag{7.3.26}$$

The turning rate in time, while travelling with the wave energy, is obtained from Eq. (7.3.25) and $dn = C_g dt$ as

$$\begin{aligned} \frac{d\theta}{dt} &= c_{\theta,ref+dif} \\ &= C_g\left(\frac{1}{k}\frac{\partial k}{\partial m} + \frac{1}{2(1+\delta_a)}\frac{\partial \delta_a}{\partial m}\right) \end{aligned} \quad \begin{array}{l}\text{stationary, spatially variable} \\ \text{depth and amplitude; no} \\ \text{currents}\end{array} \tag{7.3.27}$$

An approximation of this expression is used in the Eulerian approach of the energy-balance equation for short-crested, random waves (see Section 8.4.2). However, the established manner in which to compute combined refraction and diffraction (at least for harmonic, long-crested incident waves) is based on equations that are similar to the Helmholtz equations (Eqs. 7.3.24). Only the product of phase speed and group velocity cc_g is added, resulting in equations that are called the *mild-slope equations*[9] (Berkhoff, 1972):

$$\frac{\partial}{\partial x}\left(cc_g \frac{\partial A}{\partial x}\right) + \frac{\partial}{\partial y}\left(cc_g \frac{\partial A}{\partial y}\right) + k^2 cc_g A = 0$$
$$\frac{\partial}{\partial x}\left(cc_g \frac{\partial B}{\partial x}\right) + \frac{\partial}{\partial y}\left(cc_g \frac{\partial B}{\partial y}\right) + k^2 cc_g B = 0 \tag{7.3.28}$$

Adding cc_g to the Helmholtz equations seems trivial but the numerical techniques to solve the mild-slope equations for a given situation (finite-element methods) are considerably more complicated than the technique for solving the Helmholtz

---

[9] These equations too (like the Helmholtz equations) are usually written in complex notation, resulting in one complex differential equation of the same appearance: $\partial(cc_g \partial G/\partial x)/\partial x + \partial(cc_g \partial G/\partial y)/\partial y + k^2 cc_g G = 0$, or, in vector notation, $\nabla(cc_g \nabla G) + k^2 cc_g G = 0$, where $G = A + iB$ is a complex wave function (see the previous footnote). This equation is called '*the*' mild-slope equation.

equations (with the boundary-element method). The reason is that, with a varying water depth, the amplitude and phase information does not travel (radiate) to a point in the computational area along straight lines, as in the solution of the Helmholtz equations. The velocity potential (see Eq. 7.3.20) needs therefore to be computed at a large number of points in the computational area simultaneously. A numerically more economic approach is provided by a parabolic version of the mild-slope equation, in which the wave condition is computed line-by-line, marching in the forward direction, (only for waves travelling towards a non-reflecting coast, in a directional sector of 60°–90° on either side of some main wave direction that is constant over the computational region). More advanced models (Boussinesq-type models), in which the waves may be random, short-crested and nonlinear, are briefly introduced in Section 7.5.2.

Literature:
Beji and Nadaoka (1997), Berkhoff (1972), Booij (1981), Dalrymple and Kirby (1988), Dalrymple *et al.* (1989), Dingemans (1997a), Ebersole (1985), Holthuijsen *et al.* (2003), Kaihatu and Kirby (1995), Kirby (1984, 1986), Maa *et al.* (2002), Mei (1989), Mei *et al.* (2006), Radder (1979), Sawaragi (1995), Vincent and Briggs (1989).

7.3.5 Tides and currents

Tides and currents, or, more specifically, time-varying water depths and ambient currents, which may be due to tides, storm surges or river discharge, may change the amplitude, frequency and direction of an incoming harmonic wave. The first phenomenon (the change in amplitude) has several causes: energy bunching (as in shoaling), current-induced refraction and transfer of energy between wave and current. The second phenomenon (the change in frequency) is closely related to the well-known Doppler effect. The third phenomenon (the change in direction) is refraction, induced by current-related changes in propagation speed. All these phenomena are due to the bodily transport of the wave by the ambient current with a varying speed (horizontally and in time).

If the harmonic wave propagates in an area with *constant* depth across a *constant* ambient current (constant in space and time), the linear theory is still valid in its entirety in a frame of reference moving with the current (the wave doesn't 'know' that it moves in an ambient current, it just moves with it as if in a water tank that is carried with the ambient current). In this case, all results of the linear wave theory can therefore be applied in a frame of reference moving with the current. The *frequency* of the wave in this moving frame of reference is called the *relative* or *intrinsic* frequency, denoted as σ, and the relationship with wave number and depth (the dispersion relationship; see Eqs. 5.4.17 and 7.3.1) is retained:

$$\sigma^2 = gk \tanh(kd) \qquad\qquad (7.3.29)$$

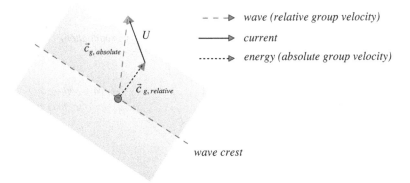

Figure 7.14 The energy propagation speed is the sum of the current vector and the vector of the group velocity (relative to the current).

In a *fixed* frame of reference (fixed to the stationary bottom), the frequency of the wave is called the *absolute* frequency and denoted as ω (as observed, for instance, with a wave pole fixed to the bottom). It is related to the relative frequency (this follows directly from the bodily transport of the wave by the current) as

$$\omega = \sigma + kU_n \tag{7.3.30}$$

where U_n is the component of the current in the wave direction (i.e., normal to the wave crest). The propagation velocity of the wave *energy* in this fixed frame of reference, i.e., relative to the bottom, $\vec{c}_{g,absolute}$, is obtained by adding as vectors the current velocity $\vec{U}$ to the group velocity relative to the current, $\vec{c}_{g,relative}$ (see Fig. 7.14):

$$\vec{c}_{g,absolute} = \vec{c}_{g,relative} + \vec{U} \tag{7.3.31}$$

The direction of wave energy transport is therefore generally not normal to the wave crest in the presence of an ambient current (some energy propagates parallel to the wave crest).

In these circumstances, of a constant current in water with a constant depth, both the relative and the absolute frequencies are constant. If, however, the water depth or the ambient current varies horizontally or in time, these frequencies will generally also vary. Determining the rate of this (relative-) frequency change in time, when travelling with the wave energy, requires a distinction amongst the wave ray (the line with co-ordinate r, along which the wave *energy* propagates), the wave orthogonal (the line with co-ordinate n *normal* to the *crest*) and the streamline of the *current* with co-ordinate s (see Fig. 7.15).

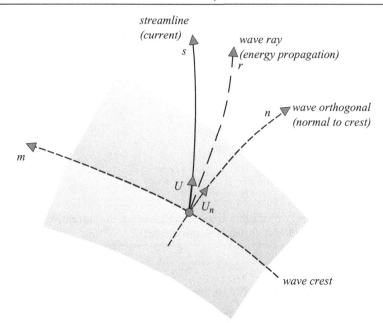

Figure 7.15 An ambient current generally deflects the propagation direction of the wave energy away from the wave direction and a distinction must be made between the wave direction (normal to the wave crest; the wave orthogonal) and the direction of energy propagation (the wave ray). In general, some wave energy travels parallel to the wave crest in the presence of an ambient current (the *m*-component of the energy transport).

With a current or water depth varying horizontally or in time, the rate of change of relative frequency, denoted as $\mathrm{d}\sigma/\mathrm{d}t = c_\sigma$, in a frame of reference moving with the wave energy along the wave ray, is given by (for the derivation, see Appendix D):

$$\frac{\mathrm{d}\sigma}{\mathrm{d}t} = c_\sigma = \frac{\partial\sigma}{\partial d}\left(\frac{\partial d}{\partial t} + U\frac{\partial d}{\partial s}\right) - c_g k\frac{\partial U_n}{\partial n} \tag{7.3.32}$$

The first term in the brackets relates to the time variation in the depth and the second term in these brackets to the effect of the current bodily moving the wave across a horizontally varying bottom. The second term on the right-hand side represents the effect of the wave moving with a horizontally varying current. The corresponding variations in *absolute* frequency ω and wave number k follow directly from the variation in the relative frequency σ with Eqs. (7.3.29) and (7.3.30), without any additional computations (the time variation of the current is thus accounted for).

In addition to changing the frequency of the waves, currents can also change the direction of the waves. This phenomenon of *current*-induced refraction is essentially the same as *depth*-induced refraction: the wave turns towards the area with lower (absolute) propagation speed of the crest (i.e., relative to the fixed bottom), which is now affected not only by the depth but also by the ambient current (the component of the current *in* the wave direction U_n). The rate of change of the wave direction (i.e., of the normal to the wave crest) due to depth- and current-induced refraction is then (for the derivation, see Appendix D)

$$c_{\theta,ref,depth+current} = -\frac{c_g}{c}\frac{\partial c}{\partial m} - \frac{\partial U_n}{\partial m} \qquad (7.3.33)$$

where obviously the term $\partial U_n/\partial m$ represents current-induced refraction and the term with $\partial c/\partial m$ represents depth-induced refraction (see Eq. 7.3.13).

The current interacts with the waves also by exchanging energy (work done by the current against the radiation stress, see footnote in Section 5.5.2; e.g., Longuet-Higgins and Stewart, 1960, 1961, 1962, 1964). This implies that the wave energy is not conserved as the wave propagates through a current field. Instead, another, closely related quantity, *action*, is conserved. It is defined as energy divided by relative frequency $A = E/\sigma$ (e.g., Bretherton and Garrett, 1969; Mei *et al.*, 2006). Wave models that account for wave–current interactions are therefore often based on an action balance equation rather than an energy balance equation (see Chapter 8).

Literature:
Briggs and Liu (1993), Christoffersen (1982), Crapper (1984), González (1984), Gutshabash and Lavrenov (1986), Johnson (1947), Jonsson (1990), Jonsson and Wang (1980), Lai *et al.* (1989), Peregrine (1976), Raichlen (1993), Sakai *et al.* (1983), Sakai and Saeki (1984), Suh *et al.* (1994), Tolman (1990), Verhagen *et al.* (1992), Yamaguchi and Hatada (1990).

7.3.6 Reflections

The coast to which the waves propagate will very probably reflect waves to some degree. A vertical cliff may well reflect 100% of the incoming wave energy, whereas a gentle beach will barely reflect energy. Computing the effect of such reflection on the wave field is generally complicated, even if the incoming wave is a simple long-crested, harmonic wave. At each point in front of the reflecting coast, the wave motion would be the sum of the incoming wave and one or more reflected waves (a geometrically intricate rocky coast may well reflect in many different directions). The type of wave model that should be used to compute the wave field under such conditions depends very much on the nature of the reflection. If the coast has many rocky outcrops with a horizontal scale roughly equal to or smaller than the wave length, the situation may well be beyond any mathematical modelling. If the reflections are somewhat less intricate, but still rather variable, for instance, a length of dyke, a breakwater or some other line-barrier, a wave diffraction model

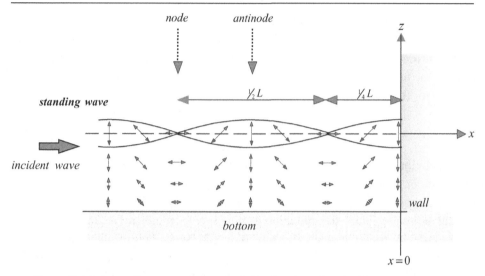

Figure 7.16 A standing wave due to the full reflection of an incident wave against a vertical wall. The short straight arrows are the trajectories of the water particles as they undergo their motion in one wave period.

may be required. If the reflection is fairly homogeneous, a wave model based on the spectral energy balance equation may be used with reflective boundary conditions. However, reflection is very often ignored, particularly near sandy, beach-like coasts, where wave reflection is often deemed to be insignificant (and certainly in wave models operating on an oceanic scale).

One of the most notable phenomena of reflecting waves is the standing wave (also called clapotis). To introduce this phenomenon, consider a one-dimensional situation with a long-crested wave at normal incidence, reflecting off a vertical wall at location $x = 0$ (see Fig. 7.16). The resulting wave profile is the summation of two waves: the incident wave, propagating towards the wall, and the reflected wave, propagating away from that wall:

$$\eta(x, t) = a_i \sin(\omega t - kx) + a_r \sin(\omega t + kx) \qquad (7.3.34)$$

with amplitudes a_r and a_i of the reflected and incident waves, respectively. In the case of 100% reflection, the two amplitudes are equal, $a_r = a_i$, and the wave may also be written (with standard trigonometric rules) as

$$\eta(x, t) = 2a_i \cos(kx)\sin(\omega t) \qquad (7.3.35)$$

The surface elevation of this standing wave (it does not propagate) fluctuates as a sine wave in time. Its amplitude is modulated horizontally with a cosine, from a minimum of 0 at locations $x = \frac{1}{4}L + (n/2)L$ (where n is an integer; these points are called 'nodes'; see Fig. 7.16) to a maximum of $2a_i$ at locations $x = (n/2)L$ (these points are called 'antinodes'). The corresponding wave-induced pressure and orbital velocities are, according to the linear wave theory,

$$p_{wave} = 2\rho g a_i \frac{\cosh[k(d+z)]}{\cosh(kd)} \cos(kx)\sin(\omega t) \tag{7.3.36}$$

$$u_x = -2\omega a_i \frac{\cosh[k(d+z)]}{\sinh(kd)} \sin(kx)\cos(\omega t) \tag{7.3.37}$$

$$u_z = 2\omega a_i \frac{\sinh[k(d+z)]}{\sinh(kd)} \cos(kx)\cos(\omega t) \tag{7.3.38}$$

In engineering practice, nonlinear estimates, not given here, are often used instead, e.g., Sainflou (1928). These expressions Eqs. (7.3.36)–(7.3.38) show that the *vertical* structure of this standing wave is the same as in a propagating wave, but that the *horizontal* structure is very different. Under a node (i.e., a point with zero surface elevation) of the standing wave, the orbital velocity is always horizontal, whereas it is always vertical in a propagating wave (under the propagating point of zero surface elevation). Under an antinode (i.e., a point with maximum elevation) of the standing wave, the orbital velocity is always vertical, whereas it is always horizontal in a propagating wave (under the propagating point of maximum surface elevation, i.e., the crest).

Very often, the reflection is (far) less than 100% and the resulting wave is not a perfectly or fully standing wave but a partially standing wave (see Fig. 7.17). If no phase shift occurs at reflection, such a wave can be written as the sum of a propagating wave travelling towards the obstacle (with an amplitude equal to that of the incident wave minus that of the reflected wave) and a fully standing wave (with a maximum amplitude equal to twice the reflected amplitude):

$$\eta(x, t) = (a_i - a_r)\sin(\omega t - kx) + 2a_r \cos(kx)\sin(\omega t) \tag{7.3.39}$$

The maximum amplitude a_{max} (at the quasi-antinodes) and the minimum amplitude a_{min} (at the quasi-nodes) of the partially standing wave are then, respectively,

$$a_{max} = a_i + a_r$$
$$a_{min} = a_i - a_r \tag{7.3.40}$$

The reflection coefficient $K_{refl} = a_r/a_i$ can readily be measured in a laboratory flume by measuring these maximum and minimum amplitudes along the flume (this is the most common way to determine reflection coefficients). In real life, the waves are almost always random and short-crested and, in addition, a coast

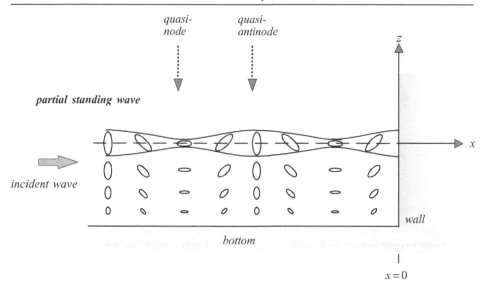

Figure 7.17 A partially standing wave due to the (partial) reflection of an incident wave against an obstacle. The ellipses are the trajectories of the water particles as they undergo their motion in one wave period.

or reflecting structure always has a finite length. This complicates the pattern of reflecting waves (e.g., diffraction will occur at the tips of a reflecting structure), but it also limits the area of large amplitudes in front of such a structure: the standing-wave patterns of the individual harmonic, long-crested wave components overlap in a random, short-crested wave field. The wave field is therefore smoothed by adding randomness and short-crestedness to the waves (maximum amplitudes are reduced and minimum amplitudes are enhanced).

The mechanisms involved in the reflection off a structure or coast are usually so complex that the reflection coefficient cannot be determined theoretically (it may well involve wave breaking). It requires observations. Such observations have been generalised to some extent for various types of coasts and coastal structures using a parameter called the Iribarren parameter or surf similarity parameter, which is defined in terms of the bottom slope and the wave steepness (see also Section 7.6):

$$\xi = \tan \alpha / \sqrt{H/L_0} \tag{7.3.41}$$

where α is the bottom slope, H is the incident wave height and L_0 is the deep-water wave length ($L_0 = gT^2/(2\pi)$, where T is the wave period). For gentle slopes or steep waves (i.e., $\xi < 2.5$) on flat, smooth, impermeable, inclined surfaces, observations show that

$$K_{refl} \approx 0.1\xi^2 \qquad \text{for } \xi < 2.5 \text{ (i.e., gentle slopes or steep waves)} \qquad (7.3.42)$$

Under the same conditions, i.e., $\xi < 2.5$, but with surfaces that are rough or permeable, the reflection is less (by as much as 50%). Under other conditions, i.e., with steep slopes or mild wave steepness (i.e., $\xi > 2.5$), the waves tend to reflect without breaking and with a higher reflection coefficient:

$$K_{refl} > 0.1\xi^2 \qquad \text{for } \xi > 2.5 \text{ (i.e., steep slopes or gentle wave steepness)}$$
$$(7.3.43)$$

Observations in real Nature off natural sandy beaches exhibit a corresponding frequency dependence: reflection practically absent for wind-sea frequencies (frequencies > 0.1 Hz; steep waves) but significant at swell frequencies (frequencies 0.05–0.1 Hz; gentle wave steepness) and strong for infra-gravity waves (frequencies <0.05 Hz; very gentle wave steepness).

Literature:
Allsop and Hettiarachchi (1988), Battjes (1974a, 1974b), Goda (2000), Herbers *et al.* (1999), Seelig and Ahrens (1981).

7.4 Wave-induced set-up and currents

7.4.1 Introduction

Waves transport not just energy; they also transport momentum. Such momentum transport is equivalent to a stress and horizontal *variations* in this stress act as *forces* on the water (body forces; gravitation is another body force) and may thus tilt the mean sea level or generate currents. In oceanic waters, these forces are generally too weak to be relevant in an engineering context, except that they may generate long waves that are bound beneath wave groups. These long waves, with a wave length equal to the wave length of the groups, are usually well outside the frequency range of wind-generated waves and they will not be considered here. In contrast to this, in coastal waters these forces can be rather large, particularly in the surf zone, where they may induce considerable changes in mean sea level and strong currents (long-shore currents, sometimes breaking out to sea as rip currents).

7.4.2 Wave momentum and radiation stress

Wave momentum is a vector property: it is the product of the mass and the wave-induced velocity of the water particles. It is represented here by its density $\rho\vec{u}$, where

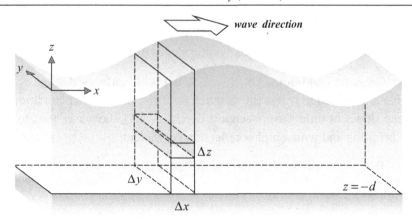

Figure 7.18 The slice of water that is used in the derivation of the horizontal momentum under a wave.

ρ is the mass density and $\vec{u}$ is the particle velocity. To estimate the total amount of momentum beneath a wave per unit horizontal area (i.e., integrated over depth; e.g., per m^2 sea surface), consider a long-crested (i.e., cylindrical) wave propagating in the positive x-direction and a column of water beneath that wave, from the bottom to the sea surface with horizontal surface area $\Delta x \Delta y$ (see Fig. 7.18). Consider next the wave-induced x-momentum in a horizontal slice in the column with thickness Δz. The amount of x-momentum in this slice is equal to $\rho u_x \Delta x \Delta y \Delta z$. The *total* amount of x-momentum in the column q_x is obtained by integrating from the bottom to the instantaneous water surface (note the upper limit of the integration):

$$q_x = \left(\int_{-d}^{\eta} \rho u_x \mathrm{d}z \right) \Delta x \Delta y \qquad (7.4.1)$$

Per unit surface area (i.e., divided by $\Delta x \Delta y$) and averaged over time, the amount of x-momentum then is

$$Q_x = \overline{\int_{-d}^{\eta} \rho u_x \mathrm{d}z} \qquad (7.4.2)$$

where the overbar denotes averaging over one wave period.

For a wave propagating in the positive x-direction, the result of this integration (to second-order accuracy; see Note 7B), with u_x taken from the linear wave theory, is

$$\boxed{Q_x = \frac{\rho a^2}{2 \tanh(kd)} \omega} \qquad (7.4.3)$$

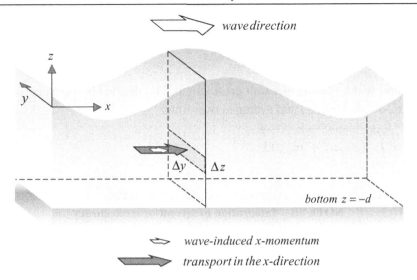

Figure 7.19 The horizontal transport of wave-induced x-momentum in the x-direction through a vertical window parallel to the wave crest with height Δz and width Δy.

For this wave, the other horizontal component of wave momentum, ρu_y, i.e., that parallel to the wave crest, is zero (the wave-induced y-momentum, i.e., directed along the crest), because the orbital velocity u_y is zero, so

$$\boxed{Q_y = 0} \qquad\qquad (7.4.4)$$

These two expressions give the total amount of horizontal momentum beneath the waves per unit horizontal surface area.

The *transport* of wave-induced momentum is equivalent to a stress (consisting of normal stresses and shear stresses) and it is called *radiation stress* (Longuet-Higgins and Stewart, 1960, 1961, 1962, 1964). To find the expressions for this stress, we first consider the horizontal transport in the wave direction of x-momentum (ρu_x), i.e., through a vertical plane, parallel to the wave crests (see Fig. 7.19). The transport through a vertical window with surface area $\Delta y \Delta z$, by bodily motion with the particle velocity u_x in a time interval Δt is $(\rho u_x) u_x \, \Delta y \Delta z \Delta t$. In addition to this, wave momentum is also transferred by the wave-induced pressure p_{wave} in the water (see Section 5.3.2 for the relation between force and momentum). This adds a transport $p_{wave} \Delta y \Delta z \Delta t$. The total transport through the vertical window $\Delta y \, \Delta z$ in a time interval Δt is therefore $(\rho u_x u_x + p_{wave}) \Delta z \Delta y \Delta t$. The transport s_{xx} through the entire vertical plane, from the bottom to the instantaneous surface, is obtained by integrating between these two limits:

$$S_{xx} = \left(\int_{-d}^{\eta} (\rho u_x u_x + p_{wave}) dz \right) \Delta y \Delta t \tag{7.4.5}$$

or, per unit width (that is, per unit crest length, i.e., divided by Δy) and per unit time (i.e., divided by Δt) and averaged over time:

$$\overline{S_{xx}} = \overline{\int_{-d}^{\eta} (\rho u_x u_x + p_{wave}) dz} \tag{7.4.6}$$

NOTE 7B   Integration to second-order accuracy

Determining the order of accuracy of an integral over the vertical beneath a harmonic wave in the linear wave theory is explained in general terms in Note 5B. In the text of Section 7.4.2. the following integrals appear:

(a) $f(z) = \rho u_x$, so the integral (averaged over time) can be written as

$$\overline{\int_{-d}^{\eta} f(z) dz} = \overline{\int_{-d}^{\eta} \rho u_x dz} = \overline{\int_{-d}^{0} \rho u_x dz} + \overline{\int_{0}^{\eta} \rho u_x dz} = \overline{\int_{0}^{\eta} \rho u_x dz} = \frac{\rho a^2}{2 \tanh(kd)} \omega$$

Since $f(z) = \rho u_x$ is of first order in amplitude (the orbital velocity is proportional to the amplitude), it follows that the first integral is of first order and that the second integral is of second order. Therefore, both need to be accounted for in a second-order approximation. However, the outcome of the first integral (averaged over time) is zero. Using the expression for u from the linear wave theory gives the result for the second integral as indicated.

(b) $f(z) = \rho u_x^2$, so the integral (averaged over time) can be written as

$$\overline{\int_{-d}^{\eta} f(z) dz} = \overline{\int_{-d}^{\eta} \rho u_x^2 dz} = \overline{\int_{-d}^{0} \rho u_x^2 dz} + \overline{\int_{0}^{\eta} \rho u_x^2 dz} \approx \overline{\int_{-d}^{0} \rho u_x^2 dz} = nE$$

Since $f(z) = \rho u_x^2$ is of second order in amplitude (the orbital velocity is proportional to the amplitude), it follows that the first integral is of second order and that the second integral is of third order. The second integral can therefore be ignored in a second-order approximation. Using the expression for u_x from the linear wave theory gives the result indicated (where E is the wave energy and n is the ratio of the group velocity over the phase speed; see Eqs. 5.5.5 and 5.5.12 or 7.3.3).

(c) $f(z) = -\rho u_z^2$, so the integral, with upper limit 0 (and averaged over time) can be written as

$$\overline{\int_{-d}^{0} f(z) dz} = -\overline{\int_{-d}^{0} \rho u_z^2 dz} = (n-1)E$$

Since $f(z) = -\rho u_z^2$ is of second order in amplitude (the orbital velocity is proportional to the amplitude), it follows that this integral is of second order in amplitude. Using the expression for u_z from the linear wave theory gives the result indicated.

(d) $f(z) = p_{wave}$, so the integral (averaged over time) can be written as

$$\overline{\int_0^\eta f(z)dz} = \overline{\int_0^\eta p_{wave}dz} = \frac{1}{2}E$$

Since $f(z) = p_{wave}$ is of first order in amplitude (the wave-induced pressure is proportional to the amplitude), it follows that this integral is of second order. Using the expression for p_{wave} from the linear wave theory gives the result indicated.

which can be split into (note the upper limits of the integrals)

$$S_{xx} = \overline{\int_{-d}^\eta \rho u_x^2 dz} + \overline{\int_{-d}^0 p_{wave}dz} + \overline{\int_0^\eta p_{wave}dz}$$

$$= \overline{\int_{-d}^\eta \rho u_x^2 dz} + \int_{-d}^0 \overline{p}_{wave}dz + \overline{\int_0^\eta p_{wave}dz} \tag{7.4.7}$$

In the linear wave theory, the average wave-induced pressure $\overline{p}_{wave}$ (the second integral on the right-hand side of Eq. 7.4.7) is zero. However, in a second-order approximation it is not, due to the *vertical* motion of the water particles; to second-order accuracy (see Note 7C) it is $\int_{-d}^0 \overline{p}_{wave}dz = -\int_{-d}^0 \overline{\rho u_z^2}dz$, so

$$S_{xx} = \overline{\int_{-d}^\eta \rho u_x^2 dz} - \int_{-d}^0 \overline{\rho u_z^2}dz + \overline{\int_0^\eta p_{wave}dz} \tag{7.4.8}$$

The outcome of the integrals in Eq. (7.4.8), to second-order accuracy is (see Note 7B) $\overline{\int_{-d}^\eta \rho u_x^2 dz} = nE$, $-\int_{-d}^0 \overline{\rho u_z^2}dz = (n-1)E$ and $\overline{\int_0^\eta p_{wave}dz} = \frac{1}{2}E$ (where n is the ratio of group velocity over phase speed, see Eq. 7.3.3; and E is the wave energy, see Eq. 5.5.5). With these results, the time-averaged *transport of x-momentum in the x-direction* per unit width and per unit time, i.e., the radiation stress component S_{xx}, is (by substitution into Eq. 7.4.8)

$$\boxed{S_{xx} = \left(2n - \frac{1}{2}\right)E}$$
if the wave direction is the direction of the positive x-axis $\tag{7.4.9}$

The double x in the subscript of S_{xx} denotes that x-momentum is transported in the x-direction (for the notation, see Note 7D). We may also say that S_{xx} is equivalent to a *normal* stress acting in the x-direction (i.e., like σ_{xx} in Note 7D).

The transport of y-momentum in the y-direction S_{yy} can be expressed as in Eq. (7.4.7), with the subscripts x replaced with y. However, the orbital motion in the y-direction is zero, so that only $\int_{-d}^{0} p_{wave} \mathrm{d}z = -\int_{-d}^{0} \overline{\rho u_z^2} \mathrm{d}z = (n-1)E$ and $\int_{0}^{\eta} \overline{p_{wave}} \mathrm{d}z = \frac{1}{2}E$ remain. The sum of these two expressions is the time-averaged *transport of y-momentum in the y-direction* per unit width and per unit time, i.e., the radiation stress component S_{yy}, which is given by

$$\boxed{S_{yy} = \left(n - \frac{1}{2}\right)E}$$

if the wave direction is the direction of the positive x-axis   (7.4.10)

NOTE 7C   The average wave-induced pressure (second-order accurate)

Consider the balance of *vertical* momentum ρu_z (z-momentum) in the column beneath the wave, from a certain level $z = z_1$ below the mean water surface, to the surface (take the x-direction in the wave direction; see illustration below).

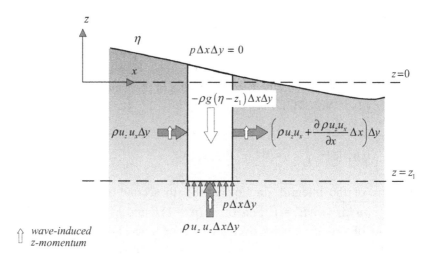

The balance of vertical momentum for a column from level z_1 to the surface.

In analogy with the other derivations of balances of properties (e.g., Section 5.3.2 but now for the column between $z = z_1$ and $z = \eta$, with horizontal surface area $\Delta x \Delta y$, so that integrals $\int_{z_1}^{\eta} \ldots \mathrm{d}z \Delta x \Delta y$ appear in the balance) and treating forces as the generation or transfer of momentum (vertical forces only: the pressure at the underside of the column and the weight of the water in the column; see the above illustration and also Section 5.3.2), we find for the balance of z-momentum ρu_z (ignoring vertical shear stresses, not only because we are considering idealised water but also because the horizontal gradients in vertical velocity are small)

$$\frac{\partial}{\partial t} \int_{z_1}^{\eta} \rho\, u_z dz \Delta x \Delta y = -\frac{\partial}{\partial x} \int_{z_1}^{\eta} \rho u_z u_x dz \Delta x \Delta y + \rho u_z u_z\, \Delta x \Delta y$$

$$-\rho g(\eta - z_1)\Delta x \Delta y + p\,\Delta x \Delta y$$

The left-hand side represents the rate of change of vertical momentum in the column. On the right-hand side; the first term represents the net import of vertical momentum through the vertical sides of the column (in this situation, with $u_y = 0$, no transport occurs in the y-direction), the second term represents the net import of vertical momentum through the underside of the column (where $z = z_1$), the third term represents the weight of the column and the fourth term represents the force at the underside of the column due to the pressure in the water. Per unit horizontal surface area this is (divide by $\Delta x \Delta y$)

$$\frac{\partial}{\partial t} \int_{z_1}^{\eta} \rho u_z dz = -\frac{\partial}{\partial x} \int_{z_1}^{\eta} \rho u_z u_x dz + \rho u_z^2 - \rho g(\eta - z_1) + p$$

If we average over time, we can write

$$\overline{\frac{\partial}{\partial t} \int_{z_1}^{\eta} \rho u_z dz} = -\overline{\frac{\partial}{\partial x} \int_{z_1}^{\eta} \rho u_z u_x dz} + \overline{\rho u_z^2} - \overline{\rho g(\eta - z_1)} + \overline{p}$$

Assuming the effect of the waves on the *mean* surface elevation to be small, so that we can take $\overline{\eta} = 0$ here, and considering stationary conditions, so that the average rate of change $\overline{\partial ../\partial t} = 0$, we find

$$0 = -\frac{\partial}{\partial x} \int_{z_1}^{\eta} \overline{\rho u_z u_x} dz + \overline{\rho u_z^2} + \rho g z_1 + \overline{p}$$

Moreover, in propagating, harmonic waves u_z and u_x are $90°$ out of phase so that $\overline{u_z u_x} = 0$. Substituting this into the above gives

$$0 = \overline{\rho u_z^2} + \rho g z_1 + \overline{p}$$

The average pressure in the water at $z = z_1$ is therefore $\overline{p} = -\rho g z_1 - \overline{\rho u_z^2}$, in which $-\rho g z_1$ is the mean hydrostatic pressure (for $\overline{\eta} = 0$; z is negative below the mean water surface) and $-\overline{\rho u_z^2}$ is the time-averaged wave-induced pressure $\overline{p}_{wave}$:

$$\overline{p}_{wave} = -\overline{\rho u_z^2}$$

S_{yy} is equivalent to a normal stress acting in the y-direction (i.e., like σ_{yy} in Note 7D). In addition to these transports of x-momentum in the x-direction and of y-momentum in the y-direction, there is also transport of x-momentum in the y-direction (S_{xy}) and of y-momentum in the x-direction (S_{yx}). The definition of these additional transports is analogous to those of S_{xx} and S_{yy} (compare with

Eq. 7.4.6):

$$S_{xy} = \overline{\int_{-d}^{\eta} (\rho u_x u_y + \tau_{xy})dz}$$ (7.4.11)

and

$$S_{yx} = \overline{\int_{-d}^{\eta} (\rho u_y u_x + \tau_{yx})dz}$$ (7.4.12)

where τ_{xy} and τ_{yx} are equivalent to the shear stresses in the water in the x- and y-directions, respectively (i.e., like σ_{xy} and σ_{yx} in Note 7D). However, in the idealised fluid that we consider here, such shear stresses are assumed to be zero. Moreover, the orbital velocities in the y-direction u_y are zero, so these transports of momentum are zero:

$\boxed{S_{xy} = 0}$ if the wave direction is the direction

of the positive x-axis   (7.4.13)

and

$\boxed{S_{yx} = 0}$ if the wave direction is the direction

of the positive x-axis   (7.4.14)

If the wave direction is *not* in the x-direction (i.e., the wave travels in a direction θ relative to the positive x-direction), then u_x needs to be replaced with $u_x \cos\theta$ and u_y by $u_y \sin\theta$ in the above expressions, so that

$$S_{xx} = \left(n - \frac{1}{2} + n\cos^2\theta\right)E$$

$$S_{yy} = \left(n - \frac{1}{2} + n\sin^2\theta\right)E \qquad \text{if the wave direction is } \theta$$ (7.4.15)

$$S_{xy} = n\cos\theta\sin\theta\,E \qquad\qquad \text{(relative to the positive } x\text{-axis)}$$

$$S_{yx} = n\sin\theta\cos\theta\,E$$

where S_{xx} and S_{yy} are equivalent to normal stresses and S_{xy} and S_{yx} to shear stresses (this set of stresses is called the radiation stress tensor, see Note 7D).

NOTE 7D   Scalar, vector and tensor

I am considering here the concepts of scalar, vector and tensor in a physical context, in Cartesian co-ordinates (with x-, y- and z-co-ordinates). I must emphasise this, since these concepts have a more general meaning in mathematics and non-Euclidian space. An excellent introduction to this subject is given by Aris (1962).

A *scalar* is a property, e.g., mass density ρ, that is characterised by only one number: the magnitude of the property. It may, but need not, be dimensionless, e.g., mass density ρ is a scalar with dimension mass per volume (in S.I. units this is [kg m^{-3}]), whereas the ratio of wave length over depth is a dimensionless scalar. A scalar may well vary with space or time but, at each location or moment in time, it is characterised by only one number, e.g., $\rho = \rho(x, y, z, t)$.

A *vector* is a property with a direction and a magnitude, e.g., the velocity of a particle $\vec{u}$. It is characterised by a set of orthogonal reference directions (unit vectors) and magnitudes (numbers). A vector too may, but need not, be dimensionless, e.g., velocity $\vec{u}$ is a vector with dimension length per time (in S.I. units [m s^{-1}]). In three spatial dimensions it is characterised by a set of three orthogonal unit vectors $\vec{e}_x$, $\vec{e}_y$ and $\vec{e}_z$ oriented in the positive x-, y- and z-directions, respectively (they define the reference directions) and a set of three numbers $\{u_x, u_y, u_z\}$. These directions and magnitudes define the three components of $\vec{u}$ in the x-, y- and z-directions, which are denoted as $\vec{u}_x = u_x \vec{e}_x$, $\vec{u}_y = u_y \vec{e}_y$ and $\vec{u}_z = u_z \vec{e}_z$. The vector $\vec{u}$ is the vector sum of these components: $\vec{u} = \vec{u}_x + \vec{u}_y + \vec{u}_z$. A vector can vary with space and time, e.g., $\vec{u} = \vec{u}(x, y, z, t)$.

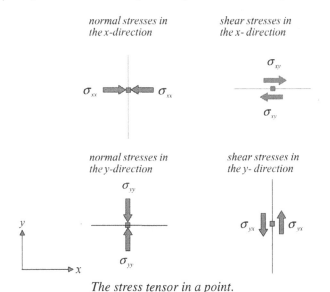

The stress tensor in a point.

A *tensor* is a property describing a state at a point with a *set* of vectors (e.g., the stress in a point). This set consists of normal vectors and tangential vectors. For instance, the *stress* in a point in a two-dimensional x, y-plane can be characterised by a tensor consisting of two *normal* stresses, σ_{xx} and σ_{yy} (loosely called tension or pressure; see the next paragraph), and two tangential (i.e., *shear*) stresses, σ_{xy} and σ_{yx} (see the above illustration). The convention for the notation of these vectors used here is that the first subscript denotes the direction of the vector, while the second subscript denotes the direction of the normal of the surface *on* which or *in* which the vector acts. Obviously, a tensor may also vary in space and time, e.g., $\sigma_{ij} = \sigma_{ij}(x, y, t)$, where $i = x$ or y and $j = x$ or y. Since these stresses are vectors, they are described with unit vectors and magnitudes.

The average of the normal stresses is usually called tension (when positive; usually in solids only, because fluids support tension poorly) or pressure (when negative; in both solids and fluids). In an ideal or stationary fluid there are no shear stresses: $\sigma_{xy} = \sigma_{yx} = 0$; and the normal stresses are equal: $\sigma_{xx} = \sigma_{yy}$.

7.4.3 Wave-induced set-up, set-down and currents

As indicated in the introduction of this chapter, horizontal variations in the transport of wave-induced momentum, i.e., horizontal variations in the radiation stress, may affect the mean sea level and generate currents, particularly in the surf zone. The reason for this is that an increase of momentum transport, i.e., an increase in radiation stress over a horizontal distance is equivalent to exerting an *opposite* force on the water body (which is similar to a situation in which an increase in water pressure in the positive x-direction induces a net force on the water body in the negative x-direction). The corresponding wave-induced radiation force *per unit horizontal surface area* in the x-direction is (see Fig. 7.20)

$$F_x = -\frac{\partial S_{xx}}{\partial x} - \frac{\partial S_{xy}}{\partial y} \qquad \text{in the } x\text{-direction} \tag{7.4.16}$$

where obviously $-\partial S_{xx}/\partial x$ represents the effect of variations in the x-directed radiation *normal* stresses and $-\partial S_{xy}/\partial y$ the effect of variations in the x-directed radiation *shear* stress.

Note that the first index of S_{xx} and S_{xy} indicates that the x-momentum is considered. The minus sign in this expression indicates that, if the radiation stress increases in the positive x-direction, the corresponding force is oriented in the negative x-direction, and also the reverse: if the radiation stress *decreases* in the positive x-direction, the corresponding force is oriented in the *positive* x-direction (e.g., in the surf zone, the wave heights generally decrease towards the shore, and the radiation stress correspondingly decreases, resulting in a force on the water body directed towards the shore; see below). Similarly for the y-direction:

$$F_y = -\frac{\partial S_{yy}}{\partial y} - \frac{\partial S_{yx}}{\partial x} \qquad \text{in the } y\text{-direction} \tag{7.4.17}$$

These forces generally cause currents and changes in the mean water level.[10]

---

[10] In numerical, hydrodynamic models, which are used to compute these currents and sea-level changes, these forces F_x and F_y are usually treated as a horizontal shear stress at the water surface (just like the wind stress, which is often used in such models to drive wind-induced currents and storm surges).

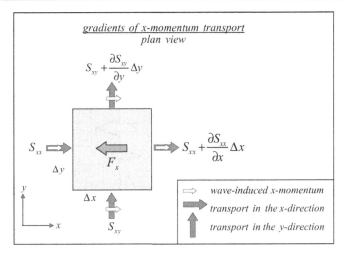

Figure 7.20 The gradients of the wave-induced x-momentum transport S_{xx} and S_{xy} (plan view of sea surface) and the corresponding force F_x in the opposite direction.

To illustrate this with a simple but important application, consider the one-dimensional situation of a long-crested harmonic wave approaching a beach at normal incidence. In a stationary situation, the (vertically averaged) current is zero,[11] because the water has piled up against the coast until the hydrostatic pressure gradient under the tilting (mean) water surface balances the driving radiation stress gradient. The corresponding change in the mean water level (indicated as $\overline{\eta}$ above still-water depth d) is readily computed with a momentum balance equation. To that end, consider a vertical water column with horizontal surface area $\Delta x \Delta y$ (see Fig. 7.21).

For the x-momentum Q_x in this column, the momentum balance equation gives the balance amongst the change of x-momentum over time interval Δt, the net import of x-momentum and the local generation of x-momentum (i.e., forces in the x-direction):

storage of momentum Q_x during time interval Δt

= net import of Q_x during time interval Δt

+ local production of Q_x during time interval Δt (7.4.18)

The amount of x-momentum in the column is $Q_x \Delta x \Delta y$. Its change during the time interval Δt is $(\partial Q_x/\partial t)\Delta x \Delta y \Delta t$. The net *import* of Q_x during that time interval is $-(\partial S_{xx}/\partial x)\,\Delta x \Delta y \Delta t$ (note that $(\partial S_{xy}/\partial y)\Delta x \Delta y \Delta t = 0$ because we

---

[11] In a situation with oblique wave incidence, a horizontal current along the shore would be generated with occasional outbreaks to sea that are called rip currents (caused by hydrodynamic instability or variations along the coast in the seabed topography or in the incident wave field).

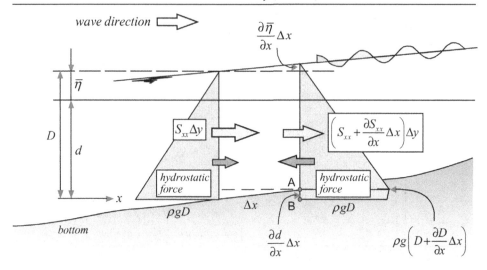

Figure 7.21 The balance of gradients of the radiation stress and the hydrostatic pressure on a vertical column under a wave with varying wave amplitude approaching a beach at normal incidence.

consider a one-dimensional case in which, by definition, all derivatives in the y-direction are zero). Ignoring bottom friction and other horizontal forces, the local production of x-momentum (per unit time) is the net hydrostatic horizontal force (hydrostatic pressure over the depth and width of the column and along the bottom; see Section 5.3.2 for the interpretation of force as a source of momentum), which is equal to the hydrostatic force on the left-hand side of the column $\frac{1}{2}\rho g D^2 \Delta y$, where $D = d + \overline{\eta}$, minus that force on the right-hand side of the column (i.e., between point A in Fig. 7.21 and the mean surface elevation, $\frac{1}{2}\rho g [D + (\partial D/\partial x)\Delta x]^2 \Delta y$), and minus the horizontal component of the hydrostatic force along the bottom (i.e., between points A and B, $\rho g[D + \frac{1}{2}(\partial D/\partial x)\Delta x](\partial d/\partial x)\Delta x \Delta y$). Ignoring second-order terms (i.e., second-order in Δx), the resulting net horizontal force is $-\rho g(d + \overline{\eta})\partial \overline{\eta}/\partial x \Delta x \Delta y$, so the balance equation, after dividing by $\Delta x \Delta y \Delta t$, is

$$\frac{\partial Q_x}{\partial t} = -\frac{\partial S_{xx}}{\partial x} - \rho g(d + \overline{\eta})\frac{\partial \overline{\eta}}{\partial x} \qquad (7.4.19)$$

For a stationary situation, i.e., when all time derivatives are zero, this reduces to (the derivatives in x may be written as d../dx since the only variations now occur in the x-dimension)

$$\frac{dS_{xx}}{dx} + \rho g(d + \overline{\eta})\frac{d\overline{\eta}}{dx} = 0 \qquad (7.4.20)$$

or, assuming that $\bar{\eta} \ll d$,

$$\frac{d\bar{\eta}}{dx} = -\frac{1}{\rho g d}\frac{dS_{xx}}{dx} \tag{7.4.21}$$

This implies that, if the radiation stress gradient is positive, $dS_{xx}/dx > 0$, the slope of the mean surface is negative, $d\bar{\eta}/dx < 0$ (resulting in a *set-down*): and, if the radiation stress gradient is negative, $dS_{xx}/dx < 0$, the slope of the mean surface is positive, $d\bar{\eta}/dx > 0$ (resulting in a *set-up*).

If no energy dissipation occurs (in this one-dimensional case of a harmonic wave at normal incidence), then the variation of the wave amplitude is due solely to shoaling. Integrating Eq. (7.4.20), taking $\bar{\eta} = 0$ in deep water, and using the energy balance equation for a shoaling wave (see Eq. 7.3.5) and the expression for S_{xx} (see Eq. 7.4.9), we find (e.g., Longuet-Higgins and Stewart, 1962)

$$\bar{\eta} = -\frac{1}{2}\frac{a^2 k}{\sinh(2kd)} \qquad \text{set-down, without energy dissipation} \tag{7.4.22}$$

Apparently, the average water level $\bar{\eta}$ depends, in this one-dimensional stationary situation, not on the bottom profile but only on the local parameters, water depth d, wave amplitude a and wave number k. The minus sign in this expression shows that shoaling lowers the mean sea level: in other words, a set-down occurs. For very shallow water (where $\sinh(2kd) \approx 2kd$), the set-down can be written, with $a = \frac{1}{2}H$, as (from Eq. 7.4.22):

$$\bar{\eta} \approx -\frac{1}{16}\frac{H^2}{d} \qquad \begin{array}{l}\text{set-down, without energy dissipation}\\ \text{in very shallow water}\end{array} \tag{7.4.23}$$

This equation shows that the set-down is proportional to the square of the wave height H (which is generally increasing towards the coast, if the waves do not break) and inversely proportional to the water depth (which is generally decreasing), so that the set-down generally increases as the wave propagates without dissipation towards the coast. At the location where the waves start to break (the point of incipient breaking, i.e., the outer edge of the surf zone), the wave height-to-depth ratio is typically $H_{br}/d_{br} \approx 0.8$ (where H_{br} and d_{br} are the wave height and local water depth at incipient breaking, respectively), so that the set-down at this point is 4%–5% of the local water depth or the local wave height (see Fig. 7.22). At the point of incipient breaking, the amplitude starts to decrease and so does the radiation stress. This implies that $dS_{xx}/dx < 0$ and the slope of the mean water surface changes sign and the character of set-down changes to that of set-up (see Fig. 7.22). In the simple case considered here, this set-up can be readily estimated as follows. Assuming that the water is shallow enough that the wave is non-dispersive (so that $n = 1$ and therefore, from Eq. 7.4.9, $S_{xx} = \frac{3}{2}E = \frac{3}{16}\rho g H^2$ and also that

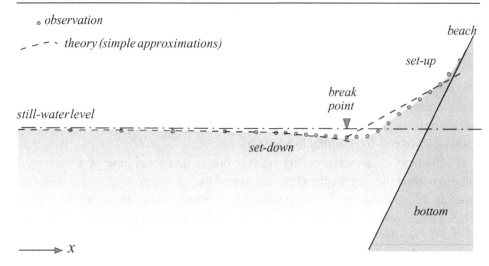

Figure 7.22 The simple theoretical approximations (see the text) of the set-down and set-up (Eqs. 7.4.22 and 7.4.25), induced by waves approaching a very steep beach, in a one-dimensional laboratory situation with a harmonic wave at normal incidence (the bottom slope is $1 : 12$; $\gamma = 1.2$), compared with observations of Bowen *et al.* (1968; shown with distorted scales).

the wave height remains equal to a fixed fraction of the local water depth (so that $H = \gamma(d + \bar{\eta})$, with constant γ, implying the assumption that the water depth decreases monotonically towards the beach), it follows from these two equations that $S_{xx} = \frac{3}{16}\rho g \gamma^2 (d + \bar{\eta})^2$. Substituting this result into the stationary momentum balance of Eq. (7.4.21), gives

$$\frac{\mathrm{d}\bar{\eta}}{\mathrm{d}x} = -\frac{3}{8}\gamma^2 \frac{\mathrm{d}(d + \bar{\eta})}{\mathrm{d}x} \qquad \text{or} \qquad \frac{\mathrm{d}\bar{\eta}}{\mathrm{d}x} = -K\frac{\mathrm{d}d}{\mathrm{d}x} \qquad (7.4.24)$$

with $K = \frac{3}{8}\gamma^2/(1 + \frac{3}{8}\gamma^2)$.

Since the water depth decreases towards the shore ($\mathrm{d}d/\mathrm{d}x < 0$), the mean surface elevation in the surf zone tilts up towards the shore ($\mathrm{d}\bar{\eta}/\mathrm{d}x > 0$). Integrating Eq. (7.4.24) from the point of incipient breaking (where $d = d_{br}$ and where the mean surface elevation is $\bar{\eta}_{br} \approx -\frac{1}{16}H_{br}^2/d_{br}$ gives the set-up as a function of the local water depth:

$$\bar{\eta} = \bar{\eta}_{br} + K(d_{br} - d) \qquad \text{set-up with energy dissipation} \qquad (7.4.25)$$

With the wave height and set-down at incipient breaking $H_{br} = \gamma d_{br}$ and $\bar{\eta}_{br} = -\frac{1}{16}H_{br}^2/d_{br}$, it follows that the set-up at the mean waterline, where $D = d + \bar{\eta} = 0$, is

$$\overline{\eta}_{waterline} = \frac{5}{16}\gamma H_{br} \qquad \text{at mean waterline waterdepth, where } D = d + \overline{\eta} = 0,$$
$$\text{relative to still-water level} \qquad (7.4.26)$$

With γ varying between 0.5 and 1.5 (approximately; see Note 8G), the proportionality coefficient in Eq. (7.4.26) varies accordingly between 0.15 and 0.45:

$$0.15 H_{br} < \overline{\eta}_{waterline} < 0.45 H_{br} \qquad \text{set-up at mean waterline,}$$
$$\text{for } 0.5 < \gamma < 1.5 \qquad (7.4.27)$$

which represents a significant set-up at the beach for high incoming waves. This rising of the mean sea level at the beach implies that, in the presence of high waves, the mean waterline moves up the beach over a considerable vertical and horizontal distance (depending on the bottom slope and the incoming wave height). Despite the fairly simple assumptions in the above (e.g., $n = 1$ and $H = \gamma(d + \overline{\eta})$), these results are realistic, as shown in Fig. 7.22 in which the theoretical results of Eqs. (7.4.22) and (7.4.25) are compared with laboratory observations.

The set-up depends on the incoming wave height and, if this wave height is stationary, then the set-up is stationary. At an actual beach, the waves tend to arrive in groups and the incoming wave height correspondingly fluctuates more or less periodically with the period of the wave groups. This causes the set-up to fluctuate accordingly, so that the surf zone moves periodically up and down as the wave groups arrive one after another, generating low-frequency waves that travel out to sea. This phenomenon is called *surf beat* (one of the forms of infra-gravity waves; see Section 1.3).

Literature:
Arcilla and Lemos (1990), Battjes (1972b), Gourlay (1992), Longuet-Higgins and Stewart (1962, 1963), Munk (1949b), Phillips (1977), Svendsen *et al.* (2003), Tucker (1950).

7.5 Nonlinear, evolving waves

7.5.1 Introduction

The three nonlinear wave theories that were introduced in Chapter 5 (Stokes, 1847; the stream-function theory of Dean, 1965; and the cnoidal theory of Korteweg and de Vries, 1895) are essentially theories for waves that do not change their characteristics horizontally or in time. In other words, these theories consider only local characteristics of permanent waves. For *evolving* waves, other nonlinear theories have been developed. Most of these are based on the same nonlinear equations as those which underlie the linear wave theory (sometimes extended to the Navier–Stokes equations; see Appendix B), but usually only in two spatial dimensions:

either the *vertical* plane or the *horizontal* plane. Because of the considerable computing power that is needed, computations based on these theories can be carried out only over short distances, typically only a dozen wave lengths or so.

For most of the theories that are formulated in the *vertical* x, z-plane (no variations in the y-dimension, i.e., the waves are assumed to be long-crested and travelling in the x-direction), the nonlinear equations, with nonlinear boundary conditions, are not solved analytically but with advanced numerical techniques, in which the free surface is tracked as part of the solution. Examples are the marker-and-cell (MAC) method, the volume-of-fluid (VOF) method, the boundary-element method and mesh-free methods (based on particle-tracking). Three-dimensional versions have also been developed (e.g., Broeze *et al.*, 1993). These methods provide realistic images of waves under a wide variety of conditions (including breaking and the presence of structures) but we will not consider such techniques here.

The most widely accepted nonlinear theory for the two-dimensional *horizontal* x, y-plane (assuming a vertical profile of the velocity potential) is the theory of Boussinesq (1872),[12] originally for one-dimensional propagation over a horizontal bottom; which later was extended to two-dimensional propagation over mildly sloping bottoms by Peregrine (1967) and further extended by many others, mostly to expand the region of applicability to deeper water.[13] The equations of this approach are essentially the shallow-water equations for the water motion (a one-layer model; see Appendix E), supplemented with corrections for vertical accelerations. A recent and promising discovery is that, instead of adding these corrections, discretising the water into two or more layers and using the Euler equations (see Appendix B) instead of the shallow-water equations gives a model with characteristics similar to those of extended Boussinesq-models (Stelling and Zijlema, 2003; Zijlema and Stelling, 2005).

Literature:
Agnon *et al.* (1993), Battjes (1994), Broeze *et al.* (1993), Eldeberky and Madsen (1998), Fenton (1999), Hirt and Nichols (1981), Kaihatu and Kirby (1995), Kirby (1990), Lin and Liu (1999), Longuet-Higgins and Cokelet (1976), Nadaoka *et al.* (1997), Peregrine (1990), Rogers and Dalrymple (2004), Yamaguchi (1986), Wei *et al.* (1995).

7.5.2 *The Boussinesq model*

If waves enter very shallow water, the particle motions become more and more horizontally oriented and eventually (when the water is very shallow) all vertical

---

[12] Valentin Joseph Boussinesq (1842–1929) was a French physicist and mathematician who contributed greatly to our understanding of hydraulics, in particular turbulence in fluids. He was a professor at the Sorbonne (Paris) and a member of the French Academy of Sciences.

[13] A nonlinear version of the mild-slope equation (see Section 7.3.3) can also be used, but that is rarely done.

accelerations may be ignored. The wave can then be described with the *shallow-water equations* (see Appendix *E*). However, before this stage has been reached, the wave motion is not yet horizontal and the shallow-water equations do not apply, but neither does the linear wave theory (the ratio of depth over amplitude is too small). The transition between these two regions of application is covered by the theory of Boussinesq. In this theory, the *vertical* structure of the velocity is not an exact solution of the basic nonlinear balance equations. Instead, it is imposed (*horizontal* velocity constant over the vertical and vertical velocity varying nearly linearly along the vertical). Substituting the corresponding velocity potential function into the nonlinear dynamic and kinematic surface boundary conditions, for a one-dimensional situation with a horizontal bottom, gives the original Boussinesq equations (see Dingemans, 1997b, his Eqs. 5.6a and 5.6b). For a non-horizontal bottom, the Boussinesq equations are (Peregrine, 1967, his Eqs. 13 and 14; Dingemans, 1997b, his Eqs. 5.73)

$$\frac{\partial \eta}{\partial t} + \frac{\partial}{\partial x}[(d + \eta)\,\bar{u}_x] = 0 \tag{7.5.1}$$

and

$$\frac{\partial \bar{u}_x}{\partial t} + \bar{u}_x \frac{\partial \bar{u}_x}{\partial x} + g \frac{\partial \eta}{\partial x} = \frac{1}{2}d\frac{\partial^3 (d\bar{u}_x)}{\partial t\, \partial x^2} - \frac{1}{6}d^2 \frac{\partial^3 \bar{u}_x}{\partial t\, \partial x^2} \tag{7.5.2}$$

where $\bar{u}_x$ is the vertically averaged, horizontal particle velocity. These equations are the one-dimensional shallow-water equations (see Eqs. E.8, E.15 and E.20), supplemented with corrections for the vertical accelerations under the wave (the terms on the right-hand side of Eq. 7.5.2). The remarkable third-order derivatives $\partial^3../\partial t\, \partial x^2$ in these terms are caused by the fact that the Laplace equation (Eq. 5.3.22) forces the (imposed) vertical structure of the velocity potential function to be expressed in terms of the horizontal structure of the velocity potential function (i.e., the second-order horizontal derivative $\partial^2../\partial x^2$). The (nonlinear) dynamic surface boundary condition takes the time derivative of this function, thus producing the mixed third-order derivatives $\partial^3../\partial t\, \partial x^2$. Note that the above Boussinesq equations are one-dimensional but they can readily be expanded to two horizontal dimensions.

Many investigators have modified the original Boussinesq equations to improve various desired characteristics of the corresponding wave (these modified versions are known as *extended* Boussinesq equations or Boussinesq-*like* equations). One of the most successful results is due to Madsen and Sørensen (1992), who extended the applicability of the Boussinesq equations to deeper water and were also able to include (empirically) the process of breaking in their Boussinesq model by imposing a separate body of fluid (also called the '*roller*') on the wave surface, simulating the breaker in real life. An important development was the introduction of spectral versions of the Boussinesq models (frequency-domain Boussinesq models), in terms of the phase and amplitude of a harmonic wave. This has led to an exchange of source terms between Boussinesq models and spectral energy balance models: surf-breaking from spectral energy balance models to Boussinesq models and triad wave–wave interactions from Boussinesq models to spectral energy balance models (see Chapter 8).

Literature:
Abbott *et al.* (1978), Battjes *et al.* (1993), Dingemans (1997b), Freilich and Guza (1984), Madsen and Schäffer (1999), Madsen and Sørensen (1993), Nwogu (1993, 1994), Svendsen (1984), Liu *et al.* (1985).

7.6 Breaking waves

The most nonlinear process affecting waves in coastal waters is depth-induced breaking, also called surf-breaking. This process is poorly understood, certainly the violent breaking of waves against rocks is beyond any theoretical modelling, although the numerical models mentioned earlier, in which the motion of small parcels of fluid is computed with two-dimensional (vertical) Navier–Stokes equations (see Section 7.5.1), produce remarkably realistic pictures of the water surface and particle motions. In practical terms, however, all that is available to the scientist and engineer on breaking waves is empirical information (occasionally supplemented with some clever speculations; see also Section 8.4.5).

If the shore is a flat beach, the type of breaker can be predicted on the basis of the Iribarren number or *surf similarity parameter* $\xi = \tan \alpha / \sqrt{H/L_\infty}$ (Iribarren and Nogales, 1949; Battjes, 1974b; see also Section 7.3.6). If we take the value of this parameter in deep water, $\xi_0 = \tan \alpha / \sqrt{H_\infty / L_\infty}$, where H_∞ is the deep-water wave height and $L_\infty = gT^2/2\pi$, or at the point of incipient breaking $\xi_{br} = \tan \alpha / \sqrt{H_{br}/L_\infty}$, where H_{br} is the wave height at the point of incipient breaking, then observations show the following types of breaking (see Fig. 7.23 and Battjes, 1974b):

spilling:	if $\xi_\infty < 0.5$	or $\xi_{br} < 0.4$
plunging:	if $0.5 < \xi_\infty < 3.3$	or $0.4 < \xi_{br} < 2.0$
collapsing or *surging:*	if $\xi_\infty > 3.3$	or $\xi_{br} > 2.0$

Battjes (1974b) shows that the value of ξ characterises not only the type of breaking, but also the reflection of waves off the beach, the run-up of waves up a beach or a dyke and the stability of the armour of a breakwater (concrete blocks or rubble).

The process of breaking limits the wave height in shallow water. This is of obvious interest to the engineer who wants to estimate the maximum wave height

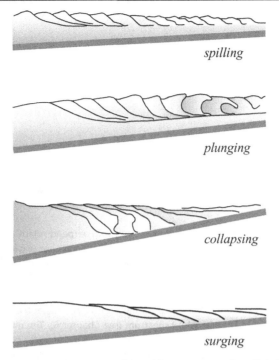

Figure 7.23 The four main types of breaking waves (after Galvin, 1968). All intermediate states may appear on a real beach.

at a certain coastal location, e.g., to formulate design conditions. The individual wave height in an irregular (!) wave field in shallow water cannot exceed some maximum H_{max}, which can be roughly estimated from the local depth. The value of H_{max} is typically a factor 0.75 times the local water depth:[14]

$$H_{max}/d \approx 0.75 \tag{7.6.1}$$

However, under exceptional conditions that factor may be as low as 0.5 or as high as 1.5 (depending on the bottom slope and wave steepness, wind etc.; see Section 8.4.5).

Literature:
Battjes (1974a, 1974b), Fenton (1999), Galvin (1968), Harlow and Welch (1965), Hirt and Nichols (1981), Iribarren and Nogales (1949), Lin and Liu (1999), Liu (2001), Nelson (1994, 1997).

---

[14] This ratio should not be confused with the ratio of maximum *significant* wave height over depth $H_{s,max}/d$, which for wind sea over an extended horizontal bottom is approximately 0.45.

8

Waves in coastal waters

8.1 Key concepts

- In this book, coastal waters are waters that are shallow enough to affect the waves, adjacent to a coast, possibly with (small) islands, headlands, tidal flats, reefs, estuaries, harbours or other features, with time-varying water levels and ambient currents (induced by tides, storm surges or river discharge).

- Under certain idealised conditions (constant wind blowing perpendicularly off a long and straight coastline, over shallow water with a constant depth), the significant wave height is determined by the wind speed, the distance to the upwind coastline (*fetch*), the time elapsed since the wind started to blow (*duration*) and the *depth*. So are the significant wave period and the energy density spectrum.

- Under these idealised conditions, the *spectrum* has a universal shape: the TMA spectrum, which is a generalised version of the JONSWAP spectrum (see Chapter 6). The directional width of this spectrum seems to be the same as in deep water (30°, one-sided width).

- Under more realistic, *arbitrary* coastal-water conditions, the spectral energy balance of the waves is used to compute the wave conditions. This shallow-water version of the energy balance is conceptually a straightforward extension of the energy balance in oceanic waters (see Chapter 6). It represents the time evolution of the wave spectrum, based on the propagation, generation, wave–wave interactions and dissipation of all spectral wave components individually.

- As in oceanic waters, an Eulerian representation (based on a computational grid projected onto the coastal region) should be used for computations with the spectral energy balance.

- Ambient *currents* can be accounted for by replacing the energy density with the action density (i.e., the energy density divided by the relative frequency) in the energy-balance equation and taking some other relatively simple (conceptually) measures.

- *Refraction* is readily accounted for in the energy or action balance equation with an extra transport term. The presentation of *diffraction* is only experimentally formulated in the energy balance equation.

- As the water depth decreases, the *processes* of wave generation by wind, quadruplet wave–wave interactions and dissipation by white-capping intensify and are joined by the process of bottom dissipation. In very shallow water (just outside and in the surf zone), triad wave–wave interactions and depth-induced breaking are added. These two processes dominate the wave evolution in the surf zone.

- In general, the combination of triad wave–wave interactions and depth-induced wave breaking seems to stabilise the shape of the spectrum in the universal shape of the TMA spectrum. However, the triad wave–wave interactions may create an additional (secondary) high-frequency peak in the spectrum under fully developed, shallow-water conditions and near the outer edge of the surf zone.

Table 8.1. *The relative importance of the various processes affecting the evolution of waves in oceanic and coastal waters (after Battjes, 1994)*

Process	Oceanic waters	Coastal waters		
		Shelf seas	Nearshore	Harbour
Wind generation	•••	•••	•	○
Quadruplet wave–wave interactions	•••	•••	•	○
White-capping	•••	•••	•	○
Bottom friction	○	••	••	○
Current refraction / energy bunching	○/•	•	••	○
Bottom refraction / shoaling	○	••	•••	••
Breaking (depth-induced; surf)	○	•	•••	○
Triad wave–wave interactions	○	○	••	•
Reflection	○	○	•/••	•••
Diffraction	○	○	•	•••

••• = dominant, •• = significant but not dominant, • = of minor importance, ○ = negligible.

8.2 Introduction

In the previous chapter it was shown how, in coastal waters, the propagation of waves is affected by a limited water depth and varying wave amplitude (shoaling, refraction and diffraction). However, a limited water depth also affects the generation, nonlinear wave–wave interactions and dissipation. Battjes (1994) has given a review of the relative importance of the various processes in deep and shallow water (see Table 8.1), which shows that modelling waves in coastal waters needs to take into account many more processes than in oceanic waters.

The processes of generation, wave–wave interactions and dissipation that are important in deep water tend to intensify in shallow water, but other processes become active that tend to be even stronger. As waves enter shallow water, they slow down, thus increasing the ratio of wind speed over wave phase speed (there is more transfer of energy from wind to waves) and the waves steepen, thus enhancing quadruplet wave–wave interactions and white-capping. These are the same processes as those which affect the waves in deep water. The additional processes in shallow water are related to propagation (shoaling and refraction), wave–wave interactions (shallow water permits near-resonance of three wave components, resulting in triad wave–wave interactions) and dissipation (bottom friction and depth-induced breaking). In very shallow water the triad interactions seem to be as important as the quadruplet wave–wave interactions are in deep water.

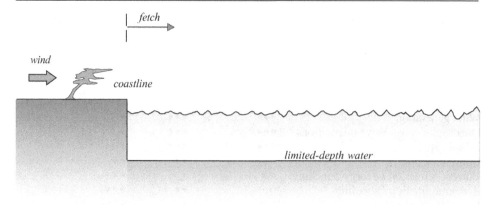

Figure 8.1 The ideal situation of wave growth in limited-depth water with a constant wind blowing over water with a constant depth, perpendicularly off a straight and infinitely long coastline.

The processes of reflection and diffraction dominate the evolution of waves in front of and behind breakwaters and other obstacles, such as rocks and (small) islands, and in harbours. Bottom friction is important only over long distances.

Literature:
Battjes (1988, 1994), WMO (1998).

8.3 Wave modelling for idealised cases (coastal waters)

The idealised case for wave growth in shallow water is essentially the same as for deep water (see Section 6.3), except that a water depth d is added as an extra parameter (see Fig. 8.1): a constant wind (constant in space and time) is blowing perpendicularly off a straight and infinitely long coastline, over water with a limited, constant depth. The waves are described with only the significant wave height and significant wave period (or peak period, i.e., the inverse of the peak frequency of the wave spectrum) or with a universal one- or two-dimensional spectrum. In this approach, the waves depend only on the distance to the upwind coastline (fetch), the time elapsed since the wind started to blow (duration), the wind speed and the water depth. Usually the duration is taken to be infinitely long (in practical applications this means sufficiently long that the precise duration is irrelevant). The transformation from duration to equivalent fetch has been treated in Section 6.3.1 and it will not be repeated here, leaving only fetch, wind speed and depth as the independent variables.

8.3.1 The significant wave

In water with a limited, constant depth, observations indicate that initially, i.e., at short fetches, the water depth has no effect on the waves. This is to be expected. The wave lengths at short fetches are so short that the depth/wave-length ratio is large and the water is *relatively* deep. As the waves grow along the fetch, the wave length becomes longer and the depth becomes more and more important (the waves start to 'feel' the bottom). At very large fetch ($F \to \infty$) the waves are fully developed but with lower values of the corresponding significant wave height and period than in deep water. Observations show that these limit values depend on the *dimensionless* water depth $\tilde{d} = gd/U_{10}^2$, where d is depth, U_{10} is the wind speed at elevation 10 m and g is the gravitational acceleration. This dependence can be approximated with tanh expressions:

$$\tilde{H}_{\infty,d} = \tilde{H}_\infty \tanh(k_3 \tilde{d}^{m_3}) \qquad (8.3.1)$$

fully developed in limited-depth water

$$\tilde{T}_{\infty,d} = \tilde{T}_\infty \tanh(k_4 \tilde{d}^{m_4}) \qquad (8.3.2)$$

where $\tilde{H}_{\infty,d}$ and $\tilde{T}_{\infty,d}$ are the limit values of the dimensionless significant wave height and period, respectively. Note the subscript ∞, d in the notation to distinguish these limit values from the deep-water limit values $\tilde{H}_\infty$ and $\tilde{T}_\infty$ (see Section 6.3.2). The coefficients k_3, k_4, m_3 and m_4 are tunable coefficients, to be determined from observations.

To obtain the shallow-water growth curves, it seems obvious to replace the deep-water limits $\tilde{H}_\infty$ and $\tilde{T}_\infty$ in the deep-water growth curves (Eqs. 6.3.8) with the shallow-water limits $\tilde{H}_{\infty,d}$ and $\tilde{T}_{\infty,d}$. However, such re-scaling would apply to *all* fetches and thus reduce the significant wave height and period at *short* fetches in the same proportion as at *long* fetches, which would not be correct: as indicated above, the waves retain their deep-water character at short fetches. To achieve this, the reduction at short fetches is compensated by the following double (!) inclusion of the depth dependence

$$\tilde{H} = \tilde{H}_\infty \tanh(k_3 \tilde{d}^{m_3}) \tanh\left(\frac{k_1 \tilde{F}^{m_1}}{\tanh(k_3 \tilde{d}^{m_3})}\right)$$

limited-depth water, all   (8.3.3)
sea states

$$\tilde{T} = \tilde{T}_\infty \tanh(k_4 \tilde{d}^{m_4}) \tanh\left(\frac{k_2 \tilde{F}^{m_2}}{\tanh(k_4 \tilde{d})^{m_4}}\right)$$

where $\tilde{F}$ is the dimensionless fetch (see Section 6.3.1). That the waves thus retain their deep-water character at short fetches is shown in Note 8A. Some recent results from observations are given in Note 8B and Fig. 8.2.

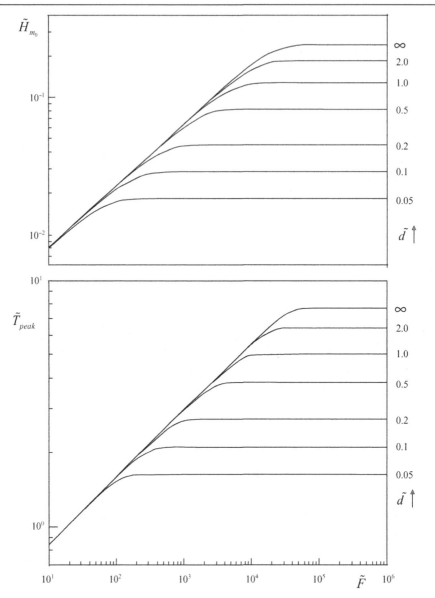

Figure 8.2 The dimensionless significant wave height and period (left-hand vertical axes; see Eqs. 6.3.4 and 6.3.5) as a function of dimensionless fetch (horizontal axes) and depth (right-hand vertical axes, with deep water $\tilde{d}$ at the top in each panel; Kahma and Calkoen, 1992, Pierson and Moskowitz, 1964, and Young and Verhagen, 1996a; modified by Breugem and Holthuijsen, 2006).

NOTE 8A   Limiting values of the depth-limited wave-growth curves

One property of the tanh function is that $\tanh x \approx x$ for $x \ll 1$. In this case, the effect is that, for short fetches, where $\tilde{F} \ll 1$, $\tanh[k_1 \tilde{F}^{m_1} / \tanh(k_3 \tilde{d}^{m_3})] \approx k_1 \tilde{F}^{m_1} / \tanh(k_3 \tilde{d}^{m_3})$. For the significant wave height, therefore,

$$\tilde{H} = \tilde{H}_\infty \tanh(k_3 \tilde{d}^{m_3}) \tanh\left(\frac{k_1 \tilde{F}^{m_1}}{\tanh(k_3 \tilde{d}^{m_3})}\right) \approx \tilde{H}_\infty k_1 \tilde{F}^{m_1} \qquad \text{for } \tilde{F} \ll 1$$

which is the *deep*-water expression for short fetches (as required; see Eq. 6.3.7). Since $\tanh x \to 1$ for $x \to \infty$, we have for large fetches, where $\tilde{F} \to \infty$,

$$\tilde{H} = \tilde{H}_\infty \tanh(k_3 \tilde{d}^{m_3}) \tanh\left(\frac{k_1 \tilde{F}^{m_1}}{\tanh(k_3 \tilde{d}^{m_3})}\right) \approx \tilde{H}_\infty \tanh(k_3 \tilde{d}^{m_3}) \qquad \text{for } \tilde{F} \to \infty$$

$$\text{(i.e., fully developed for a given value of } \tilde{d})$$

The same applies, of course, to the significant wave period.

NOTE 8B   The growth curves of the significant wave height and peak period (all depths)

Probably the best shallow-water data set that is at present available to determine the coefficients of the tanh-expressions of Eq. (8.3.3) is that obtained by Young and Verhagen (1996a) in Lake George, Australia (see Section 6.3.2). To control the transition from a young sea state to the fully developed sea state, Young and Verhagen (1996a) added two extra parameters, p and q, to these expressions:

$$\tilde{H} = \tilde{H}_\infty \left[\tanh(k_3 \tilde{d}^{m_3}) \tanh\left(\frac{k_1 \tilde{F}^{m_1}}{\tanh(k_3 \tilde{d}^{m_3})}\right)\right]^p$$

$$\tilde{T} = \tilde{T}_\infty \left[\tanh(k_4 \tilde{d}^{m_4}) \tanh\left(\frac{k_2 \tilde{F}^{m_2}}{\tanh(k_4 \tilde{d}^{m_4})}\right)\right]^q$$

They used the wind, averaged over the upwind fetch (Taylor and Lee, 1984; see Section 6.3.2). As indicated in Section 6.3.2, Breugem and Holthuijsen (2006) re-analysed the data of Young and Verhagen (1996a). The resulting coefficients[1] are summarised below with the corresponding dimensionless growth curves given in Fig. 8.2. It may be noted that Young and Babanin (2006) recently showed that replacing the expression for the limit value $\tilde{T}_\infty, d$ (Eq. 8.3.2) with an expression in terms of the dimensionless

---

[1] These coefficients relate to conditions with neutral atmospheric stability. Young (1998) gives corrections for non-neutral atmospheric conditions.

peak wave number $\tilde{k}_{peak,\infty,d} = k_{peak,\infty,d}U_{10}^2/g$ would improve the fit to observations, in particular for very shallow water ($\tilde{d} < 0.1$, say).

The coefficients representing wind-wave growth in the idealised situation
(see also Note 6B)

Deep water and finite-depth water		
Pierson and Moskowitz (1964) *fully developed sea states, deep water, Eqs. (6.3.8) and (8.3.3) and this note*	Kahma and Calkoen (1992) *young sea states, deep water, Eqs. (6.3.7)*	Young and Verhagen (1996a) modified by Breugem and Holthuijsen (2006) *all sea states, all water depths, equations of this note*
$\tilde{H} = \tilde{H}_{m_0}$ $\tilde{H}_\infty = 0.24$	$\tilde{H}_\infty = a_1\tilde{F}^{b_1}$ $a_1 = 2.88 \times 10^{-3}$ $b_1 = 0.45$	$\tilde{H}_\infty = 0.24$ $k_1 = 4.41 \times 10^{-4}$ $m_1 = 0.79$ $p = 0.572$ $k_3 = 0.343$ $m_3 = 1.14$
$\tilde{T} = \tilde{T}_{peak}$ $\tilde{T}_\infty = 7.69$	$\tilde{T}_\infty = a_2\tilde{F}^{b_2}$ $a_2 = 0.459$ $b_2 = 0.27$	$\tilde{T}_\infty = 7.69$ $k_2 = 2.77 \times 10^{-7}$ $m_2 = 1.45$ $q = 0.187$ $k_4 = 0.10$ $m_4 = 2.01$

Literature:
Bretschneider (1958), CEM (2002), SPM (1973, 1984), Thijsse and Schijf (1949), Thijsse (1948, 1952).

8.3.2 The one-dimensional wave spectrum

Under the idealised shallow-water conditions considered here, the wave spectrum evolves essentially as in deep water: from the high frequencies to lower frequencies while the area under the spectrum increases (and therefore also the significant wave height). However, in contrast to the situation in deep water, the spectrum does *not* retain its shape along the fetch. Instead, observations show that the high-frequency tail grows flatter as the waves evolve along the fetch; it changes from an f^{-5}-shape to an f^{-3}-shape. This corresponds very well to the hypothesis of Kitaigorodskii *et al.* (1975) that the shape of the spectral tail of a young sea state is more universally characterised in terms of wave number (k) than in terms of frequency (f). To show the effect of this, consider first the tail of the spectrum in deep water. The formulation in terms of wave number follows

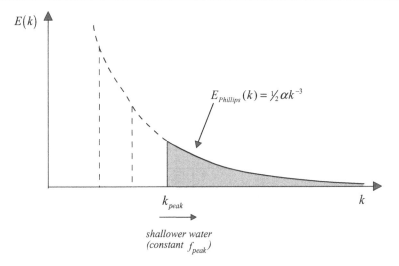

Figure 8.3 The shift of the peak wave number as the waves propagate into shallower water with constant peak frequency, resulting in a decrease of total energy and significant wave height.

directly from the (Phillips) f^{-5}-shape (see Section 6.3.3) by transforming from frequency f-space to wave-number k-space with the corresponding Jacobian $(\mathrm{d}f/\mathrm{d}k)_\infty$ for deep water (see Section 3.5.8 and the footnote in Section 3.5.5). The tail of the spectrum in k-space is then

$$E_{Phillips,\infty}(k) = \alpha g^2 (2\pi)^{-4} f^{-5} \left(\frac{\mathrm{d}f}{\mathrm{d}k}\right)_\infty \qquad \text{for deep water} \qquad (8.3.4)$$

where the subscript ∞ indicates deep water. With the dispersion relationship of the linear wave theory in deep water, this can be written as

$$E_{Phillips,\infty}(k) = \tfrac{1}{2}\alpha k^{-3} \qquad \text{for deep water} \qquad (8.3.5)$$

The hypothesis of Kitaigorodskii *et al.* (1975) implies that this expression would apply in any depth, so that the deep-water restriction may be removed (drop the subscript ∞):

$$E_{Phillips}(k) = \tfrac{1}{2}\alpha k^{-3} \qquad \text{for arbitrary-depth water} \qquad (8.3.6)$$

The corresponding *frequency* spectrum for arbitrary-depth water can now be obtained by simply transforming $E_{Phillips}(k)$ back to frequency f-space with the Jacobian for arbitrary-depth water, $\mathrm{d}k/\mathrm{d}f = 2\pi/c_g$. The result of this transformation can be written as (see Note 8C and Fig. 8.3)

$$E_{Phillips}(f) = E_{Phillips,\infty}(f)\phi(f,d) \qquad (8.3.7)$$

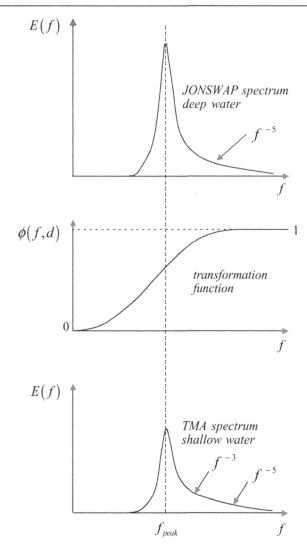

Figure 8.4 The transformation of the deep-water JONSWAP spectrum into the shallow-water TMA spectrum (the same scales are used in the upper and lower panels).

where $\phi(f, d)$ is a simple transformation function (see Fig. 8.4):

$$\phi(f, d) = \frac{1}{2n}\tanh^2(kd) \qquad (8.3.8)$$

where n is the ratio of group velocity over phase velocity (see Note 8C). In shallow water, this transformation gives $E(f) \sim f^{-3}$ (a dimensional analysis,[2] see Section 6.3.3, would give $E(f) \sim g d f^{-3}$).

---

[2] An alternative reasoning, leading to the same result, is given by Thornton (1977), who argued that breaking occurs when the forward speed of the water particles at the surface exceeds the speed of propagation of the wave

These transformations from f-space to k-space and back to f-space are conservative (no energy is lost) since the Jacobians are properly included.[3] However, with a low-frequency cut-off in deep water, as required for real waves, the transformations are no longer conservative. For instance, when the wave-number spectrum is approximated simply by cutting the k^{-3}-tail off at a (peak) wave number $k = k_{peak}$, (see Fig. 8.3) then $m_0 = \int_{k_{peak}}^{\infty} \frac{1}{2}\alpha k^{-3} dk = \frac{1}{4}\alpha k_{peak}^{-2}$, and, since $H_{m_0} = 4\sqrt{m_0}$, it follows that $H_{m_0} = 2\sqrt{\alpha}\, k_{peak}^{-1}$. With a constant α and peak wave number k_{peak}, the total energy or H_{m_0} would be constant. However, as the waves propagate into shallower water, the peak wave number is *not* constant. It increases because, in general, the peak *frequency* remains (nearly) constant. The energy at wave numbers lower than k_{peak} is correspondingly removed in the transformation (see Fig. 8.3). For a constant value of α, this implies a decreasing significant wave height.

NOTE 8C   The transformation of a deep-water f^{-5}-spectrum to shallow water

The shape of the spectral tail of waves in arbitrary-depth water suggested by Kitaigorodskii *et al.* (1975) can be derived as follows. Multiply the right-hand side of Eq. (8.3.6) by $1 = [\alpha g^2 (2\pi)^{-4} f^{-5}]/[\frac{1}{2}\alpha(k^{-3}dk/df)_\infty]$ (from Eqs. 8.3.4 and 8.3.5), so that

$$E_{Phillips}(k) = \frac{1}{2}\alpha k^{-3} \frac{\alpha g^2 (2\pi)^{-4} f^{-5}}{\frac{1}{2}\alpha (k^{-3}dk/df)_\infty}$$

Transform this expression to frequency space (include the proper Jacobian):

$$E_{Phillips}(f) = \frac{1}{2}\alpha k^{-3} \frac{\alpha g^2 (2\pi)^{-4} f^{-5}}{\frac{1}{2}\alpha (k^{-3}dk/df)_\infty} \frac{dk}{df}$$

Re-arranging the factors gives

$$E_{Phillips}(f) = \alpha g^2 (2\pi)^{-4} f^{-5} \frac{k^{-3}dk/df}{(k^{-3}dk/df)_\infty}$$

which can be written as

$$E_{Phillips}(f) = E_{Phillips,\infty}(f)\phi(f,d)$$

where

$$\phi(f,d) = \left(k^{-3}dk/df\right) / \left(k^{-3}dk/df\right)_\infty$$

---

itself. This leads to a shape of the spectral tail $E(f) \sim c^2 f^{-3}$, which, in very shallow water, with $c^2 = gd$, is $E(f) \sim gdf^{-3}$ (see also footnote in Section 6.3.3).

[3] The fact that $\phi(f,d)$ is less than unity, see Fig. 8.4, is compensated by the transformation of the k-axis to the f-axis, as implied in the use of the Jacobian.

is the function that transforms the deep-water Phillips spectrum $E_{Phillips,\infty}(f)$ into the arbitrary-depth Phillips spectrum $E_{Phillips}(f)$. This function can also be written (with phase speed $c = \omega/k$ and group velocity $c_g = d\omega/dk$) as

$$\phi(f, d) = (c^3/c_g)/(c^3/c_g)_\infty = \frac{1}{2n}\tanh^2(kd)$$

where n is the ratio of group velocity over phase velocity (see Eq. 5.4.31).

Instead of the discontinuous cut-off at the peak wave number, Bouws *et al.* (1985) proposed to use the more elegant low-frequency cut-off of the JONSWAP spectrum and also its peak-enhancement function (see Section 6.3.3). The result is called the *TMA spectrum* (thereby generalising the applicability of the JONSWAP spectrum from deep water to arbitrary-depth water; the name TMA derives from the names of the data sets TEXEL, MARSEN and ARSLOE that were used by Bouws *et al.*, 1985, to verify this idea):

$$\boxed{E_{TMA}(f) = E_{JONSWAP}(f)\,\phi(f, d)} \qquad (8.3.9)$$

This transformation is shown in Fig. 8.4. As waves move from deep water to shallow water, the f^{-3}-shape slowly replaces the f^{-5}-shape, starting at the lower frequencies (where depth effects take hold first). Apparently a universal f^{-n}-tail of the spectrum would be valid only either in deep water (with a frequency-independent value of n, e.g., $n = 5$) or in very shallow water (e.g., $n = 3$). The shape of the TMA spectrum has been verified with a large number of observations by Bouws *et al.* (1985) and independently by Young and Verhagen (1996b). Results of other studies indicate a slightly different shape (see Note 8D).

The hypothesis of Kitaigorodskii *et al.* (1975), on which the TMA spectrum is based, implies that the evolution of the waves is more universally described in wave-number space than in frequency space. This has led Young and Verhagen (1996b) to relate the values of the JONSWAP parameters[4] α, γ and σ to the dimensionless peak wave number $\tilde{k}_{peak} = k_{peak}U_{10}^2/g$ in water of arbitrary depth, rather than the dimensionless peak frequency $\tilde{f}_{peak} = f_{peak}U_{10}/g$ as was done earlier for deep water (see Section 6.3.3). From their spectra observed in Lake George, they found

$$\alpha = 0.0091\tilde{k}_{peak}^{0.24} \qquad \text{for arbitrary-depth water} \qquad (8.3.10)$$

They also showed that these values of α as a function of $\tilde{k}_{peak}$ were consistent with the values found in JONSWAP (when the transformation of $\tilde{f}_{peak}$ to $\tilde{k}_{peak}$ is carried out), thus further supporting the hypothesis of Kitaigorodskii *et al.* (1975). The values of σ were too scattered for Young and Verhagen to find any systematic

---

[4] They did not distinguish between σ_a and σ_b as defined for the JONSWAP spectrum (see Section 6.3.3).

dependence on $\tilde{k}_{peak}$ (the average value was $\sigma = 0.12$), but they suggested for γ that

$$\gamma = -5.8 \log_{10} \tilde{d} + 1.1 \qquad \text{for } 0.05 < \tilde{d} < 1 \tag{8.3.11}$$

Unfortunately, this is not consistent with deep-water observations such as those of JONSWAP (for which $\gamma = 3.3$ for young sea states, on average). For *depth-limited, fully developed* conditions, alternative relationships are given by Young and Babanin (2006); see also Note 8D.

It must be noted that, just outside and in the surf zone, the tail of the spectrum may develop a secondary, high-frequency peak due to triad wave–wave interactions (the second harmonic of the incident spectral peak; see Section 8.4.4), but such a peak seem to disappear over fairly short distances (within a few characteristic wave lengths, as the waves break and nonlinear interactions restore the smooth tail).

Literature:
Battjes (1984), Resio (1987), Smith and Vincent (2002, 2003), Smith (2004), Suh *et al.* (1994), Tucker (1994), Vincent (1985), Vincent and Hughes (1985), Zakharov (1999).

NOTE 8D   The FRF spectrum

The derivation of the TMA spectrum depends critically on the assumption that the high-frequency tail of the JONSWAP spectrum in deep water is proportional to f^{-5}. However, as shown earlier (see Section 6.3.3), a better approximation to observed spectra in deep water is obtained with an f^{-4}-tail. The corresponding tail of the wave-number spectrum in deep water would be proportional to $k^{-5/2}$. This, in turn, corresponds in shallow water to an $f^{-5/2}$-tail of the frequency spectrum. Miller and Vincent (1990) verified this with observations and suggested a corresponding adaptation of the TMA spectrum. They baptised this spectrum 'the FRF spectrum' (after the location of their observations, the Field Research Facility of the U.S. Army Engineer Research & Development Center):

$$E_{FRF}(k) = \alpha_{FRF} g^{-1/2} U_{10} k^{-5/2} \exp\left[-\frac{5}{4}\left(\frac{f}{f_{peak}}\right)^{-4}\right] \gamma_{FRF}^{\exp\left[-\frac{1}{2}\left(\frac{f/f_{peak}-1}{\sigma_{FRF}}\right)^2\right]}$$

where $\sigma_{FRF} = \sigma_{FRF,a}$ for $f \leq f_{peak}$ and $\sigma_{FRF} = \sigma_{FRF,b}$ for $f > f_{peak}$. One result of this study was that the FRF shape and the TMA shape fitted the observations equally well but the FRF shape provided a constant value $\alpha_{FRF} = 0.0029$ and a relationship between the peak enhancement coefficient γ_{FRF} and the overall wave steepness: $\gamma_{FRF} = 1.03 \times 10^4 s^{2.25}$, where the overall steepness $s = H_{m_0}/L_{peak}$. Like the authors of all other similar studies, they did not find any correlation between the values of $\sigma_{FRF,a}$ or $\sigma_{FRF,b}$ on the one hand and any other wave parameter on the other (the mean values were $\sigma_{FRF,a} = 0.115$ and $\sigma_{FRF,b} = 0.114$). The $k^{-5/2}$-shape was later found to be forced by the quadruplet wave–wave interactions in shallow water (but not so shallow that the waves break; Resio *et al.*, 2001). Later still, Smith and Vincent (2002) found, from observations in the field and in the laboratory, supported by the theory of Zakharov

(1999), that, for *very* shallow water ($kd < 1$, i.e., the lower frequencies) the wave-number spectrum is proportional to $E(k) \sim k^{-4/3}$. For higher frequencies ($kd \geq 1$), the $k^{-5/2}$-shape of the FRF spectrum would apply (see also Smith and Vincent, 2003, and Smith, 2004).

Note that a spectrum similar to the FRF spectrum may be found by applying the Kitaigorodskii scaling of Eq. (8.3.8) to the Donelan spectrum of Note 6C. It must also be noted that near the outer edge of the surf zone, a secondary, high-frequency peak may evolve due to nonlinear *triad* wave–wave interactions (see Section 8.4.4). This secondary peak disappears when the waves propagate further into the surf zone. Recent observation of Young and Babanin (2006) have shown that, also under the above idealised conditions with a constant depth, the *depth-limited, fully developed* spectrum generates a second harmonic at frequencies slightly lower than twice the peak frequency (presumably due to triad wave–wave interactions). This spectrum can be approximated as a Donelan spectrum (as in deep water) with a second, high-frequency Donelan spectrum superimposed to represent the second harmonic.

8.3.3 The two-dimensional wave spectrum

The only systematic observations of the two-dimensional frequency–direction spectrum in water of limited depth seem to be those of Young *et al.* (1996), in the same study in Lake George as referred to earlier. They report that there is no clear variation in the directional spreading σ_θ (see Section 6.3.4) as a function of $k_{peak}d$. However, they also note that (a) the scatter in their observations may be too large and the range of $k_{peak}d$ too small to detect such variation and (b) numerical experiments indicate that the quadruplet wave–wave interactions in shallow water tend to broaden the spectrum directionally.[5] Such being the uncertain state of affairs, nothing better can be concluded than that the directional width, in water of limited depth, is about equal to that in deep water as given in Section 6.3.4.

8.4 Wave modelling for arbitrary cases (coastal waters)

Modelling waves in coastal waters is conceptually as straightforward as modelling waves in deep water: we need only follow each and every single wave component from deep water to a coastal location and account for all effects of propagation, generation, wave–wave interactions and dissipation. However, as in deep water, the nonlinear character of the processes involved does not allow the *computation* of the wave field to follow this Lagrangian approach. In coastal waters there is even more reason why this approach cannot be used. In general, a harmonic wave

---

[5] This is in contrast with the observation of Young *et al.* (1996) that, at high wind speeds, the waves were distinctly long-crested, which implies a directional narrowing of the spectrum.

does not propagate in shallow water along straight wave rays, but rather along wave rays that are curved due to refraction and diffraction. When diffraction, as usual, is ignored in the computations, these wave rays often cross one another. An energy balance of the waves, based on the distance between (initially) adjacent wave rays, cannot then be used because the wave energy, being inversely proportional to this distance, would go to infinity at the crossing points. Transporting the energy *along* each individual wave ray (rather than between the rays) seems to solve this problem,[6] but (a) diffraction would still be ignored and (b) it does not solve the problem of nonlinear source terms. In coastal waters, therefore, the Eulerian approach should be used, just as in oceanic waters. Note that the difference between the Lagrangian approach and the Eulerian approach is only in the technique of computation. Both should converge to the same solution for finer and finer geographic resolution of the computational grids involved.

The fact that coastal waters cover a smaller geographic area than oceanic waters suggests that the number of equations involved would also be smaller: the horizontal scale of an ocean is of the order of 1000–10 000 km, whereas it is typically only 10 km for a coastal region. However, in coastal waters, a much higher spatial resolution is needed in the computations. It is of the order of 100 km for oceanic waters and typically only 100 m for coastal waters. This higher resolution for coastal regions thus compensates for the smaller scale of these regions. The number of equations to be integrated in coastal wave models is therefore roughly equal to that in oceanic wave models (it is of the order of 10 000 000 per time step in the integration; see Section 6.3.4).

Literature:
Ardhuin *et al.* (2001), Benoit *et al.* (1996), Booij *et al.* (1999), Bouws and Battjes (1982), Brink-Kjaer (1984), Cavaleri and Malanotte-Rizolli (1981), Cavaleri *et al* . (1989), Hasselmann *et al.* (1973), Karlsson (1969), LeMéhauté and Wang (1982), O'Reilly and Guza (1991), Piest (1965), Smith *et al.* (2000), Southgate (1984), Yamaguchi (1988).

8.4.1 The energy/action balance equation

The Eulerian spectral energy balance is formulated for coastal waters in the same manner as it is for oceanic waters. The only differences are that (1) it involves a more complicated formulation for the propagation of the wave energy, which now also needs to account for shoaling, refraction and diffraction, and (2) the number and complexity of the source terms are greater, since, in addition to the processes of

---

[6] The wave energy density formulated in $\vec{k}$-space is conserved along a wave ray. This approach can be supplemented with averaging over small regions the energy allocated to individual rays (e.g., Bouws and Battjes, 1982). Another alternative, which is not used very often, is a hybrid approach that combines the Lagrangian approach (for propagating the waves) with an Eulerian approach (to determine the source terms).

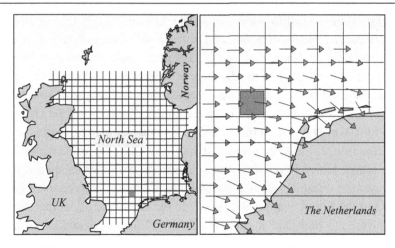

Figure 8.5 Refraction turns the waves towards the coast (the vectors in the right-hand panel represent the wave direction of a harmonic, unidirectional wave propagating across the North Sea from west to east). The turning of the waves is exaggerated here for illustrative purposes. In the Eulerian energy balance, the directional turning is considered as a transport of energy from one direction to another in *each* cell of the geographic grid (solid cell: see Fig. 8.6).

wave generation by wind, quadruplet wave–wave interactions and white-capping, we now need to represent also triad wave–wave interactions, bottom friction and depth-induced (i.e., surf-)breaking. The only simplification is that, in view of the scale of coastal regions, a wave model for coastal waters need not account for propagation on a sphere. Wave reflection off obstacles or a coastline is usually treated as a boundary condition, if it is considered at all.

In the energy balance equation for coastal waters, *shoaling* is readily accounted for by using the depth-dependent group velocity in the equation. *Refraction* and *diffraction* are not so easily dealt with. They require an additional propagation term in the equation. The essence of deriving this extra term is that, as the energy density of an individual wave component travels through the coastal region, it changes direction. In other words, while the wave energy propagates through x, y-space it simultaneously propagates through θ-space (it thus propagates through the three-dimensional x, y, θ-space; see Figs. 8.5 and 8.6). For a non-stationary situation, we need to add the time domain, so the energy balance needs to be formulated in the four-dimensional t, x, y, θ-space[7]. The derivation of the energy balance in x, y, t-space has already been given for oceanic waters (see Section 6.4.1). For coastal waters this derivation is identical except that the energy propagation speed may now not be taken out of the derivatives of Eq. (6.4.8), because in coastal

---

[7] If the ambient current or the water depth vary horizontally, or the water depth varies in time, the frequency of an individual wave component may also change, and an additional term to propagate wave energy in f-space is needed. The energy balance is then formulated in the five-dimensional t, x, y, σ, θ-space (the relative frequency σ replaces the frequency f in the presence of an ambient current; see Section 7.3.5 and Eq. 8.4.4).

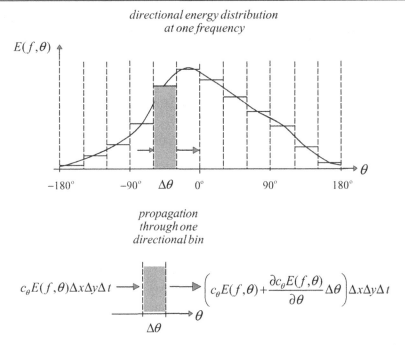

Figure 8.6 Upper panel: the directional distribution of wave energy at one frequency, at one geographic location, at one moment in time (continuous, with the solid line, and approximated with directional bins). Lower panel: refraction- or diffraction-induced turning of the wave direction, represented as propagation of wave energy through a bin in directional space. Compare with Fig. 6.12.

waters that speed generally varies with x and y. Here, we need to address only the derivation in θ-space.

For this derivation, the *directional* energy distribution at each frequency in the spectrum (see Fig. 6.7) is discretised into directional bins, each with a width $\Delta\theta$ (see Fig. 8.6). In the energy balance, the directional turning of the waves is presented as energy moving from one directional bin to the next as follows: the net import of energy into a directional bin during a time interval Δt is equal to the energy import through the left-hand side of the bin minus the energy export through the right-hand side of the bin during that interval (note that the propagation speed through the directional space is the refraction- or diffraction-induced rate of turning c_θ as derived in Section 7.3):

$$
\begin{aligned}
net\ import\ of\ energy &= c_\theta E(f,\theta)\Delta x\,\Delta y\,\Delta t \\
&\quad - \left(c_\theta E(f,\theta) + \frac{\partial c_\theta E(f,\theta)}{\partial\theta}\Delta\theta \right)\Delta x\,\Delta y\,\Delta t \\
&= -\frac{\partial c_\theta E(f,\theta)}{\partial\theta}\Delta\theta\,\Delta x\,\Delta y\,\Delta t
\end{aligned}
\tag{8.4.1}
$$

Note that, if c_θ were constant over the directions (it generally is not), the directional energy distribution would shift along the directions without changing form. Adding the result of Eq. (8.4.1) to the deep-water energy balance of Eq. (6.4.7) gives

$$
\frac{\partial}{\partial t} E(f,\theta) \Delta x \, \Delta y \, \Delta \theta \, \Delta t = -\frac{\partial c_{g,x} E(f,\theta)}{\partial x} \Delta x \, \Delta y \, \Delta \theta \, \Delta t
$$

$$
-\frac{\partial c_{g,y} E(f,\theta)}{\partial y} \Delta x \, \Delta y \, \Delta \theta \, \Delta t
$$

$$
-\frac{\partial c_\theta E(f,\theta)}{\partial \theta} \Delta x \, \Delta y \, \Delta \theta \, \Delta t
$$

$$
+ S(f,\theta) \Delta x \, \Delta y \, \Delta \theta \, \Delta t \tag{8.4.2}
$$

Dividing by $\Delta x \, \Delta y \, \Delta \theta \, \Delta t$ and moving the transport terms to the left-hand side gives the Eulerian spectral energy balance equation for arbitrary depth, which applies to all wave frequencies, all directions, all locations (geographic cells) and all points in time, including the effects of directional turning (in Cartesian co-ordinates and adding the x, y, t-dependence in the absence of an ambient current):

$$
\frac{\partial E(f,\theta;x,y,t)}{\partial t} + \frac{\partial c_{g,x} E(f,\theta;x,y,t)}{\partial x} + \frac{\partial c_{g,y} E(f,\theta;x,y,t)}{\partial y}
$$

$$
+ \frac{\partial c_\theta E(f,\theta;x,y,t)}{\partial \theta} = S(f,\theta;x,y,t) \qquad \text{shallow water} \quad (8.4.3)
$$

where c_θ is the refraction- or diffraction-induced turning rate of the individual wave components. This equation is identical to the energy balance equation for deep water (Eq. 6.4.8) except for the fourth term on the left-hand side that has been added. Note that including this term implies a horizontally variable water depth (refraction) or wave height (diffraction) and therefore a horizontally variable group velocity, so that, in an energy balance equation with this term included, the group velocity c_g may not be taken outside the derivatives (as in deep water with slowly varying wave heights).

If *ambient currents* are present, the energy balance equation needs to be supplemented with terms representing the energy transfer between waves and currents and the effects on the propagation of the waves. The effects on propagation are refraction, energy bunching (similar to shoaling, which is also a form of energy bunching, see Section 7.3.1) and frequency–shifting (related to the Doppler effect). The latter phenomenon can be accounted for by adding a propagation term in frequency space. The derivation of this extra propagation term is essentially the same as the above derivation of the refraction–diffraction term (the only difference is that direction needs to be replaced with frequency). The energy transfer between the waves and the current is not so easily represented; it involves adding terms to the energy balance equation that represent the effect of work done by the current against the radiation stresses (see Sections 5.5.2 and 7.4.2). A much simpler approach is to consider the *action balance* of the waves. The corresponding equation is identical to the energy balance equation with the energy density $E(\sigma, \theta)$ replaced with the action density $N(\sigma, \theta) = E(\sigma, \theta)/\sigma$, where σ is the relative radian frequency (the radian frequency in a system moving with the current; see Section 7.3.5). The reason is that, in contrast to wave energy, wave action is conserved in the presence of currents. The current-induced energy bunching and refraction are accounted for by using the proper expressions for the propagation speeds c_g and c_θ in the action balance equation (see Eqs. 7.3.31, 7.3.32 and 7.3.33, possibly with additional terms to account for diffraction; see below). The action balance equation, with frequency-shifting included, is then

$$
\frac{\partial N(\sigma, \theta; x, y, t)}{\partial t} + \frac{\partial c_{g,x} N(\sigma, \theta; x, y, t)}{\partial x} + \frac{\partial c_{g,y} N(\sigma, \theta; x, y, t)}{\partial y}
$$
$$
+ \frac{\partial c_\theta N(\sigma, \theta; x, y, t)}{\partial \theta} + \frac{\partial c_\sigma N(\sigma, \theta; x, y, t)}{\partial \sigma} = \frac{S(\sigma, \theta; x, y, t)}{\sigma} \tag{8.4.4}
$$

where the fifth term on the left-hand side represents the frequency-shifting of the waves.

The theories of the generation, nonlinear wave–wave interactions and dissipation of waves in finite-depth water are, as for deep water, generally rather complicated, and a full treatment is outside the scope of this book. Instead, only the essence of each theory and its results will be given. The source term $S(f, \theta)$ can again, as for deep water, be divided into terms representing generation by wind $S_{in}(f, \theta)$, nonlinear wave–wave interactions $S_{nl}(f, \theta)$ and dissipation $S_{diss}(f, \theta)$:

$$
S(f, \theta) = S_{in}(f, \theta) + S_{nl}(f, \theta) + S_{diss}(f, \theta) \tag{8.4.5}
$$

These source terms should now be subdivided to represent more processes than in deep water. The source term representing the nonlinear wave–wave interactions should represent not only quadruplet wave–wave interactions $S_{nl4}(f, \theta)$, but also triad wave–wave interactions $S_{nl3}(f, \theta)$. The source term for dissipation should represent not only white-capping $S_{wc}(f, \theta)$ but also bottom friction $S_{bfr}(f, \theta)$ and depth-induced (surf-)breaking $S_{surf}(f, \theta)$. Other processes of wave dissipation may be added. For instance, under some conditions, dissipation is caused by the penetration of water into the bottom (also called percolation, e.g., over a pebble or gravel bottom), by bottom motion (e.g., muddy bottoms), by vegetation or other marine life (e.g., mangroves, kelp, coral or mussels) or by current-induced turbulence in the water (e.g., near strong gradients in an ambient current). These processes will be ignored here because they are important only in exceptional situations.[8]

The inclusion of nonlinear processes in (very) shallow water is stretching the applicability of the linear theory of the waves even further than is the case in deep water (see Note 8E). However, the computational results obtained with the model approach described here are very reasonable when compared with observations, at least in terms of overall wave parameters such as the significant wave height, period and mean wave direction, even in the surf zone.

The processes of generation by wind, quadruplet wave–wave interactions and white-capping have been treated in Chapter 6 and need not be addressed again, except to indicate the effect of the finite depth and (where possible) ambient currents on these processes.

NOTE 8E   Nonlinear processes and the random-phase/amplitude model (coastal waters)

In Note 6E it was observed that the occurrence of nonlinear processes in deep water is, strictly speaking, at odds with the random-phase/amplitude model for ocean waves. However, such nonlinear processes are to some extent accounted for in deep-water ocean wave models. In shallow water, and in particular in very shallow water (just outside and in the surf zone), the nonlinear processes are much stronger. Spectral energy-balance models based on the random-phase/amplitude model should then perhaps be abandoned, the more so since nonlinear, phase-resolving models (such as Boussinesq-like models; see Section 7.5.2) provide excellent alternatives. Unfortunately, such phase-resolving models are generally not operationally feasible when the area of interest is large (a dozen wave lengths or more; the required computer capacity would be very considerable). It is for this reason that, in spite of these fundamental objections, spectral energy-balance models are used in shallow water, for instance when the waves approach the coast over an extended shallow foreshore (e.g., tidal flats or reefs). Engineering practice shows that this does provide reasonable estimates of the significant wave height, period and mean direction, but the interpretation of the spectrum is not a trivial matter, since phase-coupling between the wave components may be important, and requires considerable skill and care.

Literature:
Ardhuin *et al.* (2001), Bretherton and Garrett (1969), Elwany *et al.* (1995), Garrett (1967), Shemdin *et al.* (1977), Sheremet and Stone (2003), Whitham (1974).

---

[8] Another process that seems to dissipate energy is backscatter of wave energy off the bottom due to resonance with features in the seabed topography (see Section 8.4.4). It causes a decay of wave energy in the wave direction and therefore *seems* to be dissipation (i.e., the wave height decreases) but it is not, since the energy is conserved during the backscatter.

8.4.2 Wave propagation

The propagation of waves in shallow water is well understood (at least within the approximation of the linear wave theory) and also well represented in the energy balance equation, except for diffraction:

(1) shoaling is represented in the energy balance equation by the depth- and current-related variation of propagation speeds $c_{g,x}$ and $c_{g,y}$ determined with the linear wave theory (Eqs. 7.3.3 and 7.3.31),

(2) refraction is represented in the energy balance equation by the refraction-induced directional turning rate $c_\theta = c_{\theta,ref}$ from the linear wave theory (Eq. 7.3.13 or 7.3.33), but

(3) diffraction is *not* properly represented, because the expression for the diffraction-induced turning rate $c_{\theta,dif}$ (see Eq. 7.3.18) is formulated in terms of the wave amplitude of a harmonic wave, which for random waves is not defined (but an approximation is available; see below).

The effect of shoaling on a harmonic wave is generally, as noted in Section 7.3.1, to enhance the wave height as the wave approaches the coast. This is also the case for random waves. The frequency of a shoaling harmonic wave remains constant (in the absence of an ambient current) and one would therefore perhaps expect also the (mean) wave frequency of random waves to remain constant when they shoal, but that is only approximately true (even in the linear approximation). Since shoaling is stronger for the lower frequencies than for the higher frequencies, the low-frequency part of the spectrum is affected more than the high-frequency part (see Fig. 8.7). In fact, the highest frequencies might not be affected at all, because the water depth may be relatively large for these frequencies.[9] The net result of this (frequency-dependent) effect is therefore to shift the mean frequency slightly to lower values as the waves approach the coast. When random waves enter deeper water again, for instance after travelling over a sand bar, the opposite effect occurs (de-shoaling). It must be stressed that these effects of shoaling may well be dominated by other effects of propagation, generation and dissipation.

An opposing current may have the same effect as depth-induced shoaling (in addition to various other effects to be addressed later). If an opposing current increases its velocity as the waves propagate up-current, then the forward speed (relative to the fixed bottom) of the wave energy is increasingly reduced and wave energy (per unit surface area) is enhanced in the same way as in shoaling. We may therefore call this phenomenon current-induced energy bunching (just as shoaling is depth-induced energy bunching). One marked difference is that shoaling affects the low frequencies more than the high frequencies, whereas the opposite is true

---

[9] Before shoaling increases the wave height (for a harmonic wave), it slightly decreases the wave height (see Section 7.3.1). Within a certain frequency range, the spectral density may therefore slightly decrease, rather than increase, as the waves move into shallower water.

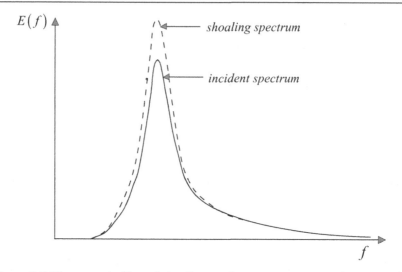

Figure 8.7 The general effect of shoaling on the wave spectrum as it approaches the coast (in the absence of other processes): the lower frequencies are enhanced more than the higher frequencies, resulting in a higher significant wave height and a slight frequency-down-shifting of the spectrum.

for current-induced energy bunching because, in the case of an ambient current, the ratio of current speed over group velocity $U_{current}/c_g$ is the controlling parameter. For frequencies propagating in an opposing current, this ratio may approach unity ($U_{current}/c_g \rightarrow 1$), resulting in blocking and reflection of the energy at the high frequencies. In addition, the waves have usually steepened enough that wave breaking occurs. One situation in which this occurs is when waves propagate into an estuary. The waves, propagating against the increasing river current in the estuary, may enhance their wave height to a point at which they will break, thus creating a zone of breaking waves at the seaward edge of the outgoing flow. Another situation in which such current-induced energy bunching with enhanced breaking may occur is in tidal eddies, for instance near intricate coastal features, or in narrow straits, or along the edges of major ocean currents such as the Agulhas Current and the Gulf Stream.

Near a smoothly curved coastline, with regular bottom topography, refraction turns all wave components to roughly the same direction at the outer edge of the surf zone. This implies that the waves there will be more long-crested than those in deeper water, i.e., the two-dimensional wave spectrum becomes directionally narrower as the waves approach the surf zone. Near such a gently varying coast, this would indeed be the case. However, near a somewhat less gently varying coast, refraction tends to create rather chaotic wave ray patterns (see Fig. 8.8 and Section 7.3.2), implying many different wave directions (cross-seas) and extreme

(a) wave rays and depth contours
  (harmonic, unidirectional incident wave)

(b) mean direction and depth contours
  (random, short-crested incident waves)

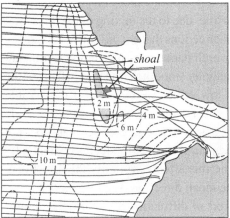 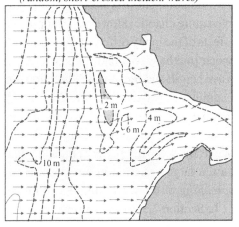

(c) directional spreading (degrees)
  (random, short-crested incident waves)

(d) normalised significant wave height $H_s / H_{s,deep}$
  (random, short-crested incident waves)

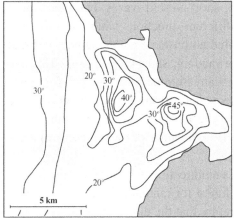 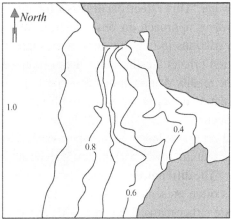

Figure 8.8 Panel a: the chaotic wave-ray pattern for a unidirectional, harmonic, wave of period 7 s from a westerly direction in the Haringvliet bay (the Netherlands). Panels b–d: the results for an incident, two-dimensional JONSWAP spectrum with the same peak period and mean wave direction and a $\cos^2\theta$-directional distribution (propagation only, all processes of generation, dissipation and wave–wave interactions are ignored). The geographic size of the area shown is approximately 15 km × 15 km, which is approximately $300\overline{L} \times 300\overline{L}$ (where $\overline{L}$ is the spatially averaged peak wave length in the wave spectrum). The reason for the decrease of the significant wave height towards the shores, in this case, is the shadowing effect of the lateral coastlines of the bay and the turning of individual wave components towards these coastlines.

variations of wave heights, even under swell conditions. For wind-sea conditions, with random, short-crested incident waves, one would expect the situation to be even more chaotic. However, the opposite is true: due to the mixing of the refraction effects of the many different frequencies and directions that are present in a wind-sea spectrum, the wave conditions in terms of overall parameters such as the significant wave height and mean wave direction are spatially smoothed: the high values of the (significant) wave height are lower and the low values are higher. It is remarkable that the *mean* wave direction in such situations is affected only by the large-scale features of the seabed topography. This is not to say that a confused sea state does not occur: the wave ray patterns indicate that *individual* wave components do generally have a large variation in direction. However, this affects primarily the directional *spreading* of the waves and not so much the mean direction (see Fig. 8.8(c)).

In Section 7.3.3 it was shown that the effect of diffraction on a harmonic, unidirectional wave is to turn the wave towards regions with lower amplitude. The wave thus turns around obstacles, such as small islands, or around the tip of a breakwater. This effect occurs also for random, short-crested waves, but, since these waves approach an obstacle from many different directions simultaneously, the variations in (significant) wave height behind an obstacle are smoothed. Diffraction effects are therefore much reduced in random, short-crested seas. This can be readily demonstrated by superimposing many Sommerfeld solutions for unidirectional, harmonic waves with various frequencies from various directions (see Section 7.3.3). The result is, as usual with the introduction of randomness and short-crestedness, a smoother wave field: the high values of the (significant) wave height are lower and the low values are higher.

The differences between the solution for a unidirectional, harmonic wave (which is often presented in diagrams) and the solution for random, short-crested waves are rather marked. Goda (2000) warns therefore that the direct application of such diagrams to real situations should be avoided, because they can lead to erroneous results. It should also be noted that, because diffraction depends on the wave length (and thus on the wave period), adding randomness to the waves changes not only the pattern of the significant wave *height*, but also that of the significant wave *period* (different frequencies turn at different rates, thus affecting the energy at different frequencies differently). Goda (2000) provides diagrams for random, short-crested waves.

Diffraction is usually not represented in operational wave models that are based on the spectral energy balance. The directional turning rate c_θ that *is* used in these models represents only refraction. This implies that these models should not be used where the wave height varies rapidly in geographic space, i.e., behind obstacles such as breakwaters and headlands and certainly not in harbours. The reason for not representing diffraction in these models is two-fold. First, the formulation of diffraction as a directional turning of the waves, as presented in Sections 7.3.3

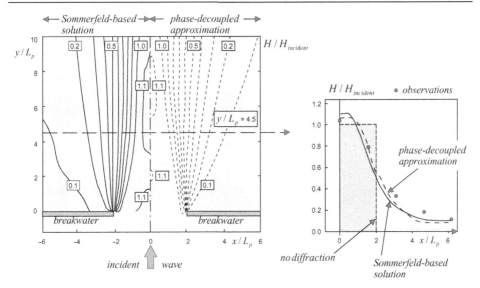

Figure 8.9 Diffraction of *random, unidirectional* waves propagating through a gap in an infinitely long, straight breakwater (JONSWAP spectrum; unidirectionality is used here only to emphasise diffraction effects; adding short-crestedness would have given more realistic results but the resulting smoothing effect on the wave field would have made diffraction effects barely visible). The left-hand panel is a plan view with the half-plane solution for each approximation. The right-hand panel gives the cross section at $y/L_p = 4.5$ (right-hand side only), where L_p is the peak wave length of the incident spectrum. Dashed lines: computed with the phase-decoupled diffraction approximation in SWAN. Solid lines: Sommerfeld-based solution (for random waves propagating through a gap between two wave breakwaters; see also Fig. 7.13). Dots: laboratory observations of Yu et al. (2000).

and 7.3.4, applies to a harmonic wave, not to random waves (it is formulated in terms of an amplitude, which cannot be defined for random waves). Second, a numerical implementation would be difficult because the inclusion of diffraction in the energy balance equation would transform that equation from a first-order differential equation into a fourth-order differential equation (Eq. 7.3.27, or an energy-based equivalent version thereof, would introduce higher-order derivatives into the energy balance equation). In addition, the tips of breakwaters or other sharp coastal features of obstacles create singularities in the wave field, which are notoriously difficult to handle in numerical models. In spite of these problems, numerical experiments have been carried out with the amplitude a in Eq. (7.3.26) replaced with the square root of the energy density $\sqrt{E(f)}$, which implies ignoring the wave phases. The results of numerical experiments with the SWAN model (see Chapter 9), obtained on the basis of such a phase-decoupled approach are reasonable, as shown in Fig. 8.9.[10] However, this approach should not be used

[10] The numerical problems mentioned here have been solved in SWAN by using under-relaxation in the iterative computational procedure of SWAN (see Section 9.5.2).

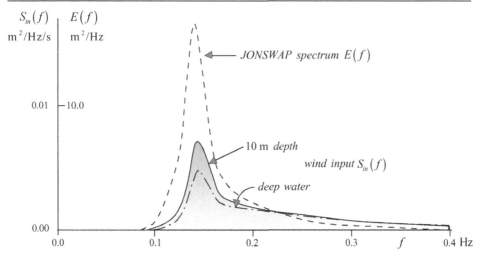

Figure 8.10 The wind input source term, for a JONSWAP spectrum in deep and shallow water (computed with the initial-growth formulation of Cavaleri and Malanotte-Rizzoli, 1981, and the feedback model of Miles, 1957; for $H_{m_0} = 3.5$ m, $T_{peak} = 7$ s and $U_{10} = 20$ m/s).

in front of reflecting obstacles where standing waves can appear, in particular in harbours, because phase information, which is not available in this approach, would be crucial.

Literature:
Booij *et al.* (1997), Brevik and Aas (1980), Briggs and Liu (1993), Briggs *et al.* (1995), Chawla and Kirby (2002), Goda (2000), Holthuijsen and Tolman (1991), Holthuijsen *et al.* (2003), Kenyon (1971), Lai *et al.* (1989), Raichlen (1993), Rivero *et al.* (1997), Sakai and Saeki (1984), Schumann (1976).

8.4.3 Generation by wind

The formulations that represent the generation of waves by wind (Section 6.4.3) show that the essential parameter in this process is the ratio of wind speed over the phase speed of the waves. Finite depth reduces the phase speed, thus increasing this ratio and, consequently, enhancing the transfer of energy to the waves (see Fig. 8.10; see also Young, 1999). In operational wave models, the effect of an ambient current is usually taken into account by replacing the absolute wind speed (i.e., that relative to the fixed bottom) with the relative wind speed (i.e., the relative to the current) in the expressions of the wind input source term (Eq. 6.4.15a or Eq. 6.4.15b).

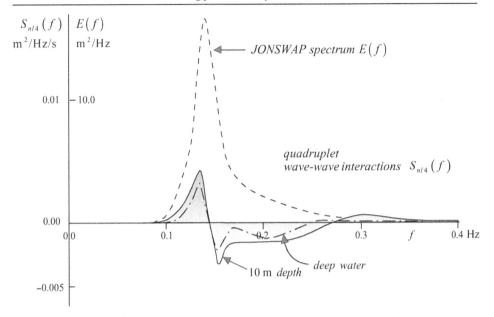

Figure 8.11 The source term for quadruplet wave–wave interactions, for a JONSWAP spectrum in deep and shallow water (near-exact computations with the Xnl code of van Vledder, 2006; for $H_s = 3.5$ m and $T_{peak} = 7$ s).

8.4.4 Nonlinear wave–wave interactions

Quadruplet wave–wave interactions

The expression for the quadruplet wave–wave interactions (see Eq. 6.4.18) is universal in the sense that it applies to any wave components that fulfil the resonance condition for four wave components, be it in deep water or in finite-depth water. In finite-depth water, the configurations of the quadruplets change and the corresponding wave–wave interactions are stronger than in deep water and their low-frequency lobe (see Fig. 6.21) shifts slightly to lower frequencies. In fact, these interactions may grow so strong that the assumptions underlying the theory of quadruplet wave–wave interactions do not hold (e.g., small corrections to the linear wave theory). The effect of a finite depth is illustrated in Fig. 8.11, which shows also the frequency up-shifting of the high-frequency lobe (white-capping will dissipate this transferred energy).

The estimate of the source term in Fig. 8.11 is based on the full expression for the quadruplet interactions (near-exact computations). In operational wave models, with a large number of geographic computational cells, this is not an economically feasible option. Instead, these models first compute the source term as for deep water with the DIA approximation (see Section 6.4.4) and then scale the results to account for the finite depth with a coefficient that is constant for all frequencies and

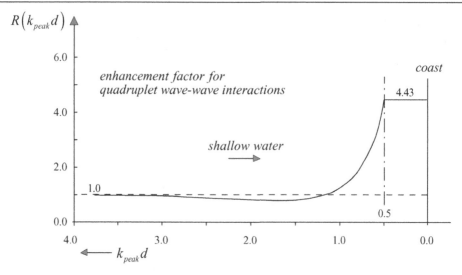

Figure 8.12 The parameterised enhancement of the quadruplet wave–wave interactions as waves enter shallow water (see also Section 9.3.3) with a shallow-water
cut-off at $k_{peak}d = 0.5$.

directions in the spectrum (which has been determined empirically for a JONSWAP
spectrum for various water depths; see Herterich and Hasselmann, 1980, Fig. 8.12
and Section 9.3.3).

Triad wave–wave interactions

In analogy with the resonance conditions for four wave components, the resonance
conditions for *three* wave components require that the sums of frequencies and
wave-number vectors of two freely propagating wave components are equal to
the frequency and wave number, respectively, of a third freely propagating wave
component (see Section 6.4.4 and Note 8F):[11]

$$f_1 + f_2 = f_3$$
$$\vec{k}_1 + \vec{k}_2 = \vec{k}_3$$

(8.4.6)

These triad resonance conditions cannot be complied with in deep water (i.e., such a
combination of wave components cannot be created with the dispersion relationship
of the linear wave theory for deep water). Triad wave–wave interactions are therefore

---

[11] An entirely different type of triad interactions occurs between two wave components and a feature in the seabed
topography (i.e., a harmonic component in this topography; see Ardhuin and Herbers, 2002). Energy is then
transferred from one water wave component to another with a harmonic in the seabed topography acting as a
sort of catalyst. The effect of this wave–wave–seabed interaction (also known as Bragg scatter) is to scatter
wave energy backwards or forwards. It may increase the directional spreading of swell propagating across
the continental shelf, depending on the spectrum of the seabed features (in the same range as the water wave
length). This Bragg scatter is ignored in operational wave models.

not relevant in deep water. The resonance conditions can be complied with only in extremely shallow water, where the waves are non-dispersive. In slightly deeper but still very shallow water, these conditions can *nearly* be complied with, so that *near-resonance* occurs, resulting in energy transfer and phase-coupling between the wave components involved. The magnitude of the energy transfer depends on the phase differences of the three wave components involved, which are quantified with the biphase $\beta_{1,2}$:

$$\beta_{1,2} = \varphi_1 + \varphi_2 - \varphi_{1+2} \tag{8.4.7}$$

in which the phases φ_1, φ_2 and $\varphi_{1+2} = \varphi_3$ are the phases of the three interacting wave components. This biphase is a characteristic of the triad and it evolves over time and distance, just as the energy of the waves evolves during propagation. In fact, the evolution of the biphase depends on the evolution of the wave energy and vice versa. Properly determining triad wave–wave interactions therefore requires coupled evolution models both for the *biphase* (a biphase evolution equation) and for the wave *energy* (the energy balance equation). This greatly complicates the modelling of waves in very shallow water.

When random waves, with a unimodal spectrum, approach very shallow water, e.g., the surf zone, the triad wave–wave interactions often generate a secondary peak in the spectrum, at twice the peak frequency, and sometimes also peaks at higher multiples of the peak frequency (see below; in addition to a low-frequency peak, which we ignore here, see Note 8F). However, these secondary peaks seem to persist only a short distance into the surf zone (several wave lengths at most). As the waves propagate further into the surf zone, the same interactions remove these peaks and force the spectral tail into a smooth, universal shape: $k^{-4/3}$ for $kd < 1$ (these components are in relatively shallow water) and $k^{-5/2}$ for $kd > 1$ (these components are in relatively deep water and $k^{-5/2}$ corresponds to f^{-4}; see Notes 6C and 8D). Observations and computations with extended Boussinesq models indicate that this forcing to a universal tail in very shallow water applies not only to a unimodal incident spectrum but also to a wide variety of other spectral shapes. In the case of initially mixed sea states (wind sea with substantial swell), the universal shape of the tail that is generated extends from the swell peak to the higher frequencies. The wind-sea peak has then completely disappeared. This smoothing effect is similar to the effect of the quadruplet wave–wave interactions in deep water, forcing the spectral tail into an f^{-4}- or $k^{-5/2}$-shape (see Note 6C; but only for wind sea in deep water). If the surf zone is narrow and followed by deeper water (which may occur when waves propagate across a narrow shoal) the energy from the secondary peak may return to the primary peak or the secondary peak may be released as freely propagating energy, at twice the peak frequency, depending on the degree of nonlinearity and the seabed topography.

**NOTE 8F   Generation of subharmonics and superharmonics
by triad wave–wave interactions**

When the wave components with subscripts 1 and 2 in the resonance conditions of Eq.
(8.4.6) are located at or near the peak of a unimodal incident spectrum, they will, in
shallow water where the triad interactions become important, transfer energy to lower
frequencies (the *difference* frequency $f_3 = f_1 - f_2$) and to higher frequencies (the *sum*
frequency $f_3 = f_1 + f_2$). Since, in a (narrow) unimodal wave spectrum, $f_1 - f_2$ is
small for the energy-rich part of the spectrum, the difference-frequency interactions
generate low-frequency waves, i.e., subharmonic or infra-gravity waves (surf beat;
see Section 1.3). They are oriented in the direction of the difference of the wave-
number vectors $\vec{k}_1$ and $\vec{k}_2$, i.e., usually at a large angle to the directions of the two
generating wave components. In the main text we do not consider these difference-
frequency interactions because infra-gravity waves fall outside the frequency range
of wind-generated waves. The sum-frequency interactions transfer energy to higher
frequencies, thus generating, in a unimodal spectrum, a high-frequency, superharmonic
peak at twice the peak frequency. A sum component is always oriented in the direction
of the sum of the wave-number vectors $\vec{k}_1$ and $\vec{k}_2$, i.e., *between* the directions of the two
generating components. This secondary high-frequency peak is therefore generated in
the same direction as the primary peak. One consequence is that the high frequencies
seem to refract more strongly than they would if Snel's law alone were applicable (their
direction follows the refracting direction of the spectral peak; this of course applies
only to the bound energy at these high frequencies). This effect has been confirmed
with measurements in the field.

Literature:
Nwogu (1994), Herbers *et al.* (1999).

A special case occurs when a unidirectional *harmonic* wave enters very shallow
water. Such a wave can interact with itself (self–self interaction: $f_1 = f_2$) to create a second harmonic, at twice its frequency, $f_3 = 2f_1$ (thus creating a nominal
triad, even if only two components are involved). The effect of this interaction is
visible as a distortion of the basic harmonic: as the wave propagates, it evolves
into a wave with sharper crests and flatter troughs (a Stokes-like wave; see Section 5.6.2). This is possible only if the second harmonic propagates with the same
phase speed as does the basic harmonic, i.e., the energy of the second harmonic
is bound to the basic harmonic. When the wave continues to propagate in ever
shallower water, it pitches forwards and breaks: it creates a steep forward face
and a gentler backward slope. This implies that the phase of the second harmonic
shifts in relation to the phase of the basic harmonic (while propagating at the same
speed).

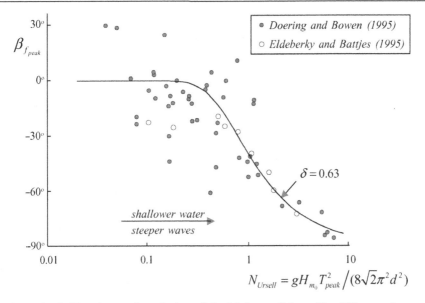

$$N_{Ursell} = gH_{m_0}T^2_{peak}\big/(8\sqrt{2}\pi^2 d^2)$$

Figure 8.13 The observed evolution of the biphase of the self–self interaction at the peak frequency of a unimodal incident frequency spectrum as the waves enter shallow water (i.e., as a function of the Ursell number N_{Ursell}, see also Sections 5.6.1 and 9.3.3). The solid line represents Eq. (8.4.8) with $\delta = 0.63$.

Similarly, when *random* waves with a *unimodal* spectrum approach the surf zone, the wave components at or near the peak of the spectrum create Stokes-like wave profiles, generating the secondary peak at twice the peak frequency referred to above. This added energy is superimposed on the freely propagating energy at this frequency, but it is bound to the primary peak. This distinction between bound and freely propagating energy is not evident in the spectrum since a variance density spectrum does not provide such a distinction. When the random waves propagate further into the surf zone, essentially the same happens as with a harmonic wave: they pitch forwards and break. Breaking reduces the energy scale of the spectrum and is therefore quite evident in the spectrum but the forward pitching is not because phase information is not provided by the spectrum. It *is* evident in, for instance, the biphase of the *self–self* interaction at the peak frequency $\beta_{f_{peak}} = 2\varphi_{f_{peak}} - \varphi_{2f_{peak}}$. Measurements in the field and under laboratory conditions show that, in *deep* water, where the waves behave reasonably linearly, this biphase is more or less uniformly distributed between $-180°$ and $+180°$ (as one may expect of freely propagating random waves) but, as the waves propagate into shallower and shallower water, the biphase concentrates more and more around a value decreasing from $0°$ to $-90°$ (see Fig. 8.13).

This implies that, also for random waves, the shape of the wave profile evolves, as observed, from a fairly symmetrical (horizontally and vertically) shape in deep water to a shape with a steeper forward pitching face and a more gently sloping backward face (the biphase of a saw-tooth-shaped wave too is $-90°$). This observed behaviour of the biphase for the self–self interaction at the peak frequency can be roughly approximated by its mean value $\overline{\beta}_{f_{peak}}$ as a function of the Ursell number (see Fig. 8.13):

$$\overline{\beta}_{f_{peak}} = -90° + 90° \tanh(\delta/N_{Ursell}) \qquad (8.4.8)$$

in which the degree of nonlinearity of the waves is represented by the Ursell number N_{Ursell} (see Section 5.6.1 and Fig. 8.13) and δ is a tunable coefficient. The value of δ varies considerably in the measurements ($\delta \approx 0.2-0.6$), illustrating the uncertainty of the parameterisation.

Most of the theoretical modelling of triad wave–wave interactions is based on the various Boussinesq equations that have been developed (see Section 7.5.2). These equations, which are fundamentally nonlinear and include the triad wave–wave interactions implicitly, are usually formulated in the time domain but they can be transformed to the frequency domain. Such a transformation gives coupled equations for the evolution of the amplitudes and biphases of the harmonic components involved. These coupled equations can, in turn, be transformed into a *biphase evolution equation* and an energy balance equation with corresponding coupling terms (which in the energy balance equation is the source term for the triad wave–wave interactions S_{nl3}). This creates a totally new approach to modelling nonlinear waves in the spectral domain, because it adds a completely new type of wave model (the biphase evolution equation). Such transformation of a Boussinesq equation, from the time domain to the spectral domain, is complicated and has successfully been carried out only for special conditions (a one-dimensional geographic situation; Herbers and Burton, 1997). In anticipation of a complete transformation, some operational wave models have been provided with a source term for the triad wave–wave interactions, $S_{nl3}(f, \theta)$ *without* a biphase evolution equation. Instead, the biphase in these models is estimated from the spectrum and the local water depth (e.g., Becq *et al.*, 1999). The simplest of these approaches is the lumped-triad approximation (LTA) of Eldeberky (1996):

$$S_{nl3}(f, \theta) = S_{nl3}^{+}(f, \theta) + S_{nl3}^{-}(f, \theta) \qquad (8.4.9)$$

with

$$S_{nl3}^{+}(f, \theta) = C_{nl3}cc_g|\sin\beta_{peak}|[E^2(f/2, \theta) - 2E(f/2, \theta)E(f, \theta)] \qquad (8.4.10)$$

and

$$S_{nl3}^{-}(f, \theta) = -2S_{nl3}^{+}(2f, \theta) \qquad (8.4.11)$$

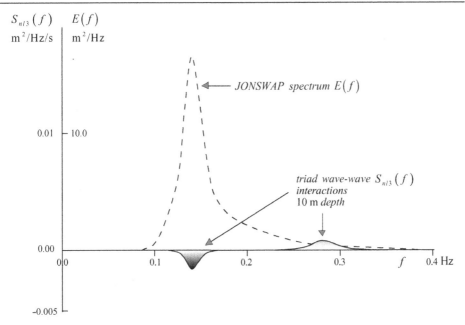

Figure 8.14 The source term for triad wave–wave interactions, for a JONSWAP spectrum in shallow water (computed with the LTA approximation of Eldeberky, 1996; for $H_s = 3.5$ m and $T_{peak} = 7$ s).

with a minimum value of $S_{nl3}^+(f, \theta) = 0$. The value of the biphase in these expressions is estimated as $\beta_{f_{peak}} = \overline{\beta}_{f_{peak}}$ of Eq. (8.4.8). The coefficient C_{nl3} is a coupling coefficient that depends on the local depth, frequency and wave number (see Section 9.3.3). The term $S_{nl3}^+(f, \theta)$ represents the energy received from frequency $f/2$ and the term $S_{nl3}^-(f, \theta)$ represents energy lost to frequency $2f$. Because $S_{nl3}^+(f, \theta)$ is always positive, each wave component in this approach receives energy from a component with half its frequency (energy: $f/2 \to f$) and loses energy to a component with double its frequency (energy: $f \to 2f$). This ensures that, in this approach, energy is always transported to higher frequencies and that no restitution of energy to lower frequencies occurs. It also implies that no subharmonics (i.e., infra-gravity waves, such as surf beat) are generated by the LTA.

The LTA describes the essential features of the energy transfer for waves entering the surf zone with a unimodal spectrum, i.e., it transfers energy from the primary peak to its superharmonic, resulting in a secondary peak at twice the peak frequency (this secondary peak, in its turn, generates *its* second harmonic at $f = 4f_{peak}$, which generates its second harmonic at $f = 8f_{peak}$ etc., but these peaks are usually outside the frequency range of numerical wave models). The corresponding source term for a JONSWAP spectrum is shown in Fig. 8.14. This rather simple approximation seems to give reasonable results near the outer edge of the surf zone (for swell *or* wind sea, but not for a mix of these two). If the waves continue to propagate in *deeper* water again (e.g., behind a shoal), the energy at the secondary peaks will propagate freely in the LTA computations and will not return to the primary peak. This too is often realistic. However, if instead the waves propagate further into *shallow* water (an extended surf area), or if the incident spectrum is multimodal, the LTA does not reproduce the transition to the smooth, universal spectral tail.

Literature:
Abreu *et al.* (1992), Agnon and Sheremet (1997), Ardhuin and Herbers (2002), Battjes and Beji (1992), Becq *et al.* (1998), Becq-Girard *et al.* (1999), Beji and Battjes (1993), Chen *et al.* (1997), Doering and Bowen (1986, 1995), Eldeberky and Battjes (1995, 1996), Elgar and Guza (1985a, 1985b, 1986), Elgar *et al.* (1990, 1993, 1995, 1997), Freilich and Guza (1984), Freilich *et al.* (1990), Hardy and Young (1996), Hasselmann (1962), Hasselmann *et al.* (1963), Herbers and Guza (1991), Herbers and Burton (1997), Herbers *et al.* (2000b, 2003), Herterich and Hasselmann (1980), Houmb and Rye (1973), Janssen *et al.* (2004), Kim and Powers (1979), Kim *et al.* (1980), Kofoed-Hansen and Rasmussen (1998), Madsen and Sørensen (1993), Mase *et al.* (1997), Masuda and Kuo (1981), Mei (1985), Mei *et al.* (2006), Norheim *et al.* (1998), Nwogu (1994), Rasmussen (1998), Sénéchal *et al.* (2001), Smith and Vincent (1992, 2002, 2003), Smith (2004), van Vledder and Bottema (2002), van Vledder (2006), Young and Eldeberky (1998).

8.4.5 Dissipation

White-capping

The proportionality coefficient in the expression for white-capping in deep water, μ (see Eq. 6.4.21), depends on the overall steepness of the waves and does not depend directly on the water depth. However, when waves enter coastal waters, shoaling tends to increase their steepness, so white-capping tends to gain importance in coastal waters (but refraction and diffraction may well induce the opposite effect). The effect may be considerable, as demonstrated in Fig. 8.15; see also Young (1999). The effects of an ambient current are equally indirect, e.g., an increasing opposing current will increase the wave steepness and thereby enhance white-capping. In fact, the steepness may thus grow much steeper than is usual for wind-generated waves and results of numerical experiments show that, under these conditions, the white-capping formulation of Komen *et al.* (1984; see Eqs. 6.4.20 and 6.4.22) underestimates the dissipation (Ris and Holthuijsen, 1996).

Bottom friction

For continental shelf seas with a sandy seabed, the dominant mechanism for bottom dissipation appears to be bottom friction. The term 'bottom friction' covers the rather complicated mechanisms in the relatively thin (compared with the water depth) turbulent boundary layer at the bottom that is created by the wave-induced

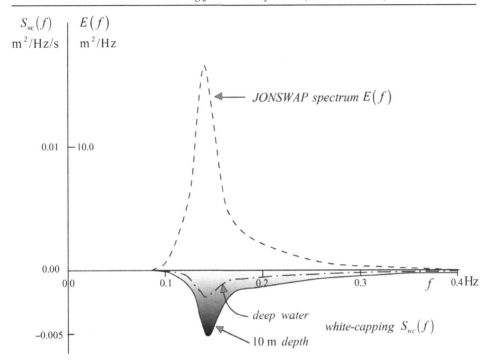

Figure 8.15 The white-capping source term, for a JONSWAP spectrum in deep and shallow water (computed with the pulse model of Hasselmann, 1974; for H_s = 3.5 m and T_{peak} = 7 s).

motion of the water particle. It is essentially a transfer of energy and momentum from the orbital motion of the water particles just above that layer to the turbulent motion in that layer. Such transfer depends therefore on the wave field itself and on characteristics of the bottom. These bottom characteristics in turn may well be affected by the waves. For instance, a smooth and flat sandy bottom may develop ripples under certain wave conditions (which would enhance the bottom friction) but these ripples may be washed away again by subsequent more severe wave conditions (which would reduce the bottom friction; the flow regime is then called 'sheet-flow').

If the *velocity* of the water particles $\vec{u}_{bottom}$ and the shear *stress* $\vec{\tau}_{bottom}$ just above the turbulent bottom boundary layer are oriented in the same direction, then the time-averaged energy-dissipation rate at the bottom $\overline{D}_{bfr}$ (per unit bottom surface area) can be written as

$$\overline{D}_{bfr} = -\overline{\tau_{bottom} u_{bottom}} \qquad (8.4.12)$$

where τ_{bottom} and u_{bottom} are the magnitudes of the (time-varying) shear stress and particle velocity, respectively. The characteristics of the particle velocity u_{bottom} are readily obtained with the linear wave theory from the wave spectrum and the local depth. Obtaining a reasonable estimate of the shear stress τ_{bottom} is not so easy. Two types of models have been developed for this. In the first type (the drag-law models), the dissipative character of the turbulent boundary layer is represented by one coefficient, which needs to be determined empirically for every wave condition and bottom condition (e.g., Collins, 1972). This type of model uses a quadratic law to estimate the magnitude of the shear stress:

$$\tau_{bottom} = \rho_{water} C_{bfr} u_{bottom}^2 \tag{8.4.13}$$

where ρ_{water} is the density of water and C_{bfr} is a bottom-friction (or drag) coefficient, so that the energy-dissipation rate becomes (from Eqs. 8.4.12 and 8.4.13)

$$\overline{D_{bfr}} = -\rho_{water} C_{bfr} \overline{u_{bottom}^2 u_{bottom}} \tag{8.4.14}$$

For random waves this is often approximated as (e.g., Collins, 1972)

$$\overline{D_{bfr}} = -\rho_{water} C_{bfr} u_{rms,bottom}^2 u_{rms,bottom} \tag{8.4.15}$$

where $u_{rms,bottom}$ is the root-mean-square orbital velocity at the bottom. Distributing $u_{rms,bottom}^2$ over frequency and direction with the linear wave theory, in other words, replacing $u_{rms,bottom}^2$ with $[2\pi f/\sinh(kd)]^2 E(f,\theta)$ and estimating $u_{rms,bottom}$ from the wave spectrum with the same theory gives the spectral distribution of the dissipation (in terms of energy density):

$$S_{bfr}^*(f,\theta) = -\rho_{water} C_{bfr} \left[\frac{2\pi f}{\sinh(kd)}\right]^2 E(f,\theta) u_{rms,bottom} \tag{8.4.16}$$

with

$$u_{rms,bottom} = \left\{ \int_0^\infty \int_0^{2\pi} \left[\frac{2\pi f}{\sinh(kd)}\right]^2 E(f,\theta)\, d\theta df \right\}^{1/2} \tag{8.4.17}$$

or, in terms of variance density (divide by $\rho_{water} g$),

$$S_{bfr}(f,\theta) = -\frac{C_{bfr}}{g} \left[\frac{2\pi f}{\sinh(kd)}\right]^2 E(f,\theta) u_{rms,bottom} \tag{8.4.18}$$

In the second type of model (eddy-viscosity models, which are based on turbulent-boundary-layer models for permanent flow), the dissipative character of the turbulent boundary layer is formulated in terms of basic bottom parameters such as the grain size of the sand (e.g., Madsen *et al.*, 1988; Weber, 1989, 1991a, 1991b). The results of the eddy-viscosity models can also be formulated as Eq. (8.4.18) but with different estimates for the bottom-friction coefficient C_{bfr}. The only parameters that seem to determine the friction (at least for sandy bottoms)

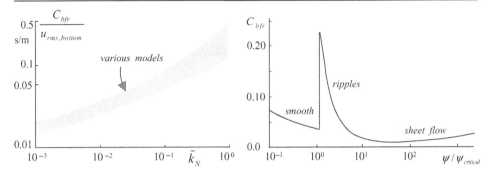

Figure 8.16 The bottom-friction coefficient C_{bfr} as a function of normalised bottom roughness $\tilde{k}_N$ and normalised Shield's parameter $\psi/\psi_{critical}$ (for swell conditions with frequency 0.1 Hz, grain diameter $D_{grain} = 0.1$ mm, $k_{N,skin} = 0.01$ m and $\psi_{critical} = 0.05$). After Tolman (1994).

are a normalised bottom roughness $\tilde{k}_N$ (defined as $\tilde{k}_N = k_N/a_{rms,bottom}$, where k_N is a bottom roughness length, also called the Nikuradse equivalent sand-grain roughness, and $a_{rms,bottom}$ is the root-mean-square amplitude of the near-bottom orbital excursion) and a parameter representing the capacity of the waves to set the bottom in motion (to create sand ripples or cause sheet-flow), called the Shields parameter ψ (e.g., Tolman, 1995):

$$\psi = C_{bfr,skin} \frac{u^2_{rms,bottom}}{(\rho_{sand}/\rho_{water} - 1) \, g \, D_{grain}} \qquad (8.4.19)$$

where ρ_{sand} and ρ_{water} are the densities of sand and water, respectively, D_{grain} is a representative grain diameter and $C_{bfr,skin}$ is the coefficient for skin friction (i.e., for a smooth bottom, when $k_N = k_{N,skin} = D_{grain}$). For a certain situation, the dependence of the friction coefficient C_{bfr} on $\tilde{k}_N$ and ψ is shown in Fig. 8.16, which shows clearly the increase of the friction coefficient with increasing bottom roughness and the sudden appearance of ripples on a smooth bottom when the Shields parameter exceeds a certain value.

A simple, often-used, alternative to the above two models is due to Hasselmann *et al.* (1973; JONSWAP, see Section 6.3.3), who represented their observations of *swell* dissipation with $C_{bfr} = \chi/(g u_{rms,bottom})$ in Eq. (8.4.18) and $\chi = 0.038 \, \text{m}^2 \, \text{s}^{-3}$. For *fully developed wind-sea* conditions in shallow water, Bouws and Komen (1983) suggest that one should use $\chi = 0.067 \, \text{m}^2 \, \text{s}^{-3}$. This approach seems to work reasonably well in operational wave models for many different situations, as long as a suitable value of χ is chosen.[12] This is also true for the other friction models

---

[12] Tolman (1994) notes that this success may be due to the fact that wave conditions dominate the dissipation (rather than movable-bed effects) and that an increase in the orbital motion is often accompanied by a decrease in friction, such that the coefficient χ is fairly constant for many wave conditions.

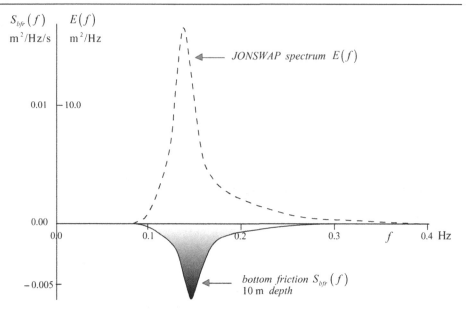

Figure 8.17 The bottom-friction source term, for a JONSWAP spectrum in shallow water ($\chi = 0.038$ m^2 s^{-3}, for $H_s = 3.5$ m and $T_{peak} = 7$ s).

mentioned above, i.e., they seem to perform reasonably well, as long as suitable values for the relevant coefficients are chosen. This implies that, given the uncertainty of this choice (information on bottom material is often rather poor), no preference can be given to any of these models from an operational point of view. However, it must be noted that the eddy-viscosity model of Weber (1989, 1991a, 1991b) has the best physical basis. The general character of the bottom-friction source term for a JONSWAP spectrum is given in Fig. 8.17.

The effect of an ambient *current* on the wave-energy dissipation due to bottom friction is often not taken into account. The reasons for this are given by Tolman (1992b), who argues that state-of-the-art expressions that nominally represent the effect of a current vary too widely in their effects to be acceptable. He also notes that the error in finding a correct estimate of the bottom roughness has a much larger impact on the dissipation than does the effect of a mean current.

Literature:
Bertotti and Cavaleri (1994), Bouws and Komen (1983), Cavaleri and Lionello (1990), Graber and Madsen (1988), Grant and Madsen (1982), Hasselmann and Collins (1968), Herbers *et al.* (2000a), Hsiao and Shemdin (1978), Hwang *et al.* (1998), Li and Mao (1992), Luo and Monbaliu (1994), Madsen *et al.* (1988), Padilla-Hernández and Monbaliu (2001), Putnam and Johnson (1949), Resio (1988), Shemdin *et al.* (1977), Tolman (1992b), Weber (1989, 1991a, 1991b).

Depth-induced (surf-)breaking

A classical and widely accepted model for depth-induced wave breaking (surf-breaking) is due to Battjes and Janssen (1978). In this model the average energy loss in a single breaking wave (per unit time, per unit horizontal bottom area) is modelled in analogy with the dissipation in a bore (a hydraulic jump) as

$$D_{surf,wave} = -\frac{1}{4}\alpha_{BJ}\,\rho g f_0 H_{br}^2 \tag{8.4.20}$$

where $\alpha_{BJ} \approx 1$ is a tunable coefficient (not related to earlier notations that use α), f_0 is the inverse of the (zero-crossing) wave period $f_0 = 1/T_0$ and H_{br} is the height of the breaking wave. In a field of random waves, $D_{surf,wave}$ is a random variable $\underline{D}_{surf,wave}$, the average of which can be estimated as an expected value:

$$\overline{D}_{surf,\,wave} = E\{\underline{D}_{surf,wave}\} = -\frac{1}{4}\alpha_{BJ}\rho g \int_0^\infty \int_0^\infty f_0 H_{br}^2\, p(H_{br},\,f_0)\,\mathrm{d}f_0\mathrm{d}H_{br} \tag{8.4.21}$$

where $p(H_{br},\,f_0)$ is the joint probability density function of the wave height and zero-crossing frequency of the *breaking* waves. The average for *all* waves (breaking and non-breaking) is then

$$\overline{D}_{surf} = Q_b\overline{D}_{surf,\,wave} \tag{8.4.22}$$

where Q_b is the fraction of breaking waves. Little is known about Q_b and $p(H_{br},\,f_0)$, although some observations of the probability density function of the breaking wave heights $p(H_{br}) = \int_0^\infty p(H_{br},\,f_0)\,\mathrm{d}f_0$ have been made (see Fig. 4.13). It is remarkable that this probability density function of the *breaking* waves overlaps the probability density function of the *non*-breaking waves, just as in deep water (see Section 6.4.5). Battjes and Janssen (1978), noting that details of $p(H_{br},\,f_0)$ are not required for estimating integral properties used a distribution in which the wave height of *all breaking* waves is equal to some maximum wave height H_{max}. They correspondingly replaced $\int_0^\infty \int_0^\infty f_0 H_{br}^2\, p(H_{br},\,f_0)\,\mathrm{d}f_0\mathrm{d}H_{br}$ with $\overline{f}_0 H_{max}^2$, so (dividing by ρg to express the result in terms of variance)

$$\overline{D}_{surf} = -\frac{1}{4}\alpha_{BJ}Q_b\overline{f}_0 H_{max}^2 \tag{8.4.23}$$

where $\overline{f}_0$ is the mean zero-crossing frequency of the breaking waves. The fraction Q_b of breakers is estimated by additionally assuming that the wave heights of *all unbroken* waves, i.e., those with heights below the maximum wave height H_{max}, are Rayleigh distributed (i.e., the Rayleigh distribution is truncated at $H = H_{max}$).[13]

---

[13] Observations have shown that assuming a full Rayleigh distribution (not truncated) for *all* waves, in which the waves above a critical value H_{cr} are breaking, may provide a better estimate of the fraction of breakers (see Baldock *et al.*, 1998): $Q_b = \int_{H_{cr}}^\infty p(H)\mathrm{d}H = \exp[-(H_{cr}/H_{rms})^2]$. The corresponding dissipation is $\overline{D}_{surf} =$

The fraction Q_b can then be estimated with

$$\frac{1 - Q_b}{\ln Q_b} = -\left(\frac{H_{rms}}{H_{max}}\right)^2 \tag{8.4.24}$$

where H_{rms} is the root-mean-square wave height $H_{rms} = \sqrt{8m_0}$, and m_0 is the zeroth-order moment of the wave spectrum. The maximum wave height H_{max} under such conditions (of depth-induced breaking)[14] is generally expressed as a fraction γ of the local water depth (including wave-induced set-up):

$$H_{max} = \gamma(d + \bar{\eta}) \tag{8.4.25}$$

where the value of the breaking index γ may depend on the wave steepness and bottom slope (see Note 8G).

NOTE 8G   The breaker index

Battjes and Stive (1985) re-analysed wave data of a number of laboratory and field exper-iments for various types of bottom topography and found values of the breaker index γ between 0.6 and 0.83 with an average of 0.73. Later, Kaminsky and Kraus (1993) found, from a compilation of data obtained in a large number of experiments, values in the range 0.6–1.59 with an average of 0.78. Obviously, γ is not a universal constant. Babanin *et al.* (2001) show that less than 10% of the breakers can be predicted to break solely on the basis of a constant wave-height-to-depth ratio. Its value seems to depend on bottom slope and incident wave steepness and even on the wind. For instance, Battjes and Stive (1985) suggest $\gamma = 0.5 + 0.4\tanh(33s_0)$, where s_0 is the *incident* wave steepness ($s_0 = H_{rms}/L_{peak,deep\ water}$, where $L_{peak,deep\ water}$ is the wave length at the peak of the incident spectrum, i.e., in deep water). A dependence on such incident wave steepness is inconvenient for a spectral wave model because these models are *locally* defined. Alternative expressions, which include a dependence on bottom slope, have been pro-posed by Bowen *et al.* (1968), Nelson (1987), Kaminsky and Kraus (1993) and Baldock *et al.* (1998). Nelson (1997) argues convincingly that, over an extended region with a horizontal bottom (bottom slope considerably less than 1 : 100), the value of γ will not exceed 0.55.

Literature:
Battjes and Janssen (1978), Battjes and Stive (1985), Douglass and Weggel (1988), Galvin (1972), Goda (1975, 2000), Goda and Morinobu (1998), Kaminsky and Kraus (1993), Nelson (1987, 1994), Smith (2001), Raubenheimer *et al.* (1996).

$\frac{1}{4}\alpha_{BJ}\bar{f}_0\exp[-(H_{cr}/H_{rms})^2](H_{cr}^2 + H_{rms}^2)$. Babanin *et al.* (2001) give explicit expressions for Q_b, derived from observations in the field, as a function of wave steepness, peak frequency, wind speed, normalised ambient surface current and the ratio of the significant wave height over depth. However, this estimate is not (yet) used in operational spectral wave models.

[14] The original model of Battjes and Janssen (1978) includes also steepness-induced breaking, i.e., white-capping.

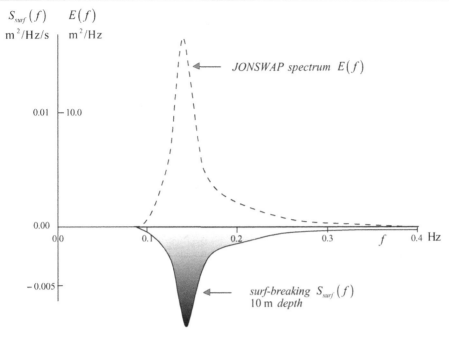

Figure 8.18 The depth-induced-breaking (surf-breaking) source term, for a JONSWAP spectrum in shallow water (computed with the spectral bore model based on Battjes and Janssen, 1978; for $H_s = 3.5$ m and $T_{peak} = 7$ s).

The above dissipation rate is the total, depth-induced dissipation rate of the waves, i.e., integrated over all frequencies and directions in the spectrum. The corresponding source term for the energy balance equation is the *spectral distribution* of this dissipation. It has been estimated from observations in laboratory flumes, which show that the shape of *unimodal* spectra of long-crested waves is barely affected by this type of breaking. Such a shape-conserving character can be achieved with a source term that is proportional to the energy density $E(f, \theta)$ itself:

$$S_{surf}(f, \theta) = \overline{D}_{surf} E(f, \theta)/m_0 \qquad (8.4.26)$$

The shape of the source term is then trivial (it is identical to the shape of the spectrum; see Fig. 8.18). However, it has been shown with observations in the field and in the laboratory that $S_{surf}(f, \theta)$ is additionally proportional to f^2. The effect of such an additional dependence is not noticeable in the evolution of the spectrum because the triad wave–wave interactions tend to compensate for errors in the precise frequency dependence (thus supporting the idea of the shape-stabilising ability of the triad wave–wave interactions).

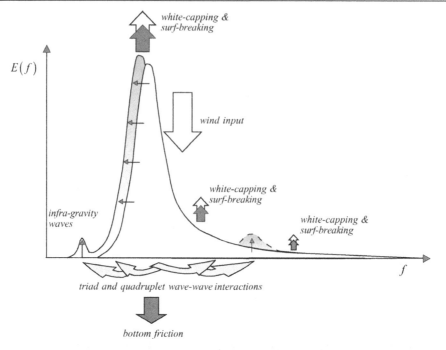

Figure 8.19 The energy flow through the spectrum in (very) shallow water: from the wind, through triad and quadruplet wave–wave interactions to (a) absorption at the lower frequencies (shifting the peak frequency and generating infra-gravity waves) and higher frequencies (possibly creating a secondary peak) and (b) dissipation by white-capping and (c) surf-breaking at all frequencies and bottom friction at the low and mid-range frequencies.

Literature:
Babanin *et al.* (2001), Baldock *et al.* (1998), Battjes (1972c), Battjes and Janssen (1978), Battjes and Beji (1992), Beji and Battjes (1993), Bertotti and Cavaleri (1985), Booij *et al.* (1999), Chen *et al.*. (1997), Collins (1970), Dally *et al.* (1984, 1985), Douglass and Weggel (1988), Elgar *et al.* (1997), Herbers *et al.* (1999, 2003), Kirby and Kaihatu (1996), Larson (1995), Le Méhauté (1962), LeMéhauté and Koh (1967), Longuet-Higgins (1969), Mase and Iwagaki (1982), Peregrine (1999), Roelvink (1993), Southgate and Nairn (1993), Svendsen (1984), Thornton and Guza (1983).

8.4.6 *Energy flow in the spectrum*

The source terms have been illustrated above for a JONSWAP spectrum, at one arbitrary moment in time. A summary of this is given in Fig. 8.19. As in deep water, the transfer of wind energy to the waves occurs mostly near the peak of the spectrum and at the mid-range frequencies. The corresponding energy gain at these frequencies is rapidly removed by wave–wave interactions (triad and quadruplet) to lower and higher frequency and by white-capping. In addition (in very shallow water), the energy at these intermediate frequencies is dissipated by bottom friction and surf-breaking. At the higher frequencies, most of the energy that is received from the mid-range frequencies is also dissipated, by white-capping and surf-breaking (high frequencies are barely affected by bottom friction), but it is not quite clear what happens additionally. Near the outer edge of the surf zone, the transfer of energy from the spectral peak to its second harmonic by triad wave–wave interactions is so strong that a secondary high-frequency peak is created, but, deeper inside the surf zone, it disappears. At the lower frequencies (below the peak frequency) the energy that is received from the mid-range frequencies is absorbed: just below the peak frequency by the quadruplet wave–wave interactions (thus down-shifting the peak frequency); and at still lower frequencies by the triad wave–wave interactions (creating infra-gravity waves or surf beat).

9

The SWAN wave model

9.1 Key concepts

- SWAN is a freely available, open-source computer model that is based on the theories that are presented in this book. It is widely used for wave research and consultancy practice by scientists and engineers.
- Since the model accounts for wave–current interactions, the basic equation of SWAN is the spectral *action balance equation*. It is formulated in Cartesian co-ordinates and optionally in spherical co-ordinates to accommodate small- and large-scale computations.
- To accommodate variable-resolution grids or boundary-fitting grids (e.g., to accommodate the results of numerical hydrodynamic flow models), *curvilinear grids* can be used as an alternative to the standard rectilinear grid. *Nested grids* permit zooming-in of computations to ever smaller regions.
- Computations with SWAN may vary from complex cases that require the full, time-dependent, two-dimensional action balance equation in spherical co-ordinates to simple one-dimensional cases that require only a stationary, one-dimensional energy balance equation in Cartesian co-ordinates.
- Bottom- and current-induced *shoaling* (energy bunching) and *refraction* are properly accounted for, but *diffraction* only approximately.
- SWAN accommodates *transmission* through, and *reflection* against, obstacles such as breakwaters and cliffs.
- Wave generation by wind is based on the feedback mechanism of Miles (Miles, 1957; Janssen 1991a). Initial waves are imposed or generated in an ad-hoc manner.
- Dissipation of wave energy is based on
  white-capping, with the pulse-model of Hasselmann (1974);
  bottom dissipation, with
  – the empirical JONSWAP model of Hasselmann *et al.* (1973), or
  – the drag-law model of Collins (1972), or
  – the eddy-viscosity model of Madsen *et al.* (1988); and
  surf-breaking, with the bore model of Battjes and Janssen (1978).
- The *quadruplet wave–wave interactions* are computed with the DIA of Hasselmann *et al.* (1985a) and the *triad wave–wave interactions* with the LTA of Eldeberky (1996).
- *Wave–induced set-up* is computed with exact computations in stationary, one-dimensional cases and with approximate computations in non-stationary or two-dimensional cases.

9.2 Introduction

The SWAN model (Booij *et al.*, 1999) is a freely available, open source computer modelthat is based on the theories that are presented in this book. It is a third-generation wave model (see Section 6.4.7 for this concept) that is used widely by scientists and engineers for research and consultancy practice (see the preface).

To support these professional users of SWAN and to demonstrate, for students, the considerations that underlie the development of such a model, the basic formulations and the numerical techniques that are used in SWAN are described here in some detail. The website from which SWAN can be downloaded can be reached through the portal website for this book at Cambridge University Press (see page vi of this book).

One of the important considerations in designing a numerical wave model for operational use is the computing time that is required for routine applications. This computing time is greatly affected by the numerical schemes that are used, particularly the schemes to propagate the waves through geographic space. These schemes are usually explicit, finite-difference schemes, which are simple, robust and economical for applications in oceanic waters. In coastal waters such schemes are not so economical, because the time step Δt in such schemes would be very small. The reason for this is that it is subject to the Courant criterion, which states that the wave energy may not travel more than one geographic cell in one time step. This implies that

$$\Delta t < \Delta x / c_{g,x} \qquad \text{and} \qquad \Delta t < \Delta y / c_{g,y} \qquad (9.2.1)$$

where Δx and Δy are the sizes of the geographic cell in the x- and y-directions, respectively (the discretisation steps in geographic space, also called the geographic resolution), and c_g is the group velocity of the lowest frequency of the waves in the model. For oceanic waters, the value of $\Delta x \approx \Delta y$ is typically 25–100 km (it is determined by the commonly used resolution for the wind fields that generate the waves in the model). This gives a time step Δt of between 20 and 80 min for a lowest wave frequency of 0.04 Hz. This is operationally acceptable. However, for applications in coastal waters, the value of $\Delta x \approx \Delta y$ is often as small as 100 m and sometimes as small as 10 m (it is determined by the scale of the features in the seabed topography or the coastline), whereas the lowest wave frequency is the same as in oceanic waters. For such low frequencies, the value of Δt would be 1.5–15 s (in a water depth of 5 m, say). This is operationally unacceptable and another numerical approach has to be chosen.[1] The SWAN model is therefore based on *implicit* propagation schemes, which are always numerically stable, independently of the values of Δt, Δx and Δy.[2] This does not imply that these values can be chosen arbitrarily. Although the computations will be stable for any value of Δt,

---

[1] If the required resolution is larger than about 1 km, the third-generation open source wave models WAM (WAMDI group, 1988; Komen *et al.*, 1994) and WAVEWATCH (Tolman, 1991; Tolman and Chalikov, 1996) can be used. They are both based on explicit propagation schemes but they do account for such shallow-water processes as depth- and current-induced refraction, bottom friction and depth-induced breaking (e.g., Monbaliu *et al.*, 2000).

[2] An alternative approach, which is based on a Lagrangian approach for propagating wave energy and an Eulerian approach for computing the source terms, is used in the third-generation wave model TOMAWAC of Benoit *et al.* (1996). Recently, another alternative, which is based on an unstructured finite-volume approach, has been developed (Hsu *et al.*, 2005 for a SWAN version and Sørensen *et al.*, 2004).

Δx or Δy, in order to obtain accurate results it is required that these values are much smaller than the time and space scales of the phenomena to be computed. For instance, the space scale of the surf zone is of the order of 100–1000 m, and so, the value of Δx or Δy must be of the order of 10–100 m (in the surf zone).

The formulations in SWAN representing the processes of generation by wind, quadruplet wave–wave interactions, white-capping and bottom friction are identical to those of the WAM model (of which there are two versions: WAM Cycle III, the WAMDI group, 1988; and WAM Cycle IV, Günther *et al.*, 1992, and Komen *et al.*, 1994). They are supplemented in SWAN with the processes of depth-induced breaking and triad wave–wave interactions.

9.3 Action balance

9.3.1 The action balance equation

Since SWAN accounts for wave–current interactions, it is based on the action balance equation rather than the energy balance equation and in terms of relative radian frequency σ rather than absolute radian frequency ω. For small-scale computations, the formulation in Cartesian co-ordinates (repeated from Eq. 8.4.4) is therefore

$$\frac{\partial N(\sigma, \theta; x, y, t)}{\partial t} + \frac{\partial c_{g,x} N(\sigma, \theta; x, y, t)}{\partial x} + \frac{\partial c_{g,y} N(\sigma, \theta; x, y, t)}{\partial y}$$
$$+ \frac{\partial c_\theta N(\sigma, \theta; x, y, t)}{\partial \theta} + \frac{\partial c_\sigma N(\sigma, \theta; x, y, t)}{\partial \sigma} = \frac{S(\sigma, \theta; x, y, t)}{\sigma} \qquad (9.3.1)$$

which reduces to the energy balance equation in the absence of an ambient current (no frequency-shifting; repeated from Eq. 8.4.3, with f replaced by ω):

$$\frac{\partial E(\omega, \theta; x, y, t)}{\partial t} + \frac{\partial c_{g,x} E(\omega, \theta; x, y, t)}{\partial x} + \frac{\partial c_{g,y} E(\omega, \theta; x, y, t)}{\partial y}$$
$$+ \frac{\partial c_\theta E(\omega, \theta; x, y, t)}{\partial \theta} = S(\omega, \theta; x, y, t) \qquad (9.3.2)$$

where $N(\sigma, \theta)$ is the *action* density spectrum and $E(\omega, \theta)$ is the energy density spectrum. The first term on the left-hand side of each of these equations represents the local rate of change of action (or energy) density in time, the second and third terms represent propagation of action (or energy) in geographic space (with propagation velocities $c_{g,x}$ and $c_{g,y}$ in x- and y-space, respectively, thus accounting for shoaling; see Sections 5.5.2 and 7.3.5). The fourth term represents depth-induced and current-induced refraction (with propagation velocity c_θ in θ-space; diffraction is optionally included; see Sections 7.3.2 – 7.3.5). The fifth term (Eq. 9.3.1 only) represents shifting of the relative frequency due to variations in depth and currents (with propagation velocity c_σ in σ-space; see Section 7.3.5). The term $S(\sigma, \theta)$ or $S(\omega, \theta)$ is the source term in terms of energy density (both in the action balance equation and in the energy balance equation; the division by σ makes $S(\sigma, \theta)/\sigma$ the source term for the action density). It represents the effects of generation, nonlinear wave–wave interactions and dissipation.

For large-scale computations, including global scales, the spectral action balance equation in SWAN is optionally formulated in terms of spherical co-ordinates (analogously to Eq. 6.4.11, for arbitrary depth and ambient currents):

$$\frac{\partial \, N(\sigma, \theta; \lambda, \varphi, t)}{\partial t} + \frac{\partial c_{g,\lambda} N(\sigma, \theta; \lambda, \varphi, t)}{\partial \lambda}$$
$$+ (\cos \varphi)^{-1} \frac{\partial c_{g,\varphi} \cos \varphi N(\sigma, \theta; \lambda, \varphi, t)}{\partial \varphi} + \frac{\partial c_\theta N(\sigma, \theta; \lambda, \varphi, t)}{\partial \theta}$$
$$+ \frac{\partial c_\sigma N(\sigma, \theta; \lambda, \varphi, t)}{\partial \sigma} = \frac{S(\sigma, \theta; \lambda, \varphi, t)}{\sigma} \tag{9.3.3}$$

with longitude λ and latitude φ (see Section 6.4.1).

In stationary situations, time is removed from the formulations (the first term on the left-hand side of each of the above balance equations) and, in one-dimensional situations,[3] the variation in the y-direction is removed. For such situations, the computations in SWAN are carried out with the much reduced one-dimensional energy balance equation:

$$\frac{\partial c_{g,x} \, E(\omega, \theta; x, y, t)}{\partial x} + \frac{\partial c_\theta \, E(\omega, \theta; x, y, t)}{\partial \theta} = S(\omega, \theta; x, y, t)$$
$$\text{stationary, one-dimensional, no currents} \tag{9.3.4}$$

which saves considerably on computer requirements. Computations with SWAN may therefore vary from large-scale, time-dependent computations with the full, two-dimensional action balance equation in spherical co-ordinates (see Eq. 9.3.3) to small-scale, stationary computations with the one-dimensional energy balance equation in Cartesian co-ordinates (see Eq. 9.3.4).

9.3.2 Generation by wind

The wind speed that is used to drive SWAN (user input) is the wind speed at 10-m elevation U_{10}, but, in the actual computations, it is converted into the friction

---

[3] Here, a one-dimensional situation is by definition a situation without variations in the y-direction, in other words, without variations along the coastline if the x-axis is taken normal to the coastline. This implies that the seabed topography may vary in the direction normal to the coast but not along the coast (the coastline and the depth contours must be parallel straight lines). It also implies that oblique incidence and short-crestedness of the waves are allowed, as long as the incident mean wave direction and short-crestedness are constant along the coast (they may vary normal to the coast).

velocity u_* with

$$u_*^2 = C_D U_{10}^2 \qquad (9.3.5)$$

in which C_D is the wind-drag coefficient (see Section 6.2). For the WAM Cycle III formulations in SWAN, the value of C_D is determined with an expression due to Wu (1982):

$$C_D = \begin{cases} 1.2875 \times 10^{-3} & \text{for } U_{10} < 7.5 \,\text{m/s} \\ (0.8 + 0.065 U_{10}) \times 10^{-3} & \text{for } U_{10} \geq 7.5 \,\text{m/s} \end{cases} \qquad (9.3.6)$$

For the WAM Cycle IV formulations in SWAN, the computation of u_* is an integral part of computing the wave generation by wind (see below).

The wave generation by wind is described with the feedback mechanism of Miles, supplemented with initial wave growth (see Section 6.4.3):

$$S_{in}(\sigma, \theta) = \alpha + \beta E(\sigma, \theta) \qquad (9.3.7)$$

For the initial wave growth (the term α), the empirical expression of Cavaleri and Malanotte-Rizzoli (1981) is used, with a cut-off to avoid growth at frequencies lower than the Pierson–Moskowitz frequency (Tolman, 1992a; compare with Eq. 6.4.13):

$$\alpha = \begin{cases} \dfrac{1.5 \times 10^{-3}}{g^2 2\pi} [u_* \cos(\theta - \theta_{wind})]^4 \, G & \text{for } |\theta - \theta_{wind}| \leq 90° \\ 0 & \text{for } |\theta - \theta_{wind}| > 90° \end{cases} \qquad (9.3.8)$$

where the cut-off function G is

$$G = \exp[-(\sigma/\sigma_{PM}^*)^{-4}] \qquad \text{with } \sigma_{PM}^* = 2\pi \frac{0.13g}{28u_*} \qquad (9.3.9)$$

θ_{wind} is the wind direction and σ_{PM}^* is the peak frequency of the Pierson and Moskowitz (1964) spectrum, reformulated in terms of friction velocity. The effects of currents are accounted for in SWAN by using the relative local wind speed and direction (i.e., the wind-speed vector minus the current vector). The user can also choose to impose initial wave conditions with a JONSWAP spectrum and a $\cos^2\theta$ directional distribution centred on the local wind direction (the significant wave height and peak frequency are obtained from the local wind and a fictitious fetch equal to the numerical grid steps Δx and Δy averaged over the computational domain used in the growth curve of Kahma and Calkoen, 1992, see Eqs 6.3.7).

For the WAM Cycle III formulations in SWAN, the coefficient β for exponential wave growth is taken from Snyder *et al.* (1981) and Komen *et al.* (1984; compare with Eq. 6.4.15b):

$$\beta = \max\left\{0, 0.25\frac{\rho_{air}}{\rho_{water}}\left[28\frac{u_*}{c}\cos[\theta - \theta_{wind}) - 1\right]\right\}\sigma \qquad (9.3.10)$$

where c is the phase velocity and ρ_{air} and ρ_{water} are the densities of air and water, respectively. For the WAM Cycle IV formulations in SWAN, the value of the coefficient β is taken from Komen *et al.* (1994; compare with Eq. 6.4.15a):

$$\beta = \max\left\{0, \gamma\frac{\rho_{air}}{\rho_{water}}\left(\frac{u_*}{c}\right)^2\cos^2(\theta - \theta_{wind})\right\}\sigma \qquad (9.3.11)$$

where γ is due to Janssen (1991a):

$$\gamma = \frac{1.2}{\kappa^2}\lambda\ln^4\lambda \qquad (9.3.12)$$

where

$$\lambda = \frac{gz_e}{c^2}\exp[\kappa c/|u_*\cos(\theta - \theta_{wind})|] \qquad \text{for } \lambda \le 1 \qquad (9.3.13a)$$

and

$$\beta = 0 \qquad \text{for } \lambda > 1 \qquad (9.3.13b)$$

and κ is the von Kármán constant, equal to 0.41, and z_e is the effective surface roughness (see next). The friction velocity u_* is computed with (Janssen, 1991a; see Mastenbroek *et al*, 1993)

$$U_{10} = \frac{u_*}{\kappa}\ln\left(\frac{10 + z_e - z_0}{z_e}\right) \qquad (9.3.14)$$

in which z_0 is the surface-roughness length and z_e is the effective surface-roughness length:

$$z_0 = 0.01\frac{u_*^2}{g} \qquad \text{and} \qquad z_e = \frac{z_0}{\sqrt{1 - \tau_{wave}/\tau}} \qquad (9.3.15)$$

where τ is the total surface stress ($\tau = \rho_{air}u_*^2$; see Section 6.2) and τ_{wave} is the wave-induced stress, which is given by

$$\tau_{wave} = \rho_{water}\int_0^{2\pi}\int_0^\infty \sigma\beta E(\sigma, \theta)d\theta\, d\sigma \qquad (9.3.16)$$

For a given wind speed U_{10} and a given wave spectrum $E(\sigma, \theta)$, the value of u_* can thus be determined. In SWAN, the iterative procedure of Mastenbroek *et al.* (1993) is used.

9.3.3 Nonlinear wave–wave interactions

Quadruplet wave–wave interactions

The computations of the quadruplet wave–wave interactions are carried out in SWAN with the discrete-interaction approximation (DIA) of Hasselmann *et al.* (1985a; see Section 6.4.4).[4] The computations of the *triad* wave–wave interactions are carried out with the lumped-triad approximation (LTA) of Eldeberky (1996; see Section 8.4.4).

Two configurations of quadruplets of wave numbers are considered in the DIA, each with the following frequencies:

$$\begin{aligned}
\sigma_1 &= \sigma_2 = \sigma \\
\sigma_3 &= \sigma(1 + \lambda) = \sigma^+ \\
\sigma_4 &= \sigma(1 - \lambda) = \sigma^-
\end{aligned} \qquad (9.3.17)$$

where $\lambda = 0.25$ is a constant (not related to λ in the previous paragraph). To satisfy the resonance conditions for quadruplet wave–wave interactions in deep water, the wave-number vectors with frequencies σ_3 and σ_4 lie at angles of $\theta_1 = -11.5°$ and $\theta_2 = 33.6°$ to the other two wave-number vectors that are identical to each other in frequency, wave number and direction. The second quadruplet is the mirror of this first quadruplet in the sense that $\theta_1 = 11.5°$, $\theta_2 = -33.6°$ and $\lambda = 0.25$. The corresponding source term in deep water for the quadruplet wave–wave interactions $S_{nl4}(\sigma, \theta)$ is

$$S_{nl4}(\sigma, \theta) = S^*_{nl4}(\sigma, \theta) + S^{**}_{nl4}(\sigma, \theta) \qquad (9.3.18)$$

where $S^*_{nl4}(\sigma, \theta)$ refers to the first quadruplet configuration and $S^{**}_{nl4}(\sigma, \theta)$ to the second. The contribution of each of these quadruplets can be written as

$$S^*_{nl4}(\sigma, \theta) = 2\delta S_{nl4}(\alpha_1\sigma, \theta) - \delta S_{nl4}(\alpha_2\sigma, \theta) - \delta S_{nl4}(\alpha_3\sigma, \theta) \qquad (9.3.19)$$

where each term is

$$\begin{aligned}
\delta S_{nl4}(\alpha_i\sigma, \theta) = {}& C_{nl4}(2\pi)^2 g^{-4}\left(\frac{\sigma}{2\pi}\right)^{11} \\
& \times \left\{ E^2(\alpha_i\sigma, \theta)\left[\frac{E(\alpha_i\sigma^+, \theta)}{(1+\lambda)^4} + \frac{E(\alpha_i\sigma^-, \theta)}{(1-\lambda)^4}\right]\right. \\
& \left. - 2\frac{E(\alpha_i\sigma, \theta)E(\alpha_i\sigma^+, \theta)E(\alpha_i\sigma^-, \theta)}{(1-\lambda^2)^4}\right\} \qquad \text{for } i = 1, 2, 3
\end{aligned} \qquad (9.3.20)$$

---

[4] The near-exact Xnl code of the WRT approach developed by van Vledder (2006; see Section 6.4.4) is optionally available in SWAN from version 40.41 onwards for research purposes (for operational use it is too demanding in terms of computer requirements).

in which $\alpha_1 = 1$, $\alpha_2 = (1 + \lambda)$, $\alpha_3 = (1 - \lambda)$ and the constant $C_{nl4} = 3 \times 10^7$. The expressions for $S_{nl4}^{**}(\sigma, \theta)$ are identical to those for $S_{nl4}^{*}(\sigma, \theta)$ for the mirror directions.

Following Hasselmann and Hasselmann (1981), the source term for the quadruplet interactions in finite water depth is taken to be identical to that in deep water, multiplied by a scaling factor R (see Fig. 8.12):[5]

$$S_{nl4, \text{ finite depth}} = R(k_{\text{peak,JONSWAP}}d)S_{nl4, \text{infinite depth}} \qquad (9.3.21)$$

where R is given by

$$
\begin{aligned}
R(k_{\text{peak,JONSWAP}}d) &= 1 + \frac{C_1}{k_{\text{peak,JONSWAP}}d}(1 - C_2 k_{\text{peak,JONSWAP}}d) \\
&\quad \times \exp(C_3 k_{\text{peak,JONSWAP}}d)
\end{aligned} \qquad (9.3.22)
$$

in which $k_{\text{peak,JONSWAP}}$ is the peak wave number of the JONSWAP spectrum for which the original computations were carried out. In SWAN with arbitrarily shaped spectra, the peak wave number $k_{\text{peak,JONSWAP}}$ is replaced with 0.75 times the *mean* wave number: $k_{\text{peak,JONSWAP}} \to k_p = 0.75\tilde{k}$ (Komen *et al.*, 1994; where $\tilde{k}$ is the mean wave number, see Eq. 9.3.28). The values of the coefficients are $C_1 = 5.5$, $C_2 = \frac{6}{7}$ and $C_3 = -1.25$. To avoid unrealistically high values of R, a maximum value of $R = 4.43$ is imposed (as in WAM Cycle IV).

Triad wave–wave interactions

The lumped-triad approximation (LTA) of Eldeberky (1996) is applied to all wave components in each of the spectral wave directions separately, with the expressions of Section 8.4.4 in terms of σ, and

$$C_{nl3} = \alpha_{EB} 2\pi J^2 \qquad (9.3.23)$$

in which α_{EB} is a tunable coefficient (the default value in SWAN is $\alpha_{EB} = 0.1$) and the biphase β is approximated with Eq. (8.4.8) using Ursell number (see Note 5C) $N_{\text{Ursell}}^{***} = gH_{m_0}\overline{T}^2/(8\sqrt{2}\pi^2 d^2)$, with $\overline{T} = m_0/m_1$, where m_0 and m_1 are the zeroth and first moments of the variance density spectrum $E(f)$ and $\delta = 0.2$. The triad wave–wave interactions are calculated only for $N_{\text{Ursell}}^{***} > 0.1$. The interaction coefficient J is taken from Madsen and Sørensen (1993):

$$J = \frac{k_{\sigma/2}^2 (gd + 2c_{\sigma/2}^2)}{k_\sigma d \left(gd + \dfrac{2}{15}gd^3 k_\sigma^2 - \dfrac{2}{5}\sigma^2 d^2 \right)} \qquad (9.3.24)$$

---

[5] The near-exact Xnl code, mentioned earlier, accommodates finite-depth water conditions properly, i.e., without scaling.

where the subscripts σ and $\sigma/2$ of wave number k and phase speed c refer to the wave number and phase speed at frequencies $\sigma/2$ and σ, respectively.

9.3.4 Dissipation

Dissipation is represented in SWAN by white-capping $S_{wc}(\sigma, \theta)$, bottom friction $S_{bfr}(\sigma, \theta)$ and depth-induced breaking $S_{surf}(\sigma, \theta)$. In addition, partial absorption (which implies partial reflection) by line structures (such as breakwaters, a group of sub-grid islands, and coastlines) is also represented in SWAN by numerically blocking energy propagating through such structures (obstacles).

White-capping

White-capping is represented by the pulse-based model of Hasselmann (1974; see Section 6.4.5), as suggested by the WAMDI group (1988):

$$S_{wc}(\sigma, \theta) = -\mu k \, E(\sigma, \theta) \tag{9.3.25}$$

where

$$\mu = C_{wc}\left((1 - n) + n\frac{k}{\tilde{k}}\right)\left(\frac{\tilde{s}}{\tilde{s}_{PM}}\right)^{p}\frac{\tilde{\sigma}}{\tilde{k}} \tag{9.3.26}$$

where the overall wave steepness $\tilde{s}$ (Janssen, 1991a; Günther *et al.*, 1992)[6] is defined as $\tilde{s} = \tilde{k}\sqrt{m_0}$ and $\tilde{s}_{PM}$ is the value of $\tilde{s}$ for the Pierson–Moskowitz spectrum (1964; $\tilde{s}_{PM} = \sqrt{3.02 \times 10^{-3}}$). The coefficients C_{wc}, n and p are tunable coefficients. The mean frequency and mean wave number, $\tilde{\sigma}$ and $\tilde{k}$, respectively, are defined (WAMDI group, 1988) as

$$\tilde{\sigma} = \left[m_0^{-1}\int_0^{2\pi}\int_0^{\infty}\sigma^{-1}E(\sigma, \theta)\mathrm{d}\theta\mathrm{d}\sigma\right]^{-1} \tag{9.3.27}$$

$$\tilde{k} = \left[m_0^{-1}\int_0^{2\pi}\int_0^{\infty}k^{-1/2}E(\sigma, \theta)\mathrm{d}\theta\mathrm{d}\sigma\right]^{-2} \tag{9.3.28}$$

For the WAM Cycle III formulations in SWAN, $n = 0$ and, for the WAM Cycle IV formulations, $n = 0.5$. The values of the coefficient C_{wc} and exponent p were obtained by Komen *et al.* (1984) by closing the energy balance of the waves in idealised deep-water wave-growth conditions. The result for the WAM Cycle III formulations is (Komen *et al.*, 1984) $C_{wc} = 2.36 \times 10^{-5}$ and $p = 4$, whereas for the WAM Cycle IV formulations it is (Günther *et al.*, 1992) $C_{wc} = 4.10 \times 10^{-5}$ (assuming $p = 4$). These tuning results depend critically on the high-frequency

---

[6] One remarkable effect of using such an overall steepness is that adding some swell to wind sea reduces the overall wave steepness and thus also the white-capping of the wind sea. This enhances the net growth of the wind sea in the presence of swell which does not seem to be very realistic (e.g., Booij *et al.*, 2001).

cut-off that was used for the spectra (Young and Banner, 1992; Banner and Young, 1994) and, since this cut-off is different in SWAN from that in WAM (see Section 9.5.1), differences in the overall wave growth rates between the WAM and SWAN models are to be expected.

Bottom friction

Bottom friction is the dominant bottom dissipation mechanism for continental shelf seas with sandy bottoms and the corresponding source term may generally be represented (see Section 8.4.5) as

$$S_{bfr}(\sigma, \theta) = -\frac{C_{bfr}}{g} \left[\frac{\sigma}{\sinh(kd)} \right]^2 E(\sigma, \theta) u_{rms,bottom} \qquad (9.3.29)$$

in which C_{bfr} is a bottom-friction coefficient and $u_{rms,bottom}$ is the root-mean-square orbital bottom velocity. Considering the large variations in bottom conditions in coastal areas (bottom material, bottom roughness length, ripple height, etc.), there is no field data evidence indicating that one should give preference to a particular model with which to estimate C_{bfr} (see Luo and Monbaliu, 1994). Movable-bed effects are ignored in SWAN and three bottom-friction models have been implemented: the drag-law model of Collins (1972), the eddy-viscosity model of Madsen *et al.* (1988) and the empirical JONSWAP model of Hasselmann *et al.* (1973).

The JONSWAP model is simply $C_{bfr} = C_{JONSWAP} = 0.038/u_{rms,bottom}$ for swell conditions (Hasselmann *et al.*, 1973) and $C_{bfr} = C_{JONSWAP} = 0.067/u_{rms,bottom}$ for wind-sea conditions (Bouws and Komen, 1983). The coefficient in the drag model of Collins (1972) is $C_{bfr} = C_{Collins} = 0.015$. The model of Madsen *et al.* (1988) gives

$$C_{bfr} = C_{Madsen} = \frac{f_w}{\sqrt{2}} \qquad (9.3.30)$$

where f_w is a non-dimensional friction factor estimated with the formulation of Jonsson (1966, 1980; cf. Madsen *et al.*, 1988):

$$f_w = 0.30 \qquad \text{for } a_b/k_N < 1.57$$
$$\text{(hydraulic rough bottom)} \qquad (9.3.31)$$

$$\frac{1}{4\sqrt{f_w}} + \log_{10}\left(\frac{1}{4\sqrt{f_w}}\right) = m_f + \log_{10}\left(\frac{a_b}{k_N}\right)$$
$$\text{for } a_b/k_N \geq 1.57$$
$$\text{(hydraulic smooth bottom)} \qquad (9.3.32)$$

where $m_f = -0.08$ (Jonsson and Carlsen, 1976) and a_b is a representative near-bottom excursion amplitude:

$$a_b^2 = 2 \int_0^\infty \int_0^{2\pi} \frac{1}{[\sinh(kd)]^2} E(\sigma, \theta) d\theta \, d\sigma \qquad (9.3.33)$$

and k_N is the bottom-roughness length scale, which of course depends on the actual bottom condition.

Depth-induced (surf-)breaking

The total dissipation (i.e., integrated over the spectrum) due to depth-induced wave-breaking (surf-breaking) can be modelled well with the dissipation of a bore, applied to the breaking waves in a random field in shallow water (Battjes and Janssen, 1978; see Section 8.4.5). The mean zero-crossing frequency in that model, $\overline{f}_0$, is replaced in SWAN with $\overline{f} = m_1/m_0$ (where m_0 and m_1 are the zeroth and first moments of the variance density spectrum $E(f)$), and the maximum possible wave height in the local water depth H_{max} is determined from $H_{max} = \gamma D$, where D is the total water depth, including the wave-induced set-up, and γ is the breaker parameter (a tunable coefficient, with a default value of 0.73 in SWAN).

Reflection, transmission and absorption

To accommodate situations with line structures such as breakwaters (i.e., dimensions in one direction smaller than the geographic resolution in the model), SWAN can *reflect* wave energy off and *transmit* wave energy through or over such structures. The difference between the incident energy on the one hand and the sum of reflected and transmitted energy on the other, is *absorbed* by the structure (or coast). This option can also be used to simulate the absorbing effect of (a group of) sub-grid islands. Reflection is modelled as specular reflection (angle of incidence equals angle of reflection) and it is assumed that the wave frequencies remain unchanged during transmission (only the energy scale of the spectrum is affected, not the spectral shape).

9.4 Wave-induced set-up

The gradients in the wave-induced radiation stresses, which generally cause currents and set-up, are standard output of SWAN and can be used for further computations with a separate hydrodynamic model (see Appendix E). The result of such computations can be returned to SWAN as input for the wave computations to achieve the feedback between waves on the one hand and set-up and currents on the other. For stationary cases, this may be iterated; for non-stationary cases, the computational results can be exchanged between the models at regular intervals. As an alternative to computations with a separate hydrodynamic model, the wave-induced *set-up* can also be *estimated* with SWAN. The corresponding SWAN computations are exact for stationary, one-dimensional cases ('exact' in the sense that conventional equations and numerical techniques are used) and approximate for non-stationary or two-dimensional cases (see below). SWAN cannot compute wave-induced currents.

In *one*-dimensional situations,[7] the computation of the wave-induced set-up in SWAN is based on the vertically integrated momentum balance equation, which represents a balance between the wave-induced force (the radiation stress gradient normal to the coast) and the vertically integrated hydrodynamic pressure gradient, as given by (repeated from Eq. 7.4.20; the x-axis is taken normal to the coastline)

$$\frac{d\bar{\eta}}{dx} = -\frac{1}{\rho g(d + \bar{\eta})} \frac{d\tilde{S}_{xx}}{dx} \tag{9.4.1}$$

The radiation stress is defined for random, short-crested waves as (see Eqs. 7.4.15, but now integrated over the spectrum)

$$\tilde{S}_{xx} = \int_0^\infty \int_0^{2\pi} \left(n - \frac{1}{2} + n \cos^2\theta \right) E(\sigma, \theta) d\theta \, d\sigma \tag{9.4.2}$$

This approach is exact in the context of the linear wave theory for *stationary* conditions. The same equation is used for *non-stationary*, one-dimensional conditions, making such computations a quasi-stationary approximation.

For stationary, *two*-dimensional situations, Dingemans *et al.* (1987) have shown that the wave-induced *set-up* is mainly due to the rotation-free part of the wave-induced forces,[8] i.e., given by the divergence $\partial F_x/\partial x + \partial F_y/\partial y$, whereas the wave-induced *currents* are mainly driven by the divergence-free part, i.e., by the rotation $\partial F_x/\partial y - \partial F_y/\partial x$, where

$$F_x = -\frac{\partial \tilde{S}_{xx}}{\partial x} - \frac{\partial \tilde{S}_{xy}}{\partial y} \quad \text{in the } x\text{-direction}$$

$$\tag{9.4.3}$$

$$F_y = -\frac{\partial \tilde{S}_{yy}}{\partial y} - \frac{\partial \tilde{S}_{yx}}{\partial x} \quad \text{in the } y\text{-direction}$$

---

[7] Here, a one-dimensional situation is by definition a situation without variations along a straight coastline. This implies that the seabed topography may vary in the direction normal to the coast but not along the coast (the coastline and the depth contours must be parallel straight lines). It also implies that oblique incidence and short-crestedness of the waves are allowed, as long as the incident mean wave direction and short-crestedness are constant along the coast (they may vary normal to the coast).

[8] Dingemans *et al.* (1987) also show (on the basis of Longuet-Higgins, 1973) that, under certain conditions, a numerically more robust estimate of the radiation stress gradient can be obtained from the dissipation source term in the energy balance equation of the waves.

with the radiation stresses for random, short-crested waves defined as (see Eqs. 7.4.15, but now averaged over the spectrum)

$$
\begin{aligned}
\tilde{S}_{xx} &= \int_0^\infty \int_0^{2\pi} \left(n - \frac{1}{2} + n\cos^2\theta \right) E(\sigma, \theta)\mathrm{d}\theta\,\mathrm{d}\sigma \\
\tilde{S}_{yy} &= \int_0^\infty \int_0^{2\pi} \left(n - \frac{1}{2} + n\sin^2\theta \right) E(\sigma, \theta)\mathrm{d}\theta\,\mathrm{d}\sigma \\
\tilde{S}_{xy} &= \int_0^\infty \int_0^{2\pi} (n\cos\theta\sin\theta) E(\sigma, \theta)\mathrm{d}\theta\,\mathrm{d}\sigma \\
\tilde{S}_{xy} &= \int_0^\infty \int_0^{2\pi} (n\sin\theta\cos\theta)\, E(\sigma, \theta)\mathrm{d}\theta\,\mathrm{d}\sigma
\end{aligned}
\tag{9.4.4}
$$

Computations that are based only on the divergence of the vertically integrated momentum balance equations (the shallow-water equations) would therefore give a reasonable estimate of the set-up. If, correspondingly, this divergence is balanced by the hydrostatic forces, then the following Poisson equation applies (if the divergence of the acceleration terms is ignored; see Appendix E):

$$
\frac{\partial F_x}{\partial x} + \frac{\partial F_y}{\partial y} + \frac{\partial}{\partial x}\left[\rho g(d + \overline{\eta})\frac{\partial\overline{\eta}}{\partial x} \right] + \frac{\partial}{\partial y}\left[\rho g(d + \overline{\eta})\frac{\partial\overline{\eta}}{\partial y} \right] = 0
\tag{9.4.5}
$$

which is used in SWAN to compute the wave-induced set-up for random, short-crested waves (it reduces to Eq. 9.4.1 in one-dimensional situations). Note that this is only an approximation, even for stationary conditions.

9.5 Numerical techniques

9.5.1 Introduction

For small-scale computations (i.e., sufficiently small that Cartesian co-ordinates can be used), the geographic space is discretised in SWAN with a rectangular grid, with constant resolutions Δx and Δy in the x- and y-directions, respectively (Δx may differ from Δy). For large-scale computations, the resolution is constant in terms of longitude and latitude, with resolutions $\Delta\lambda$ and $\Delta\varphi$, respectively ($\Delta\lambda$ may differ from $\Delta\phi$). Time is discretised with a constant time step Δt for the simultaneous integration of the propagation terms and the source terms (for stationary computations, time is removed from the equations). The spectral space is discretised with a constant directional resolution $\Delta\theta$ and a constant *relative* resolution for the radian frequency $\Delta\sigma/\sigma$ (which gives a logarithmic frequency distribution). When a coastal region is considered, with waves propagating only towards the coast, then (for reasons of economy) the option of computing only wave components travelling in a pre-defined (user-provided) directional sector ($\theta_{min} < \theta < \theta_{max}$) is available. The frequencies are defined between a fixed (user-provided) low-frequency cut-off f_{min} and a fixed (user-provided) high-frequency cut-off f_{max} (this range $f_{min} < f < f_{max}$ is called the prognostic range of the spectrum; typically $f_{min} = 0.04$ Hz and $f_{max} = 1$ Hz for conditions at sea). In this range, the spectral density is free to develop (i.e., without *a-priori*-imposed restraints). Outside this range, the spectrum is imposed: below the low-frequency cut-off, the spectrum is set to zero; above the high-frequency cut-off, an f^{-m}-tail is imposed (this range of frequencies $f < f_{min}$ and $f > f_{max}$ is called the diagnostic range of the spectrum; it is used to compute nonlinear wave–wave interactions at the high frequencies and to compute integral wave parameters). SWAN uses $m = 4$ if the WAM Cycle III wind-generation formulation of Komen *et al.* (1984) is used, and $m = 5$ if the WAM Cycle IV wind-generation formulation of Janssen (1991a) is used. The reason for using a fixed high-frequency cut-off rather than a dynamic cut-off frequency that depends on the wind speed or on the mean frequency, as in the WAM model, is that, in coastal regions, mixed sea states with rather different characteristic frequencies may occur. For instance, a local wind may generate a very young wind sea behind an island, which is totally unrelated to, but superimposed on, a simultaneously occurring swell. In such cases, a dynamic cut-off frequency may be too low to account properly for the locally generated sea state.

9.5.2 Propagation

Wave energy (or wave action) always propagates down-wave (by definition), even in the presence of an ambient current, so that the state at a geographic grid point in SWAN is determined by the state at the up-wave geographic grid points. This is also the case in spectral space. The most robust numerical propagation scheme would therefore be an implicit up-wind scheme.[9] The adjective 'implicit' is used here to indicate that, in such a scheme, all derivatives of the action density (in time t and horizontal co-ordinates x and y) are formulated[10] at one and the same computational level, i_t or i_x or i_y, *except* the derivative in the integration dimension in which also the previous or up-wave level is used: $i_t - 1$ or $i_x - 1$ or $i_y - 1$ (see Fig. 9.1). Implicit schemes are always unconditionally stable and the values of the discrete steps Δx, Δy and Δt in space and time can be chosen independently, allowing relatively large time steps (larger than the Courant criterion of Eq. 9.2.1 would allow) in the computations. However, for reasons of computational accuracy,

---

[9] Up-wind is the common term in numerical analysis, but 'up-wave' would be more appropriate in the case of wave models.
[10] I will use Cartesian co-ordinates in treating the numerical techniques of SWAN, but these techniques apply also when spherical co-ordinates are used.

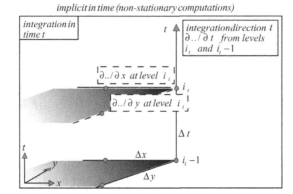

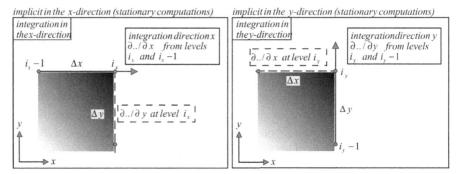

Figure 9.1 The definition of the adjective 'implicit' in time (upper panel) and in horizontal space (lower panels), using all (up-wave) derivatives at the same level, except in the integration direction.

they must be much smaller than the space and time scales of the phenomena to be computed (see Section 9.2). For small-scale computations (coastal regions, smaller than 25 km, say), an implicit first-order up-wind difference scheme in geographic space seems to be accurate enough (on the basis of several years of experience with the second-generation HISWA shallow-water wave model; Holthuijsen *et al.*, 1989). For large-scale computations, higher-order implicit schemes are needed (to reduce diffusion effects). In directional space, the same experience with HISWA shows that a first-order scheme is not suitable and a higher-than-first-order scheme is required.

The use of geographic up-wind schemes calls for the direction space to be decomposed into four quadrants at each grid point of the geographic space (see Fig. 9.2). In each of these quadrants, the computations can be carried out independently of the other quadrants, except for energy or action that is moving across the directional boundaries between the quadrants (where $\theta = n \times 90°$, with $n = 0, 1, 2$ and 3) due to refraction or diffraction (turning of wave direction) and nonlinear wave–wave

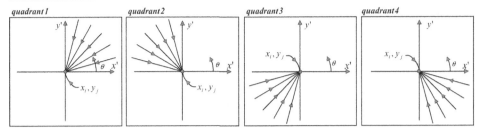

Figure 9.2 The waves propagate in all directions, divided into four quadrants. The local x- and y-co-ordinates for grid point x_i, y_i are indicated by x' and y'.

interactions (transfer of energy or action between wave components from different directions). Such exchange between quadrants is formulated in terms of corresponding conditions at the boundary directions between the quadrants. To account properly for these boundary conditions, the computations are carried out iteratively (also for each time step in the case of non-stationary computations).

Three alternative up-wind schemes for propagation in geographic space can be used in SWAN, each in a rotating sequence of four forward-marching sweeps (one sweep per quadrant) using the following schemes:

(A) for stationary and non-stationary cases, the first-order BSBT scheme (backward space, backward time) with considerable diffusion (diffusion is *first* order in Δx and Δy);
(B) for non-stationary cases, the second-order S&L scheme (Stelling and Leendertse, 1992), with very little diffusion (diffusion is *third* order in Δt, Δx and Δy); and
(C) for stationary cases, the second-order SORDUP scheme (Rogers *et al.*, 2002), with little diffusion (diffusion is *second* order in Δx and Δy).

Near open *boundaries* (e.g., the open ocean), *coastlines* and *obstacles* (e.g., breakwaters), the computations revert to the BSBT scheme (for the SORDUP scheme in the last two grids adjoining the corresponding boundary, coast or obstacle grid points and for the S&L scheme, the last three grids adjoining such grid points). This scheme has a larger numerical diffusion but that is usually acceptable over the small distances involved.

Numerical schemes

For *non-stationary conditions*, the integration in time is carried out with a simple backward finite-difference scheme. The corresponding discretisation of the action balance equation for the BSBT scheme (for positive propagation speeds; including the propagation in spectral space and the computation of the source terms but ignoring their discretisation) is

$$
\left[\frac{N^{i_t} - N^{i_t-1}}{\Delta t} \right]^n_{i_x,i_y,i_\sigma,i_\theta}
$$

$$
+ \left[\frac{(c_{g,x}N)_{i_x} - (c_{g,x}N)_{i_x-1}}{\Delta x} \right]^{i_t,n}_{i_y,i_\sigma,i_\theta}
+ \left[\frac{(c_{g,y}N)_{i_y} - (c_{g,y}N)_{i_y-1}}{\Delta y} \right]^{i_t,n}_{i_x,i_\sigma,i_\theta}
$$

$$
+ \left[\frac{(1-\nu)(c_\sigma N)_{i_\sigma+1} + 2\nu(c_\sigma N)_{i_\sigma} - (1+\nu)(c_\sigma N)_{i_\sigma-1}}{2\,\Delta\sigma} \right]^{i_t,n}_{i_x,i_y,i_\theta}
$$

$$
+ \left[\frac{(1-\eta)(c_\theta N)_{i_\theta+1} + 2\eta(c_\theta N)_{i_\theta} - (1+\eta)(c_\theta N)_{i_\theta-1}}{2\,\Delta\theta} \right]^{i_t,n}_{i_x,i_y,i_\sigma} = \left[\frac{S}{\sigma} \right]^{i_t,n^*}_{i_x,i_y,i_\sigma,i_\theta}
$$

$$
(9.5.1)
$$

where i_t is the time-level index, i_x, i_y, i_σ and i_θ are grid counters and Δt, Δx, Δy, $\Delta\sigma$ and $\Delta\theta$ are the increments in time, geographic space and spectral space, respectively. The iterative nature of the computation is indicated with the iteration index n (the iteration index for the source terms n^* is equal to n or $n-1$, depending on the source term, see below). Because of the iterations, the scheme is approximately implicit for the source terms. For negative propagation speeds, appropriate $+$ and $-$ signs are required in Eq. (9.5.1). The coefficients ν and η in this equation determine the degree to which the scheme in spectral space is up-wind or central (the signs of these coefficients depend on the signs of the corresponding propagation speeds). These coefficients thus control the numerical diffusion in frequency space and direction space, respectively. A value of $\nu = 0$ or $\eta = 0$ corresponds to central schemes, which have the largest accuracy (numerical diffusion ≈ 0). Values of $|\nu| = 1$ or $|\eta| = 1$ correspond to up-wind schemes, which are somewhat more diffusive and therefore less accurate but more robust. If large gradients of the action density in frequency space or direction space are present (i.e., discontinuities in the two-dimensional spectrum), numerical oscillations can arise (especially with the central-difference schemes), resulting in negative values of the action density. In each sweep, such negative values are removed from the two-dimensional spectrum by setting these values equal to zero and by re-scaling the remaining positive values such that the frequency-integrated action density per spectral direction is conserved.

The depth derivatives and current derivatives required to compute c_σ and c_θ (see Eqs. 7.3.32 and 7.3.33) are calculated with a first-order up-wind scheme.

Diffraction is implemented in SWAN (from version 40.41 onwards) by adding a diffraction parameter δ_E to the expressions for the group velocity components, $c_{g,x}$ and $c_{g,y}$ and to the turning rate c_θ in the above propagation schemes. This parameter is equal to δ_a of Eq. (7.3.26) with the amplitude a replaced with the square root of the energy density $\sqrt{E(\sigma)}$ (see Section 8.4.2). This option should not be used in front of reflecting obstacles where standing waves can appear (for instance, in harbours; phase information is then required, but is not available; see Section 7.3.3). In each of the iterations of the propagation scheme, the second-order derivative $\nabla(cc_g \nabla \sqrt{E(\sigma)})$ in the expression for δ_E is obtained with a simple, second-order central scheme based on the results of the previous iteration. For the x-dimension, the estimation is

$$\left\{ \frac{\partial}{\partial x}\left(CC_g \frac{\partial \sqrt{E}}{\partial x}\right)\right\}^n \approx \frac{1}{2\Delta x^2}\left\{\left[(CC_g)_i + (CC_g)_{i-1}\right]\sqrt{E_{i-1}}\right.$$
$$-\left[(CC_g)_{i-1} + 2(CC_g)_i + (CC_g)_{i+1}\right]\sqrt{E_i}$$
$$+\left.\left[(CC_g)_i + (CC_g)_{i+1}\right]\sqrt{E_{i+1}}\right\}^{n-1} \qquad (9.5.2)$$

where $E = E(\sigma)$, i is a grid counter in the x-dimension and n is the iteration number. For the y-dimension, the expression is identical, with y replacing x. The estimation of δ_E is thus based on the values of the energy density E obtained from the preceding iteration of the geographic propagation (the value of δ_E is cut off at the low side at -1 to avoid having imaginary propagation speeds; there is no upper bound). The problem of the singularity at the tips of breakwaters (see Holthuijsen et al., 2003, and Section 8.4.2) has been solved by using an optional frequency-dependent under-relaxation in the iterative procedure of the computations (see Zijlema and van der Westhuysen, 2005).

For *more accurate* computations under non-stationary conditions, the two terms that represent propagation in geographic space in Eq. (9.5.1) are replaced with the Stelling and Leendertse (S&L) scheme with the following discretisation in geographic space:

$$\left[\frac{\frac{5}{6}(c_{g,x}N)_{i_x} - \frac{5}{4}(c_{g,x}N)_{i_x-1} + \frac{1}{2}(c_{g,x}N)_{i_x-2} - \frac{1}{2}(c_{g,x}N)_{i_x-3}}{\Delta x}\right]^{i_t,n}_{i_y,i_\sigma,i_\theta}$$
$$+\left[\frac{\frac{5}{6}(c_{g,y}N)_{i_y} - \frac{5}{4}(c_{g,y}N)_{i_y-1} + \frac{1}{2}(c_{g,y}N)_{i_y-2} - \frac{1}{2}(c_{g,y}N)_{i_y-3}}{\Delta y}\right]^{i_t,n}_{i_x,i_\sigma,i_\theta}$$
$$+\left[\frac{\frac{1}{4}(c_{g,x}N)_{i_x+1} - \frac{1}{4}(c_{g,x}N)_{i_x-1}}{\Delta x}\right]^{i_t,n-1}_{i_y,i_\sigma,i_\theta}$$
$$+\left[\frac{\frac{1}{4}(c_{g,y}N)_{i_y+1} - \frac{1}{4}(c_{g,y}N)_{i_y-1}}{\Delta y}\right]^{i_t,n-1}_{i_x,i_\sigma,i_\theta} \qquad (9.5.3)$$

The numerical diffusion of the S&L scheme is so small that the so-called garden-sprinkler effect (GSE) is noticeable in wave propagation over large distances (see the footnote in Section 6.4.2). It can be counteracted by using the following diffusion terms in the action balance equation (see Booij and Holthuijsen, 1987):

$$D_{xx}\frac{\partial^2 N(\sigma, \theta)}{\partial x^2} + D_{yy}\frac{\partial^2 N(\sigma, \theta)}{\partial y^2} + D_{xy}\frac{\partial^2 N(\sigma, \theta)}{\partial x \partial y} \qquad (9.5.4)$$

The values of the diffusion coefficients D_{xx}, D_{yy} and D_{xy} depend on the spectral *resolution* and the *propagation time* of the waves. *In* the propagation direction and *normal* to that direction, the diffusion coefficients are, respectively,

$$D_{ss} = \Delta c^2 W/12$$
$$D_{nn} = c^2 \Delta\theta^2 W/12 \qquad (9.5.5)$$

where W is the wave age (the time elapsed since the generation of the wave energy for the frequency and direction considered, not to be confused with the ratio of phase speed over wind speed c/U_{10}, which is also called wave age). In terms of the x- and y-directions, the coefficients are

$$D_{xx} = D_{ss} \cos^2\theta + D_{nn} \sin^2\theta$$
$$D_{yy} = D_{ss} \sin^2\theta + D_{nn} \cos^2\theta$$
$$D_{xy} = (D_{ss} - D_{nn}) \cos\theta \sin\theta \qquad (9.5.6)$$

The numerical scheme used to compute these diffusion terms at the time level i_{t-1} is a simple, central, first-order, finite-difference scheme. This (explicit) finite-differencing is fast (having little impact on computation time) but it is only conditionally stable. With a mathematical analysis (not given here) it can be shown that a likely stability condition for the one-dimensional S&L scheme with this GSE correction is $D\Delta t/(\Delta r^2) \leq 0.5$ (where D is the maximum of D_{xx}, D_{yy} and D_{xy} and Δr is the minimum of Δx and Δy). It can also be shown that the S&L scheme with this GSE correction is stable for typical ocean cases. For shelf-sea and also for small-scale computations the GSE tends to be small and this form of GSE correction should not be used, in order to retain unconditional numerical stability.

For *stationary* conditions, SWAN can be run in stationary mode with the BSBT scheme (see Eq. 9.5.1). Time is then removed as a variable but the integration (in geographic space) is still carried out iteratively. The propagation scheme is still implicit (see Fig. 9.1) and the values of Δx and Δy can therefore still be chosen independently of each other. For *more accurate* computations under stationary conditions, the SORDUP scheme with the following discretisation in geographic space is used, instead of the BSBT scheme:

$$
\left[\frac{1.5(c_{g,x}N)_{i_x} - 2(c_{g,x}N)_{i_x-1} + 0.5(c_{g,x}N)_{i_x-2}}{\Delta x} \right]_{i_y,i_\sigma,i_\theta}^{i_t,n}
$$

$$
+ \left[\frac{1.5(c_{g,y}N)_{i_y} - 2(c_{g,y}N)_{i_y-1} + 0.5(c_{g,y}N)_{i_y-2}}{\Delta y} \right]_{i_x,i_\sigma,i_\theta}^{i_t,n} \tag{9.5.7}
$$

which replaces the two corresponding terms in Eq. (9.5.1).

Solvers, grids and boundaries

To explain the above numerical solution techniques in terms of matrix solutions, first ignore the decomposition into quadrants. The propagation of the waves in both geographic and spectral space would then be described with one large basic matrix. Removing refraction, frequency-shifting and nonlinear source terms from this basic matrix permits a matrix solution with a Gauss–Seidel technique (e.g., Golub and van Loan, 1986) in which the matrix is decomposed into four sections (the above four quadrants), which are each solved in one step (super-convergence). Restoring refraction and frequency-shifting to the matrix requires the solution of a sub-matrix for each geographic grid point. If no currents are present and the depth is stationary, this is readily done with a Thomas algorithm (e.g., Abbott and Basco, 1989; Ferziger and Perić, 2002), because $c_\sigma = 0$ and the sub-matrix is a simple tri-diagonal matrix. If currents *are* present or the depth is *not* stationary, the sub-matrix is a band matrix. It is solved with an iterative ILU-CGSTAB method (see Vuik, 1993; van der Vorst, 1992) or the strongly implicit procedure (SIP; see Ferziger and Perić, 2002; from SWAN version 40.20 onwards). Restoring refraction and frequency-shifting also introduces coefficients in each quadrant that cause a mutual dependence of matrix sections. The same happens when nonlinear source terms are added. The basic matrix as a whole needs therefore to be solved iteratively until some break-off criteria are met (see Zijlema and van der Westhuysen, 2005). To reduce the number of iterations in stationary mode with wind generation, SWAN starts with a reasonable first guess of the wave field (which is based on the second-generation source terms of Holthuijsen and De Boer, 1988, adapted for shallow water). It reduces the number of iterations typically by a factor of two. In non-stationary mode, a very reasonable first guess is available for each time step from the previous time step and the required number of iterations is usually small (one or two). If in this mode no iterations are used, the computations of propagation are still implicit and therefore still unconditionally stable.

The above descriptions are based on a rectangular grid in all five dimensions (x, y, t, σ and θ). Actually, the propagation scheme in SWAN for geographic space is formulated on a *curvilinear* grid (irregular, quadrangular, not necessarily orthogonal; such a grid may be used as a boundary-fitting grid to accommodate the water levels and currents computed with a hydrodynamic model). The above regular grid is only a special case of this curvilinear grid. The finite differences for the curvilinear grid are based on approximating the geographic distribution of the energy (action) density in the area enclosed by three neighbouring grid points with a flat triangle. The gradient at each grid point at location x_i, y_j is then readily approximated from the up-wind grid points. For the x-direction this is for grid point i, j (the grid points are ordered in x, y-space with labels i and j, respectively):

$$\frac{\partial c_x N}{\partial x} \approx \frac{(c_{g,x} N)_{i,j} - (c_{g,x} N)_{i-1,j}}{\Delta \tilde{x}_1} + \frac{(c_{g,x} N)_{i,j} - (c_{g,x} N)_{i,j-1}}{\Delta \tilde{x}_2} \qquad (9.5.8)$$

where $\Delta \tilde{x}_1 = \Delta x_1 - (\Delta y_1/\Delta y_2)\Delta x_2$ and $\Delta \tilde{x}_2 = \Delta x_2 - (\Delta y_2/\Delta y_1)\Delta x_1$. The increments are $\Delta x_1 = x_{i,j} - x_{i-1,j}$, $\Delta x_2 = x_{i,j} - x_{i,j-1}$, $\Delta y_1 = y_{i,j} - y_{i-1,j}$ and $\Delta y_2 = y_{i,j} - y_{i,j-1}$. The gradient in the y-direction is similarly estimated. This curvilinear option operates both in Cartesian and in spherical co-ordinates. In addition to these options of the propagation schemes, SWAN permits zooming-in of the computations to ever smaller areas with geographically nested computational grids.

The boundary conditions in geographic space are fully absorbing for wave energy that is *leaving* the computational domain (to open sea) or crossing a coastline (unless that coastline has been defined as a reflecting obstacle). The wave energy *entering* the computational domain along open geographic boundaries needs to be prescribed by the user. For coastal regions such incoming energy is usually provided along the deep-water boundary (which is usually taken more or less parallel to the coastline). Along the lateral boundaries (which are usually taken more or less perpendicular to the coastline), the spectral densities are usually set equal to zero by the user. Such erroneous lateral boundary conditions are practically unavoidable but they propagate into the computational area, affecting the computational results in triangular areas with the apex of each triangle at the corners between the deep-water boundary and the lateral boundaries. The angle of the apex is $30°$–$45°$ (for wind-sea conditions) on either side of the deep-water mean wave direction (the angle is less for swell conditions; it is essentially equal to the one-sided width of the directional distribution of the incoming wave spectrum). For this reason the lateral boundaries should be sufficiently far away from the area where the computational results need to be reliable.

9.5.3 Generation, wave–wave interactions and dissipation

The numerical estimations of the source terms in SWAN are essentially implicit, i.e., based on the wave conditions at the time level at which the wave conditions are being computed. This is achieved with implicit or iterative explicit schemes to compute the source terms, which, in the limit of a large number of iterations, always result in implicit estimates. In actual computations, final convergence is never achieved and the schemes used to compute the source terms are therefore, strictly speaking, only approximately implicit. In the following the adjectives 'explicit' and 'implicit' refer to the approximations of the source terms *within each iteration of the propagation scheme* (i.e., at one location, at one moment in time).

The linear term α in the source term that represents initial wave generation by wind (see Eq. 9.3.8) is independent of the spectrum, so it can be readily computed from other information than the spectrum, such as the wind speed and direction. All other source terms depend on energy density and can be written as (quasi-)linear terms: $S = \phi E$, in which ϕ is a coefficient that generally depends on (integral) wave parameters (e.g., m_0, $\tilde{\sigma}$, $\tilde{k}$, σ, k etc.) and action densities of other spectral components. In SWAN, the coefficient ϕ is *always* determined at the previous iteration level: $\phi = \phi^{n-1}$. Whether the scheme for computing the source term is explicit (with the energy density for $S = \phi E$ taken from the previous iteration) or implicit (with the energy density estimated from the previous and the present iteration) depends on whether the source term is positive or negative.

Positive source terms

For positive source terms (wave generation by wind and the triad and quadruplet wave–wave interactions when they are positive), the integration is generally more stable if an explicit scheme is used (i.e., the source term depends on the energy density in the previous iteration E^{n-1} rather than on the energy density in the present iteration E^n) rather than an implicit scheme (in which the source term would depend also on E^n). The explicit scheme for these source terms is therefore

$$S^n \approx \phi^{n-1} E^{n-1} = S^{n-1} \quad \text{for all source terms when they are positive} \quad (9.5.9)$$

In SWAN versions prior to version 40.41, this explicit scheme is also used for the formulation of the quadruplet wave–wave interactions when they are negative. This is considered reasonable since Tolman (1992a) has shown that using an explicit scheme in combination with a limiter (see below) gives results similar to those obtained with the following more expensive implicit scheme (such an implicit formulation is optionally available in SWAN).

Negative source terms

For negative source terms (white-capping, bottom friction, surf-breaking and the triad and quadruplet wave–wave interactions when they are negative, but see above) the integration is generally more stable if an implicit scheme is used. Two versions are used, depending on whether the source term is *strongly* nonlinear or *weakly* nonlinear. If a negative source term is strongly nonlinear (surf-breaking only), it is estimated at iteration level n with a linear extrapolation from the previous *iteration* $(n - 1)$:

$$S^n \approx S^{n-1} + \left(\frac{\partial S}{\partial E}\right)^{n-1} (E^n - E^{n-1}) \qquad \text{for surf-breaking} \qquad (9.5.10)$$

However, to achieve even more stable computations, the term $S^{n-1} = \phi^{n-1} E^{n-1}$ in this formulation is replaced with $S^{n-1} = \phi^{n-1} E^n$ (making the scheme somewhat more implicit and thus more robust; note that in the limit the solution is the same). Since this process of surf-breaking has been formulated such that $S = a S_{tot}$ and $E = a E_{tot}$, the derivative $\partial S / \partial E$ is analytically determined as $\partial S_{tot} / \partial E_{tot}$ (where a is identical in these two expressions, E_{tot} and S_{tot} are the integrated spectrum and source term, respectively; see Eq. 8.4.26).

If the negative source term is *weakly* nonlinear (white-capping, bottom friction and negative triad and quadruplet wave–wave interactions), a similar accuracy of estimating S^n can be achieved by replacing $(\partial S / \partial E)^{n-1}$ in Eq. (9.5.10) with $(S/E)^{n-1}$. This gives the following simpler and therefore more economical scheme:

$$S^n \approx \phi^{n-1} E^{n-1} + \left(\frac{S}{E}\right)^{n-1} (E^n - E^{n-1}) \qquad \begin{array}{l}\text{all source terms when they are} \\ \text{negative (except surf-breaking)}\end{array}$$

$$(9.5.11)$$

With $S = \phi E$, this reduces to

$$S^n \approx \phi^{n-1} E^n \qquad\qquad \begin{array}{l}\text{for all source terms when they are} \\ \text{negative (except surf-breaking)}\end{array}$$

$$(9.5.12)$$

These estimates of the source terms are added to the elements of the matrix for the propagation.

Numerical stability

Although the computation schemes for propagation in SWAN are inherently stable, the integration of the source terms is not. This may lead to numerical instabilities, which are suppressed or avoided either with a limiter (default) or with under-relaxation (optional). The limiter suppresses the development of numerical instabilities by limiting the maximum total change of action density *per iteration* of the propagation scheme at each discrete wave component to a fraction (0.1) of the Phillips (1957) equilibrium level, reformulated in terms of wave number (see

Section 8.3.2) in order to be applicable in water of arbitrary depth:

$$|\Delta N(\sigma, \theta)|_{max} = 0.1\left(\tfrac{1}{2}\alpha k^{-3}\right)J/\sigma \qquad (9.5.13)$$

where $\tfrac{1}{2}\alpha k^{-3}$ is the Phillips equilibrium level, formulated in wave number k-space (see Section 8.3.2), $J = \partial k/\partial \sigma = c_g^{-1}$ is the Jacobian used to transform from k-space to σ-space and $\alpha = \alpha_{PM} = 0.0081$ is the Phillips 'constant' of the Pierson–Moskowitz (1964) spectrum (this does not impose a shape of the spectral tail; it merely dampens the change from one iteration to the next). In versions prior to version 40.31 of SWAN, this limiter is not applied in the surf zone (in SWAN: $H_{rms}/H_{max} < 0.2$ with $H_{rms} = \sqrt{8E_{tot}}$, which implies a fraction of breakers $Q_b > 10^{-5}$). From version 40.31 of SWAN onwards, the limiter is always used, irrespective of whether the waves break or not (giving a slightly slower, but more robust convergence). When the optional under-relaxation is sufficiently strong (the relaxation coefficient and effect are to be determined by the user with trial computations), no limiter is needed (see Zijlema and van der Westhuysen, 2005).

9.5.4 Wave-induced set-up

For geographically *one*-dimensional situations, the wave-induced set-up is computed in SWAN with the momentum balance equation in which the gradient of the radiation stress balances the gradient of the hydrostatic pressure (see Eq. 7.4.20). The integration in the x-direction in this one-dimensional situation is carried out with a simple trapezoidal rule. In geographically *two*-dimensional cases, the Poisson equation of the divergence-free force field (see Eq. 9.4.5) is used. It is solved with the same technique as that which is used for wave propagation with ambient currents (Vuik, 1993; van der Vorst, 1992, prior to version 40.41 of SWAN; or Ferziger and Perić, 2002, from that version onwards). The boundary conditions for this elliptic partial differential equation are (1) at the deepest boundary point the set-up is set at zero; (2) at open boundaries, equilibrium between the radiation stress gradient and the hydrodynamic pressure gradient normal to the model boundary; and (3) at the coastline, equilibrium between the radiation stress gradient and the hydrodynamic pressure gradient normal to the coastline. The coastline in SWAN moves as dictated by the wave-induced set-up (i.e., drying and flooding are accounted for).

Appendix A

Random variables

1 One random variable

1a Characterisation

The surface elevation in the presence of waves, at any *one* location and at any *one* moment in time, will be treated as a random variable (a variable the exact value of which cannot be predicted). For instance, consider a laboratory flume with water in which a wind generates waves (Fig. A.1). Somewhere in the flume, at location A, a wave gauge measures the surface elevation as a function of time. At some moment in time t_1 (after the wind has started from zero), the surface elevation at that location has a value $^1\eta(t_1)$. The superscript 1 indicates the experiment number (more experiments will follow).

If the experiment were repeated, this value (at the same location and same moment in time after the wind has started to blow) would be $^2\eta(t_1)$. If the experiment were repeated again, the surface elevation would be $^3\eta(t_1)$, and so on and so forth. This value of the surface elevation at time t_1 obviously cannot be predicted and is therefore a random variable. It will be denoted as $\underline{\eta}(t_1)$ (underscored to show that it is a random variable). Of course, the surface elevations at *other* times t_2, t_3, ... etc. are equally unpredictable and therefore also random variables $\underline{\eta}(t_2)$, $\underline{\eta}(t_3)$, ... This is just an example with the surface elevation as the random variable. In the following, an unspecified random variable will be denoted as $\underline{x}$.

A random variable $\underline{x}$ is fully characterised by its *probability density function* $p(x)$, which is defined such that the probability of $\underline{x}$ attaining a value between x and $x + \mathrm{d}x$ is given by (see Fig. A.2)

$$\Pr\{x < \underline{x} \leq x + \mathrm{d}x\} = \int_x^{x+\mathrm{d}x} p(x)\mathrm{d}x \qquad (\text{A.1})$$

It follows that the probability of $\underline{x}$ being less than or equal to x (the probability of non-exceedance; see Fig. A.3) is

$$\Pr\{\underline{x} \leq x\} = \int_{-\infty}^x p(x)\mathrm{d}x = P(x) \qquad (\text{A.2})$$

$P(x)$ is called the *(cumulative) distribution function* of $\underline{x}$.

The above probability *density* function can be obtained as the derivative of this *distribution* function: $p(x) = \mathrm{d}P(x)/\mathrm{d}x$. Obviously, the value of a random variable is always less than infinity, so the probability that the value of x is less than ∞ is 1. This implies that the maximum value of the distribution function $P(x)$ is 1 and that the surface area of a probability density function $p(x)$ is always 1:

$$\Pr\{\underline{x} \leq \infty\} = P(\infty) = \int_{-\infty}^{\infty} p(x)\mathrm{d}x = 1 \qquad (\text{A.3})$$

Note that statisticians express probabilities in terms of fractions, rather than in terms of percentages, as you would do when talking to a friend. The inverse of the distribution

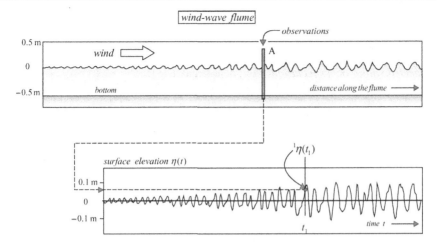

Figure A.1 One value of the surface elevation $^1\eta(t_1)$ at one location, one moment in time, in one experiment, in a wind-wave flume.

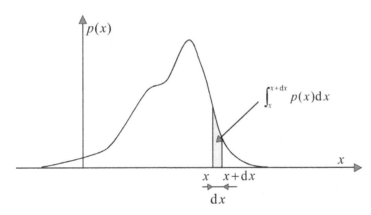

Figure A.2 The probability density function $p(x)$ of a random variable $\underline{x}$.

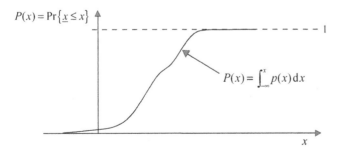

Figure A.3 The (cumulative) distribution function $P(x)$ of a random variable $\underline{x}$ (the probability of non-exceedance).

function, i.e., the function that gives the value of the random variable x for a given probability of non-exceedance, is written as $x(P) = P^{-1}(x)$ and is called the quantile function.

The average or mean value of x can be defined in terms of the probability density function $p(x)$ as the first-order moment,[1] divided by the zeroth-order moment of $p(x)$, and it is called the '*expected value*' or '*expectation*' of x, denoted as $E\{x\}$:

$$\text{expected value of } x = E\{\underline{x}\} = \mu_x = m_1/m_0 = \int_{-\infty}^{+\infty} xp(x)\mathrm{d}x \Big/ \int_{-\infty}^{+\infty} p(x)\mathrm{d}x \qquad \text{(A.4)}$$

Since $\int_{-\infty}^{+\infty} p(x)\mathrm{d}x = 1$, it follows that

$$E\{\underline{x}\} = \mu_x = \int_{-\infty}^{+\infty} xp(x)\mathrm{d}x \qquad \text{(A.5)}$$

This average may be interpreted as the location of the probability density function on the x-axis. The probability density function may be further characterised by increasingly higher-order moments. The second-, third- and fourth-order moments are thus used to define the width, skewness and kurtosis of the function respectively. For instance,

$$\sigma_x^2 = E\{(\underline{x} - \mu_x)^2\} = \int_{-\infty}^{+\infty} (x - \mu_x)^2 p(x)\mathrm{d}x = E\{\underline{x}^2\} - \mu_x^2 = m_2 - m_1^2 \qquad \text{(A.6)}$$

σ_x^2 is called the variance and σ_x the standard deviation of x, which represents the width of the probability density function. *Alternative* measures of the mean, width, skewness and kurtosis can be based on the moments of the quantile function $\beta_r = \int_0^1 P^r x(P)\mathrm{d}P$ (the probability-weighted moments). These measures β_r are called L-moments.

Averages of *functions* of x can also be defined as expected values. For instance, the above variance σ_x^2 of x can be seen as the average of the function $f(x) = (x - \mu_x)^2$. In general, the expected value of a function $f(x)$ is defined as

$$E\{f(x)\} = \int_{-\infty}^{+\infty} f(x)p(x)\mathrm{d}x \qquad \text{(A.7)}$$

1b Gaussian probability density function

Many processes in Nature behave in such a way that the well-known *Gaussian* probability density function applies:

$$p(x) = \frac{1}{\sqrt{2\pi}\sigma_x} \exp\left[-\frac{(x - \mu_x)^2}{2\sigma_x^2}\right] \qquad \text{(A.8)}$$

A theoretical explanation of this almost universal applicability is provided by the *central limit theorem*, which, simply formulated, says that the *sum* of a *large* number of *independent* random variables (not necessarily Gaussian distributed, and without one being dominant) is Gaussian distributed. Since many natural phenomena result from a large number of causes, it is reasonable to find the Gaussian probability density function to apply so often. The Gaussian probability density function is also called the *normal* probability density function. It is by no means the only one that applies to natural phenomena. Many others may also apply, e.g., the Rayleigh, exponential and Weibull probability density functions, to

---

[1] The nth-order moment of a function $h(x)$ is by definition $m_n = \int_{-\infty}^{\infty} x^n h(x)\mathrm{d}x$. The function $h(x)$ may be any function, not necessarily a probability density function.

name only a few. Note that the definition of the standard deviation σ_x is independent of the specific probability density function and therefore independent of the Gaussian probability density function.

1c Estimation

An average of a random variable $\underline{x}$ is often not determined from the probability density function $p(x)$, but *estimated* from a set of sample values of $\underline{x}$ (called *realisations*, e.g., observations in an experiment). Such a set of sample values is called an *ensemble*, and the average is called an ensemble average, denoted as $\langle . \rangle$. For instance,

$$\text{mean} = \mu_x \approx \langle \underline{x} \rangle = \frac{1}{N} \sum_{i=1}^{N} x_i \tag{A.9}$$

$$\text{variance} = \sigma_x^2 \approx \langle (\underline{x} - \langle \underline{x} \rangle)^2 \rangle$$

$$= \frac{1}{N} \sum_{i=1}^{N} (\underline{x}_i - \langle \underline{x} \rangle)^2 = \frac{1}{N} \sum_{i=1}^{N} \langle \underline{x}_i \rangle^2 - \langle \underline{x} \rangle^2 \tag{A.10}$$

where N is the number of samples. Note that these are only estimates, which will always differ from the expected values. These differences are called (statistical) sampling errors.

2 Two random variables

2a Characterisation

A *pair* of random variables $(\underline{x}, \underline{y})$ is fully characterised by its *joint* probability density function: $p(x, y)$. This two-dimensional function is defined, in analogy with the above, such that the probability of $\underline{x}$ attaining a value between x and $x + dx$ and of $\underline{y}$ (simultaneously) attaining a value between y and $y + dy$ is given by

$$\Pr\{x < \underline{x} \le x + dx \text{ and } y < \underline{y} \le y + dy\} = \int_{x}^{x+dx} \int_{y}^{y+dy} p(x, y) dy \, dx \tag{A.11}$$

The two random variables may be unrelated to one another. They are then called independent. Alternatively, they may well be related. One variable is then said to be dependent on the other. When they are *linearly* related they are said to be *correlated* (they cluster around a straight line, when one is plotted against the other). Note that two random variables can be related but uncorrelated (see Fig. A.4). Unrelated variables are obviously uncorrelated.

The degree of correlation (the degree to which the pairs of $(\underline{x}, \underline{y})$ cluster around a straight line) is quantified with the correlation coefficient $\gamma_{x,y}$, which is defined as the normalised covariance $C_{x,y}$ of the two variables:

$$\gamma_{x,y} = C_{x,y}/(\sigma_x \sigma_y) \quad \text{with} \quad -1 \le \gamma_{x,y} \le 1 \tag{A.12}$$

where the covariance is the average product of $\underline{x}$ and $\underline{y}$, each taken relative to its mean:

$$C_{x,y} = E\{(\underline{x} - \mu_x)(\underline{y} - \mu_y)\} \tag{A.13}$$

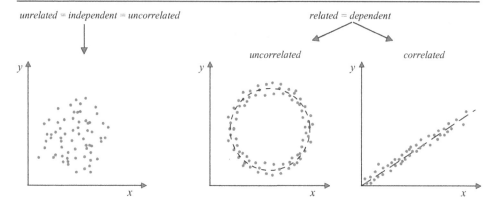

Figure A.4 (In)dependent, (un)related and (un)correlated random variables.

2b Two-dimensional Gaussian probability density function

The two-dimensional *Gaussian* probability density function for two random variables $\underline{x}$ and $\underline{y}$ is

$$p(x, y) = \frac{1}{2\pi\sigma_x\sigma_y \left(1 - \gamma_{x,y}^2\right)^{1/2}}$$

$$\times \exp\left\{-\frac{1}{1 - \gamma_{x,y}^2}\left[\frac{(x - \mu_x)^2}{2\sigma_x^2} - \gamma_{x,y}\frac{(x - \mu_x)(y - \mu_y)}{\sigma_x\sigma_y} + \frac{(y - \mu_y)^2}{2\sigma_y^2}\right]\right\}$$

$$(A.14)$$

3 Stochastic processes

3a Characterisation

Random variables may not only be dependent, related or correlated. They may also be *ordered* in some sense, i.e., the variables exist in some kind of sequence. This is a useful notion when (many) more than two variables are considered. For instance, bottles produced by a machine appear one by one from that machine. Their exact length (which is a random variable) is therefore ordered in their sequence of appearance. In this example, the ordering is one-dimensional, but that need not always be the case. For instance, the length of a student (a random variable!) in a lecture room is ordered by the (two-dimensional, horizontal) position of the students in the room. The weight of leaves on a tree is ordered in three-dimensional space. Such an *ordered set of random variables* is called a stochastic *process*.

A stochastic *process* in one-dimension can readily be visualised with the wind-generated waves in the flume of Section 1a of this appendix. The measurement starts at $t = 0$ when the wind starts to blow over still water and the subsequent (very large) set of surface elevations η observed at location A is a function of time. The values are unpredictable and this set is therefore an example of an ordering of very many random variables in *time*. (Note that, in a time sequence $\underline{x}(t_i)$, the random variable $\underline{x}$ at time t_1 is another random variable than $\underline{x}$ at time t_2, which is another random variable than $\underline{x}$ at time t_3, etc.). One such experiment is *one* realisation of the stochastic process $\eta(t_1), \eta(t_2), \eta(t_3), \ldots, \eta(t_i), \ldots$. Obviously, when the surface elevation at some moment in time is large, then, a fraction of a second

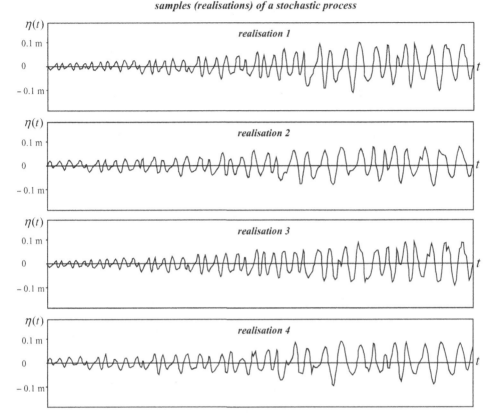

Figure A.5 A set of four realisations of the surface elevation as a function of time, at location A, in the laboratory flume of Fig. A.1 (statistically identical experiments, but with the same wind speed etc.). The waves grow as time increases until some sort of equilibrium (in a statistical sense) is reached (i.e., stationarity).

later, the surface elevation will also be large: the surface elevations at short time intervals are related and even correlated. Only after some lapse of time will the relation between the two surface elevations $\underline{\eta}(t_i)$ and $\underline{\eta}(t_j)$ be lost, i.e., when the time interval $\underline{t}_j - \underline{t}_i$ is large (compared with a characteristic wave period). The experiment can be repeated at will: the wind machine is turned off and, after the water surface has returned to its still level, the wind machine is turned on again (at $t = 0$), which starts the next experiment. Obviously, there are as many realisations of the stochastic process $\underline{\eta}(t_1)$, $\underline{\eta}(t_2)$, $\underline{\eta}(t_3)$, . . ., $\underline{\eta}(t_i)$, . . . as there are experiments (see Fig. A.5).

Like any random variable, $\underline{\eta}(t_i)$ is fully characterised by its probability density function. This implies that, to characterise the surface elevation statistically *at that moment in time*, this probability density function is required at each moment in time t_i. To characterise the surface elevations as a *process*, we need additionally, at that moment in time t_i, all joint probability density functions, i.e., the joint probability $p(\eta(t_i), \eta(t_j))$ for all t_j. There is an infinite number of moments in time t_i, each requiring an infinite number of such functions, since there are infinitely many moments t_j!

3b Stationary processes

If, after some time, the surface elevation at location A in the flume is constant in some *statistical* sense (see Fig. A.5), then all statistical characteristics of the waves are independent of time and the process is said to be *stationary* (but the statistical characteristics may still depend on time *intervals* $t_j - t_i$). The stationarity of a process greatly simplifies the description, since only the statistical characteristics for one moment in time are required (including the relationships with the random variables at all time *intervals*). The analogous condition for variables that are ordered in space is called *homogeneity*. If only the averages and the variances of the variables are constant in time or space, the process is called weakly stationary or weakly homogeneous.

3c Gaussian processes

If all (joint) probability density functions of a process (stationary or not) are *Gaussian*, the process is called a Gaussian process. A Gaussian process is relatively simple to describe, since only the averages of each pair of variables and their covariance are required. Writing the variables of one such pair (see Eq. A.13) as $\underline{x} = \underline{x}(t_1) = \underline{x}(t)$ and $\underline{y} = \underline{x}(t_2) = \underline{x}(t + \tau)$, we may write the covariance as $C_{x,x} = E\{[\underline{x}(t) - \mu_x(t)] \times [\underline{x}(t + \tau) - \mu_x(t + \tau)]\} = C(t, \tau)$. The covariance may therefore also be seen as a function of time t and time interval τ: the covariance function. Since the two variables are from the same process, $C(t, \tau)$ is also called the auto-covariance function.

3d Stationary, Gaussian processes

A stationary, Gaussian process is even simpler to describe: only the mean and the covariances for *one moment in time* are required (because they are identical at all other times). The auto-covariance is then (only) a function of the time interval τ and, if the average of the variable is taken to be zero (as usual for the surface elevation of waves), it can be written as

$$C(\tau) = E\{x(t)x(t + \tau)\} \qquad\qquad \text{if } \mu_x(t) = \mu_y(t) = 0 \qquad (A.15)$$

Note that the auto-covariance for $\tau = 0$ is the variance of this process: $E\{x^2(t)\} = C(0)$.

3e Ergodic processes

If averaging over time (or space) gives the same results as averaging over an ensemble of realisations, the process is said to be ergodic. The mean and variance of a zero-mean ergodic process can then be estimated as

$$\mu_x \approx \langle\underline{x}(t_i)\rangle = \frac{1}{D}\int_D \underline{x}(t)dt \qquad\qquad \text{(the mean)} \qquad\qquad (A.16)$$

$$\sigma_x^2 \approx \langle[\underline{x}(t_i)]^2\rangle = \frac{1}{D}\int_D [\underline{x}(t)]^2 dt \qquad\qquad \text{(the variance)} \quad \text{if } \mu_x = 0 \quad (A.17)$$

and the auto-covariance as

$$C(\tau) \approx \langle x(t)x(t + \tau)\rangle = \frac{1}{D}\int_D [\underline{x}(t)\underline{x}(t + \tau)]dt \quad \text{if } \mu_x = 0 \qquad\qquad (A.18)$$

where the brackets $\langle . \rangle$ denote ensemble averaging and D is the length of the time interval (duration) over which the time average is taken. More generally, for the average of a function $f[\underline{x}(t_i)]$

$$\langle f[\underline{x}(t_i)] \rangle = \frac{1}{D} \int_D f[\underline{x}(t)]dt \qquad \text{for an ergodic process} \tag{A.19}$$

It follows that such a process is stationary.[2] The surface elevation of random, wind-generated waves under stationary conditions happens to be ergodic (in the linear approximation of these waves), so all averages needed to describe waves can be estimated as time-averages. This is fortunate because we would not be able to obtain an ensemble in Nature it would require Nature repeating, over and over again, identical conditions, in a statistical sense, at sea.

4 The sea-surface elevation

The surface elevation of wind-generated waves as a function of time is often treated as a Gaussian process. Measurements have shown this to be very reasonable (see Section 4.3), but there are also theoretical grounds: the surface elevation at any one moment in time t_i can be seen as the sum of the elevations at that time of a large number of harmonic wave components that have been generated independently of each other (by a turbulent wind, possibly at very different locations) and that have travelled independently of each other across the sea surface (in the linear approximation of waves). The central limit theorem (see Section 1b of this appendix) shows that therefore the sea-surface elevation *should* be Gaussian distributed – but not always. Steep waves, or high waves in shallow water, do interact and are therefore not independent. Deviations from the Gaussian model do therefore occur at sea, particularly in the surf zone.

---

[2] The reverse is not true: not all stationary stochastic processes are ergodic, for instance, switching on an electric circuit that produces an unpredictable but constant current (which may well be Gaussian distributed, i.e., the constant value is drawn from a Gaussian distribution every time the circuit is activated) gives a stationary, stochastic process (after all, the values are unpredictable and the statistical characteristics are constant in time). However, it is not an ergodic process because the time-average in each realisation is different from the time-average in another realisation, and is therefore (in general) not equal to the ensemble average.

Appendix B

Linear wave theory

1 Introduction

Here we continue with the linear theory of surface gravity waves at the point where in Chapter 5 the basic equations are linearised (Section 5.3), to see what terms are actually removed by the linearisation. The end results, in terms of the linearised equations and linearised boundary conditions, are identical to those in Chapter 5.

2 Conservation equations (1)

The basic equations, in their nonlinear form, are the continuity equation and the momentum balance equations (for constant density; see Section 5.3.2):

$$\frac{\partial u_x}{\partial x} + \frac{\partial u_y}{\partial y} + \frac{\partial u_z}{\partial z} = 0 \qquad \text{continuity} \qquad (B.1)$$

$$\frac{\partial u_x}{\partial t} + \frac{\partial (u_x u_x)}{\partial x} + \frac{\partial (u_y u_x)}{\partial y} + \frac{\partial (u_z u_x)}{\partial z} = -\frac{1}{\rho}\frac{\partial p}{\partial x} \qquad \text{momentum in the } x\text{-direction}$$

$$(B.2)$$

$$\frac{\partial u_y}{\partial t} + \frac{\partial (u_x u_y)}{\partial x} + \frac{\partial (u_y u_y)}{\partial y} + \frac{\partial (u_z u_y)}{\partial z} = -\frac{1}{\rho}\frac{\partial p}{\partial y} \qquad \text{momentum in the } y\text{-direction}$$

$$(B.3)$$

$$\frac{\partial u_z}{\partial t} + \frac{\partial (u_x u_z)}{\partial x} + \frac{\partial (u_y u_z)}{\partial y} + \frac{\partial (u_z u_z)}{\partial z} = -\frac{1}{\rho}\frac{\partial p}{\partial z} - g \quad \text{momentum in the } z\text{-direction}$$

$$(B.4)$$

If we apply the chain rule of differentiation to the momentum equation for the x-direction (Eq. B.2) and subsequently subtract from the result the equation of continuity (Eq. B.1), multiplied by u_x:

$$\frac{\partial u_x}{\partial t} \boxed{+ u_x\frac{\partial u_x}{\partial x}} + u_x\frac{\partial u_x}{\partial x} \boxed{+ u_x\frac{\partial u_y}{\partial y} + u_y\frac{\partial u_x}{\partial y}} \boxed{+ u_x\frac{\partial u_z}{\partial z} + u_z\frac{\partial u_x}{\partial z}} = -\frac{1}{\rho}\frac{\partial p}{\partial x} \quad (B.5)$$

minus

$$\boxed{u_x\frac{\partial u_x}{\partial x}} \boxed{+ u_x\frac{\partial u_y}{\partial y}} \boxed{+ u_x\frac{\partial u_z}{\partial z}} = 0 \qquad (B.6)$$

318

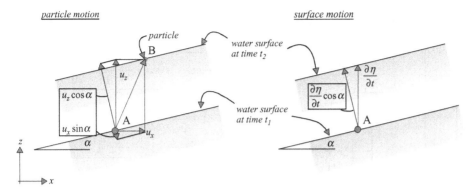

Figure B.1 A particle in the water surface remains in the surface, when it moves from a position A at time t_1 to a position B at time t_2 if the velocity component of the particle *normal* to the surface $u_z \cos \alpha - u_x \sin \alpha$ (left-hand panel) is equal to the velocity of the surface in that direction $(\partial \eta / \partial t) \cos \alpha$ (right-hand panel).

then the result is the equation of motion for the x-direction:

$$\frac{\partial u_x}{\partial t} + u_x \frac{\partial u_x}{\partial x} + u_y \frac{\partial u_x}{\partial y} + u_z \frac{\partial u_x}{\partial z} = -\frac{1}{\rho} \frac{\partial p}{\partial x} \qquad (B.7)$$

Performing the same operation on the momentum balance equations for the y- and z-directions gives the other two equations of motion. These three equations of motion are also called *the Euler equations*. If we had included viscosity (internal friction in the water) by adding the terms $v(\partial^2 u_x/\partial x^2 + \partial^2 u_x/\partial y^2 + \partial^2 u_x/\partial z^2)$, $v(\partial^2 u_y/\partial x^2 + \partial^2 u_y/\partial y^2 + \partial^2 u_y/\partial z^2)$ and $v(\partial^2 u_z/\partial x^2 + \partial^2 u_z/\partial y^2 + \partial^2 u_z/\partial z^2)$ to the right-hand side of each of these equations, respectively, where v represents the viscosity of water (called the kinematic viscosity coefficient and closely related to the molecular viscosity coefficient $\mu = \rho v$), the equations would be called the *Navier–Stokes* equations.

3 Boundary conditions (1)

The *kinematic surface boundary condition* relates the velocity of a particle in the surface to the motion of the surface. If we consider, for the sake of simplicity, a (vertically) two-dimensional situation, then it is readily seen from the geometry of the situation (Fig. B.1) that, if a particle is to remain in the surface, then the velocity component of that particle, in the direction normal to the surface $u_z \cos \alpha - u_x \sin \alpha$, should be equal to the velocity *of* the surface in that direction $(\partial \eta / \partial t) \cos \alpha$ (the particle may move *in* the surface):

$$u_z \cos \alpha - u_x \sin \alpha = \frac{\partial \eta}{\partial t} \cos \alpha \qquad \text{at } z = \eta \qquad (B.8)$$

where η is the surface elevation above some reference level and α is the slope of the surface. This can be re-written as

$$u_z = \frac{\partial \eta}{\partial t} + u_x \tan \alpha = \frac{\partial \eta}{\partial t} + u_x \frac{\partial \eta}{\partial x} \qquad \text{at } z = \eta \qquad (B.9)$$

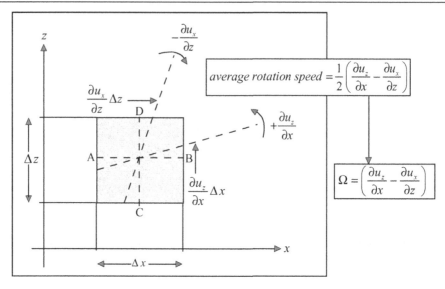

Figure B.2 Rotation Ω is twice the average rotation speed of the two main axes of an infinitesimally small rectangle (in two dimensions).

For a three-dimensional situation, this kinematic boundary condition at the surface can be written similarly as

$$u_z = \frac{\partial \eta}{\partial t} + u_x \frac{\partial \eta}{\partial x} + u_y \frac{\partial \eta}{\partial y} \qquad \text{at } z = \eta \qquad (B.10)$$

The kinematic *bottom* boundary condition is simply that the water may not penetrate the fixed, horizontal bottom, so that

$$u_z = 0 \qquad\qquad \text{at } z = -d \qquad (B.11)$$

The *dynamic* surface boundary condition (pressure is zero) is simply

$$p = 0 \qquad\qquad \text{at } z = \eta \qquad (B.12)$$

4 Rotation or vorticity

Consider an infinitesimally small rectangle in the vertical plane with dimension Δx, Δz (representing a small volume of water; Fig. B.2). Along the right-hand side of the rectangle, the vertical velocity (i.e., the velocity in the z-direction; near point B in Fig. B.2) will generally differ from the vertical velocity along the left-hand side (near point A) by $(\partial u_z / \partial x)\Delta x$. This difference makes the horizontal main axis AB rotate at a rotation speed of $+\partial u_z / \partial x$ (counter-clockwise for positive $\partial u_z / \partial x$). The horizontal velocity (i.e., the velocity in the x-direction) along the upper side of the rectangle (near point D) will likewise differ from the horizontal velocity along the lower side (near point C) by $(\partial u_x / \partial z)\Delta z$. This difference will make the vertical main axis CD rotate at a rotation speed of $-\partial u_x / \partial z$ (i.e., clockwise for positive $\partial u_x / \partial z$). The average rotation speed of the rectangle is then $\frac{1}{2}(\partial u_z / \partial x - \partial u_x / \partial z)$. *Vorticity*, i.e., the property of having rotating water particles, is expressed in terms of this concept of rotation. In the two-dimensional situation considered here, rotation is defined as twice the rotation speed of the rectangle (a small volume of water, or a water particle):

$$\Omega = \frac{\partial u_z}{\partial x} - \frac{\partial u_x}{\partial z} \tag{B.13}$$

The motion is called irrotational (or vorticity-free) if

$$\Omega = 0 \tag{B.14}$$

For a three-dimensional situation, rotation is expressed in terms of a vector $\vec{\Omega} = (\Omega_x, \Omega_y, \Omega_z)$ with

$$\Omega_x = \frac{\partial u_z}{\partial y} - \frac{\partial u_y}{\partial z}, \qquad \Omega_y = \frac{\partial u_x}{\partial z} - \frac{\partial u_z}{\partial x} \quad \text{and} \quad \Omega_z = \frac{\partial u_y}{\partial x} - \frac{\partial u_x}{\partial y} \tag{B.15}$$

The motion in three dimensions is called irrotational (or vorticity-free) if

$$\Omega_x = 0, \quad \Omega_y = 0 \quad \text{and} \quad \Omega_z = 0 \quad \text{or} \quad \vec{\Omega} = \vec{0} \quad \text{in vector notation} \tag{B.16}$$

5 The velocity potential function

The *velocity potential function* ϕ is defined as a function in x, y, z and t having the property that its spatial derivatives are the particle velocities in x, y, z-space (this is possible only for irrotational motions):

$$\phi = \phi(x, y, z, t) \qquad \text{defined such that } u_x = \frac{\partial \phi}{\partial x}, \ u_y = \frac{\partial \phi}{\partial y} \text{ and } u_z = \frac{\partial \phi}{\partial z}$$

or, in vector notation,

$$\vec{u} = \nabla \phi \tag{B.17}$$

6 Conservation equations (2)

We can express the *continuity equation* in terms of the velocity potential by substituting the expressions for u_x, u_y and u_z of Eqs. (B.17) into Eq. (B.1):

$$\frac{\partial^2 \phi}{\partial x^2} + \frac{\partial^2 \phi}{\partial y^2} + \frac{\partial^2 \phi}{\partial z^2} = 0 \qquad \text{or, in vector notation,} \qquad \nabla^2 \phi = 0 \tag{B.18}$$

We can also express the *equations of motion* in terms of the velocity potential function. To that end, we first invoke the condition that the water motion is irrotational, so that

$$\Omega_x = \frac{\partial u_z}{\partial y} - \frac{\partial u_y}{\partial z} = 0 \to \frac{\partial u_y}{\partial z} = \frac{\partial u_z}{\partial y} \tag{B.19}$$

$$\Omega_y = \frac{\partial u_x}{\partial z} - \frac{\partial u_z}{\partial x} = 0 \to \frac{\partial u_x}{\partial z} = \frac{\partial u_z}{\partial x} \tag{B.20}$$

$$\Omega_z = \frac{\partial u_y}{\partial x} - \frac{\partial u_x}{\partial y} = 0 \to \frac{\partial u_x}{\partial y} = \frac{\partial u_y}{\partial x} \tag{B.21}$$

The equation of motion in the x-direction (Eq. B.7) may then be written as

$$\frac{\partial u_x}{\partial t} + u_x \frac{\partial u_x}{\partial x} + u_y \frac{\partial u_y}{\partial x} + u_z \frac{\partial u_z}{\partial x} = -\frac{1}{\rho} \frac{\partial p}{\partial x} \tag{B.22}$$

or, moving u_x, u_y and u_z behind the differentiations,

$$\frac{\partial u_x}{\partial t} + \frac{\partial}{\partial x}\left(\frac{1}{2}u_x^2 + \frac{1}{2}u_y^2 + \frac{1}{2}u_z^2\right) = -\frac{1}{\rho} \frac{\partial p}{\partial x} \tag{B.23}$$

This equation of motion can be expressed in terms of the velocity potential merely by substituting into it the expressions for u_x, u_y and u_z of Eq. (B.17), with the following result:

$$\frac{\partial}{\partial t}\left(\frac{\partial \phi}{\partial x}\right) + \frac{\partial}{\partial x}\left\{\frac{1}{2}\left[\left(\frac{\partial \phi}{\partial x}\right)^2 + \left(\frac{\partial \phi}{\partial y}\right)^2 + \left(\frac{\partial \phi}{\partial z}\right)^2\right]\right\} = -\frac{1}{\rho} \frac{\partial p}{\partial x} \tag{B.24}$$

Changing the order of differentiation in the first term and moving the term on the right-hand side to the left-hand side allows us to write this equation as

$$\frac{\partial}{\partial x}\left\{\frac{\partial \phi}{\partial t} + \frac{1}{2}\left[\left(\frac{\partial \phi}{\partial x}\right)^2 + \left(\frac{\partial \phi}{\partial y}\right)^2 + \left(\frac{\partial \phi}{\partial z}\right)^2\right] + \frac{p}{\rho}\right\} = 0 \tag{B.25}$$

The corresponding equations for the momentum in the y- and z-directions may similarly be written as

$$\frac{\partial}{\partial y}\left\{\frac{\partial \phi}{\partial t} + \frac{1}{2}\left[\left(\frac{\partial \phi}{\partial x}\right)^2 + \left(\frac{\partial \phi}{\partial y}\right)^2 + \left(\frac{\partial \phi}{\partial z}\right)^2\right] + \frac{p}{\rho}\right\} = 0 \tag{B.26}$$

$$\frac{\partial}{\partial z}\left\{\frac{\partial \phi}{\partial t} + \frac{1}{2}\left[\left(\frac{\partial \phi}{\partial x}\right)^2 + \left(\frac{\partial \phi}{\partial y}\right)^2 + \left(\frac{\partial \phi}{\partial z}\right)^2\right] + \frac{p}{\rho} + gz\right\} = 0 \tag{B.27}$$

Note that the equation for the z-direction contains the term gz and the other two equations do not. The reason is obvious: gravitation works only in the z-direction. We can add the term gz to the other two equations without altering the meaning of these equations because this term would disappear when the derivative in the x- or y-direction is taken:

$$\frac{\partial}{\partial x}\left\{\frac{\partial \phi}{\partial t} + \frac{1}{2}\left[\left(\frac{\partial \phi}{\partial x}\right)^2 + \left(\frac{\partial \phi}{\partial y}\right)^2 + \left(\frac{\partial \phi}{\partial z}\right)^2\right] + \frac{p}{\rho} + gz\right\} = 0 \tag{B.28}$$

$$\frac{\partial}{\partial y}\left\{\frac{\partial \phi}{\partial t} + \frac{1}{2}\left[\left(\frac{\partial \phi}{\partial x}\right)^2 + \left(\frac{\partial \phi}{\partial y}\right)^2 + \left(\frac{\partial \phi}{\partial z}\right)^2\right] + \frac{p}{\rho} + gz\right\} = 0 \tag{B.29}$$

$$\frac{\partial}{\partial z}\left\{\frac{\partial \phi}{\partial t} + \frac{1}{2}\left[\left(\frac{\partial \phi}{\partial x}\right)^2 + \left(\frac{\partial \phi}{\partial y}\right)^2 + \left(\frac{\partial \phi}{\partial z}\right)^2\right] + \frac{p}{\rho} + gz\right\} = 0 \tag{B.30}$$

The sum of terms between the curly brackets appears in all three equations, in each case expressing the fact that this sum of terms is not a function of x, y or z. This sum can therefore be only an (arbitrary) function of time t, $f(t)$, for which we take the simplest possible: $f(t) = 0$. We therefore find, from the equations of motion,

$$\frac{\partial\phi}{\partial t} + \frac{1}{2}\left[\left(\frac{\partial\phi}{\partial x}\right)^2 + \left(\frac{\partial\phi}{\partial y}\right)^2 + \left(\frac{\partial\phi}{\partial z}\right)^2\right] + \frac{p}{\rho} + gz = 0$$

or, in vector notation,

$$\frac{\partial\phi}{\partial t} + \frac{1}{2}|\nabla\phi|^2 + \frac{p}{\rho} + gz = 0 \tag{B.31}$$

This is the *Bernoulli equation* for unsteady motion. Removing the quadratic terms makes this equation the *linearised Bernoulli equation* for unsteady motion (see also Eq. 5.3.28):

$$\boxed{\frac{\partial\phi}{\partial t} + \frac{p}{\rho} + gz = 0} \tag{B.32}$$

7 Boundary conditions (2)

The *kinematic boundary condition* at the *surface* (see Eq. B.10), in terms of the velocity potential function (obtained by merely substituting the expressions for u_x, u_y and u_z of Eq. B.17), is

$$\frac{\partial\phi}{\partial z} = \frac{\partial\eta}{\partial t} + \frac{\partial\phi}{\partial x}\frac{\partial\eta}{\partial x} + \frac{\partial\phi}{\partial y}\frac{\partial\eta}{\partial y} \qquad \text{at } z = \eta \tag{B.33}$$

Removing the nonlinear terms (the last two terms on the right-hand side) makes this boundary condition linear (see also Eq. 5.3.23):

$$\boxed{\frac{\partial\phi}{\partial z} = \frac{\partial\eta}{\partial t} \qquad \text{at } z = 0} \tag{B.34}$$

The *dynamic boundary condition* at the *surface*, expressed in terms of the velocity potential, is obtained by taking the Bernoulli equation (Eq. B.31) at the surface $z = \eta$, with $p = 0$, so that

$$\frac{\partial\phi}{\partial t} + \frac{1}{2}\left[\left(\frac{\partial\phi}{\partial x}\right)^2 + \left(\frac{\partial\phi}{\partial y}\right)^2 + \left(\frac{\partial\phi}{\partial z}\right)^2\right] + g\eta = 0 \qquad \text{at } z = \eta$$

or, in vector notation,

$$\frac{\partial\phi}{\partial t} + \frac{1}{2}|\nabla\phi|^2 + g\eta = 0 \tag{B.35}$$

Removing the quadratic terms makes this equation linear (see Eq. 5.3.29):

$$\boxed{\frac{\partial\phi}{\partial t} + g\eta = 0 \qquad \text{at } z = 0} \tag{B.36}$$

The *kinematic boundary condition* at the *bottom*, in terms of the velocity potential function, is (see also Eq. 5.3.24)

$$\boxed{\frac{\partial\phi}{\partial z} = 0 \qquad \text{at } z = -d} \tag{B.37}$$

The above linearised equations and boundary conditions are the equations and boundary conditions that are used in the linear theory of surface gravity waves (see Chapter 5).

Appendix C
Spectral analysis

1 Introduction

Measurements of the sea-surface elevation are almost always obtained with an electrical current in some instrument. This *analogue* signal can be transformed into an estimate of the variance density spectrum of the waves, using *analogue* systems, such as electronic circuits or optical equipment. However, with today's small and fast computers the analogue signal can also be transformed into a *digital* signal for a subsequent *numerical* analysis. The latter option has been accepted widely and it will be treated here.

The numerical analysis depends on the type of measurement. The most common and simplest measurement in this respect is a record of the sea-surface elevation at one location as a function of time (i.e., a one-dimensional record). Records like these are produced by instruments such as a heave buoy, a wave pole or a low-altitude altimeter. These can be analysed with a one-dimensional Fourier transform.[1] Other types of measurements generate multivariate signals (i.e., several, simultaneously obtained, time records), e.g., the two slope signals of a pitch-and-roll buoy. Such signals require a cross-spectral analysis (e.g., Tucker and Pitt, 2001), or some other, advanced method (e.g., Hashimoto, 1997; Young, 1994; Pawka, 1983; Lygre and Krogstad, 1986 and many others). Two-dimensional images, e.g., from a surface-contouring radar, require a two-dimensional Fourier transform (e.g., Singleton, 1969) and moving images (e.g., those produced by a ship's radar) require a three-dimensional Fourier transform. Here, we consider only the simplest possible measurement: the sea-surface elevation at one location as a function of time.

The estimation of the wave spectrum from such a measurement can be based on two numerical approaches. In the first approach the auto-covariance function of the surface elevation is computed and then Fourier-transformed (see Section 3.5.5). This method was commonly used in the 1950s and 1960s. The second approach is to Fourier-transform directly the wave record itself. This is the preferred technique today. It is usually carried out with the numerical Fast Fourier Transform (FFT), which was introduced about 1970 (it is far more efficient, i.e., faster than the 'old' technique based on the auto-covariance function). This approach will be treated here without mathematical proofs and ignoring all mathematical details. Excellent references for the spectral analysis of random signals are

---

[1] An alternative type of analysis that is slowly gaining popularity as a supplement to the Fourier analysis is the *wavelet analysis* (e.g., Farge, 1992; Foufoula-Georgiou and Kumar, 1994; Mallat, 1998). It is essentially a variation on the Fourier analysis that is considered here. The main difference is that the wavelet analysis provides a time-varying spectral estimate, on a time scale that depends on the phenomenon to be described. It is therefore particularly suited for identifying events or transient phenomena against a random background (e.g., breaking waves or freak waves in a wave record; Liu, 1994; Liu and Mori, 2000; Liu and Babanin, 2004). Similar information can also be obtained with conventional Fourier analysis by dividing the time record into short, overlapping segments of equal duration, and Fourier-analysing these segments in sequence. The wavelet analysis is more subtle: the duration of the (overlapping) segments is equal to a fixed number of wave periods (typically only a few) of the harmonic component that is being analysed. This implies that, at each moment at which the wavelet spectrum is computed, a segment of different duration is used for each wave period separately. In addition, each of these segments is multiplied by a standard function that has the same shape for all wave periods. This function is called the *mother wavelet*, after which this technique is named.

Blackman and Tukey (1958), Jenkins and Watts (1968), Bendat and Piersol (1971), Goda (2000) and Tucker and Pitt (2001).

2 Basic analysis

The spectrum has been defined in the main text in terms of amplitudes of harmonic components (see Section 3.5.3):

$$E(f) = \lim_{\Delta f \to 0} \frac{1}{\Delta f} E\{\tfrac{1}{2}\underline{a}^2\} \tag{C.1}$$

where $\underline{a}$ is the amplitude of the harmonic component and Δf is an arbitrarily chosen frequency band. The spectral analysis of a wave record is essentially the elaboration of this definition. It involves estimating the amplitude $\underline{a}$, determining the expectation $E\{\tfrac{1}{2}\underline{a}^2\}$, dividing this expectation by the frequency interval Δf and then taking the limit $\Delta f \to 0$. However, as always in real life, estimations replace these exact definitions.

The estimation of the amplitude per frequency requires the sea-surface elevation to be written as a Fourier series with unknown amplitudes and phases:[2]

$$\eta(t) = \sum_{i=1}^{N} a_i \cos(2\pi f_i t + \alpha_i) \qquad \text{with } f_i = \frac{i}{D} \text{ so that } \Delta f = \frac{1}{D} \tag{C.2}$$

where $\eta(t)$ is the record of the surface elevation. This is a non-random version of the random-phase/amplitude model that underlies the definition of the spectrum (for a given wave record, the phases and amplitudes are not random because they can be computed from the record). Using trigonometric identities, Eq. (C.2) may also be written as[3]

$$\eta(t) = \sum_{i=1}^{N} [A_i \cos(2\pi f_i t) + B_i \sin(2\pi f_i t)] \tag{C.3}$$

with amplitude a_i and phase α_i:

$$\boxed{a_i = \sqrt{A_i^2 + B_i^2}} \tag{C.4}$$

and

$$\tan \alpha_i = -\frac{B_i}{A_i} \tag{C.5}$$

The amplitudes A_i and B_i can be determined from the record with *Fourier integrals* (see Note C1):

$$\boxed{A_i = \frac{2}{D} \int_D \eta(t)\cos(2\pi f_i t)\,dt \qquad \text{for } f_i = i/D} \tag{C.6}$$

$$\boxed{B_i = \frac{2}{D} \int_D \eta(t)\sin(2\pi f_i t)\,dt \qquad \text{for } f_i = i/D} \tag{C.7}$$

---

[2] Usually a Fourier series starts with $i = 0$ to include the mean value of the time series, but that has been taken to be zero here.

[3] The harmonic $x(t) = a \cos(\omega t + \alpha)$ may be written (standard trigonometry) as $x(t) = a[\cos(\omega t)\cos\alpha - \sin(\omega t)\sin\alpha]$, so that, if we write $x(t) = A \cos(\omega t) + B \sin(\omega t)$, then $A = a \cos\alpha$ and $B = -a \sin\alpha$. From this we find $a^2 = A^2 + B^2$ and $\tan\alpha = -B/A$.

This operation on the wave record to obtain the amplitudes is called the *Fourier* transform (A_i and B_i are called the Fourier coefficients). By applying this operation to all frequencies (i.e., all i), all values of A_i and B_i can be computed, and subsequently all values of amplitude a_i and phase α_i. Now, the next steps would be to estimate the expectation $E\{\frac{1}{2}\underline{a}^2\}$, divide this by the frequency interval Δf and then take the limit of $\Delta f \to 0$. However, apart from the division by $\Delta f = 1/D$, this is not possible because of some practical problems.

NOTE C1   The Fourier transform (1)

The two Fourier integrals of Eqs. (C.6) and (C.7) that determine the values of the amplitudes A_i and B_i from a wave record $\eta(t)$ can be interpreted as filters, which filter, from the wave record, the one component with frequency $f_i = i/D$ (this implies that the duration D is a multiple of the wave period T_i because $D = i/f_i = iT_i$). The user who performs the Fourier transform chooses $i = 1, 2, 3, \ldots$ in sequence, to filter all components f_i. In this manner, all amplitudes (A_i and B_i) are obtained and therefore all amplitudes a_i and phases α_i. This filtering is readily demonstrated by replacing the time series $\eta(t)$ with its representation as a Fourier *series* in the integral of Eq. (C.6). It is shown here for the integral for A_i only; essentially the same is true for the integral for B_i.

The integral of Eq. (C.6) is

$$\boxed{\mathrm{FT}(f_i) = \frac{2}{D}\int_D \eta(t)\cos(2\pi f_i\, t)\,\mathrm{d}t}$$

With the Fourier series representation $\eta(t) = \sum_{p=1}^{P}[A_p\cos(2\pi f_p t) + B_p\sin(2\pi f_p t)]$ of Eq. (C.3) substituted, this becomes

$$\mathrm{FT}(f_i) = \frac{2}{D}\int_D \sum_{p=1}^{P}\big[A_p\cos(2\pi f_p t) + B_p\sin(2\pi f_p t)\big]\cos(2\pi f_i t)\,\mathrm{d}t$$

Since the contributions of $\sin(2\pi f_p t)\cos(2\pi f_i t)$ and $\cos(2\pi f_p t)\cos(2\pi f_i t)$ to this integral are zero for all p and i, except for $\cos(2\pi f_p t)\cos(2\pi f_i t)$ when $p = i$, the integral reduces to

$$\mathrm{FT}(f_i) = \frac{2}{D}\int_D A_i\cos^2(2\pi f_i t)\,\mathrm{d}t = \frac{2A_i}{D}\int_D \cos^2(2\pi f_i t)\,\mathrm{d}t$$

Since the duration of the wave record is a multiple of the period T_i, the outcome of the integral $\int_D \cos^2(2\pi f_i t)\,\mathrm{d}t = D/2$ so that

$$\mathrm{FT}(f_i) = A_i$$

Therefore

$$\boxed{A_i = \frac{2}{D}\int_D \eta(t)\cos(2\pi f_i t)\,\mathrm{d}t}$$

which is identical to Eq. (C.6) in the main text of this appendix.

The Fourier transform (2)

The structure of the integral in the Fourier transforms of Eqs. (C.6) and (C.7) is very simple to remember: the integral is twice the (time-)averaged product of the surface elevation $\eta(t)$ and the cosine (or sine) for each frequency f_i. To show this, define the time-averaged product of $x(t)$ and $y(t)$ as

$$\overline{xy} = \frac{1}{D} \int_D x(t)y(t)\,dt$$

Then, taking $x(t) = \eta(t)$ and $y(t) = \cos(2\pi f_i t)$, the Fourier transform to compute A_i can be written as

$$A_i = 2\,\overline{\eta(t)\cos(2\pi f_i t)}$$

and similarly

$$B_i = 2\,\overline{\eta(t)\sin(2\pi f_i t)}$$

3 Practical problems

An actual wave record differs in several respects from the surface elevation in the definition of the spectrum that underlies the analysis:

the duration of the wave record is finite;

there is usually only one record;

the wave record is discretised in time; and

the observation of the surface elevation is contaminated with instrument and processing errors.

3a The finite duration of the wave record

In the Fourier transform of a wave record, the frequency interval Δf has a constant value determined by the given duration D of the record: $\Delta f = 1/D$ (see Eq. C.2). Taking the limit $\Delta f \to 0$ (as in the definition of the spectrum, implying $D \to \infty$) is therefore not possible with the finite duration of a given wave record. The estimation of the spectrum, of necessity, therefore becomes

$$E(f) = \lim_{\Delta f \to 0} \frac{1}{\Delta f} E\{\tfrac{1}{2}\underline{a}^2\} \approx \boxed{\frac{1}{\Delta f} E\{\tfrac{1}{2}\underline{a}_i^2\}} \qquad \text{with } \Delta f = \frac{1}{D} \qquad (C.8)$$

The finite frequency interval $\Delta f = 1/D$ implies that details of the spectrum within this spectral interval Δf cannot be seen. In other words, details on a frequency scale $\Delta f = 1/D$ are lost (see Fig. C.1). The duration D should therefore be chosen long enough that details that are relevant can be seen. The capacity to resolve such spectral details is called *frequency resolution* and it is quantified with the frequency bandwidth:

$$\boxed{\Delta f = \frac{1}{D}} \qquad (C.9)$$

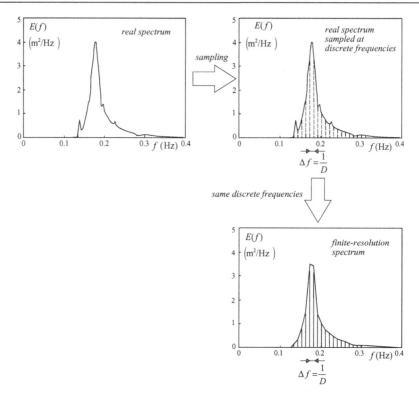

Figure C.1 The finite frequency resolution, due to the finite duration of the wave record, removes details from the spectrum.

(Resolution is actually defined as the frequency interval between independent estimates of the spectral density, which in more advanced spectral-analysis techniques may differ slightly from $1/D$.)

The frequency resolution can be improved only by taking a longer duration of the wave record. However, the wave records should also be stationary to give the spectrum any meaning. The actual duration is therefore always a compromise. On the one hand it should be sufficiently short that the assumption of a stationary situation is reasonable, on the other hand it should be sufficiently long that the frequency resolution is adequate. In addition, it should be long enough to allow one to obtain statistically reliable estimates (see below).

3b One wave record

The fact that usually only one wave record is available for the spectral analysis (at least for measurements at sea) means that the variance density must be estimated (at least initially) from just one amplitude, i.e., from $\frac{1}{2}a_i^2$, rather than from $E\{\frac{1}{2}\underline{a}_i^2\}$. This gives the so-called 'raw' estimate of $E(f)$:

$$E(f) \approx \frac{1}{\Delta f} E\{\tfrac{1}{2}\underline{a}_i^2\} \rightarrow \boxed{\frac{1}{\Delta f}\left(\tfrac{1}{2}a_i^2\right)} \qquad \text{with resolution } \Delta f = \frac{1}{D} \qquad (\text{C.10})$$

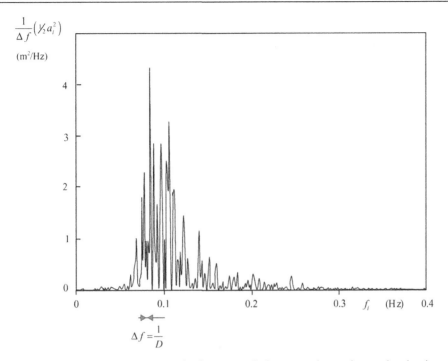

Figure C.2 The raw spectrum looks 'grassy', because the variance density is estimated from only one amplitude per frequency (the error is of the order of 100%).

This raw estimate would be acceptable if the error (the difference between the expected value $E\{\frac{1}{2}a^2\}$ and the computed value $\frac{1}{2}\underline{a}_i^2$) were relatively small, but that is not the case;[4] it is of the order of 100% (we can therefore *not* say $E(f) \approx \frac{1}{2}\underline{a}_i^2/\Delta f$). This large error is obvious from the rather 'grassy' look of the raw spectrum (see Fig. C.2). This poor reliability is unacceptable and it therefore needs to be improved (for the *given* wave record). This can be achieved only at the cost of something else.

There are several techniques to do this, but they all come at the expense of the spectral resolution. One of the simplest is to divide the time record into a number (p) of non-overlapping segments, each with a duration $D^* = D/p$. Each of these segments is then Fourier-analysed (as above) to obtain values of $\frac{1}{2}a_i^2$ with a resolution δf determined by the duration of the segment: $\delta f = 1/D^* = 1/(D/p) = p\,\Delta f$. The expectation $E\{\frac{1}{2}\underline{a}_i^2\}$ is then estimated as the average of these values (for each frequency separately; this is called the quasi-ensemble average, indicated with $\langle . \rangle$):

$$E\{\tfrac{1}{2}\underline{a}_i^2\} \approx \langle \tfrac{1}{2}a_i^2 \rangle \tag{C.11}$$

[4] The amplitude $\underline{a}_i$ is Rayleigh distributed, so the distribution of $\frac{1}{2}a_i^2$ is an exponential distribution with a mean value $E\{\frac{1}{2}\underline{a}_i^2\}$ and a width equal to $E\{\frac{1}{2}\underline{a}_i^2\}$. In other words, the error associated with estimating $E\{\frac{1}{2}\underline{a}_i^2\}$ as $\frac{1}{2}\underline{a}_i^2$ is of the same order as the mean.

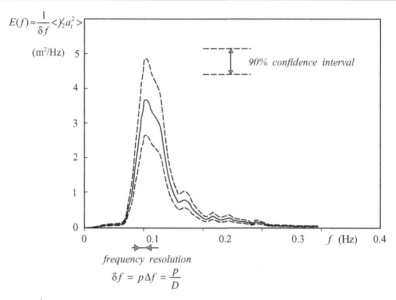

$$E(f) \approx \frac{1}{\delta f} < \frac{1}{2}a_i^2 >$$

(m²/Hz)

frequency resolution

$$\delta f = p\Delta f = \frac{p}{D}$$

Figure C.3 The (quasi-)ensemble-averaged spectrum of Fig. C.2 and its 90% confidence interval.

By this quasi-ensemble averaging, the error[5] is reduced by a factor $\sqrt{p}$:

$$\boxed{E(f) \approx \frac{1}{\delta f}\langle \tfrac{1}{2}a_i^2\rangle} \qquad \text{with resolution } \boxed{\delta f = p\Delta f} \text{ and } \textit{error} \boxed{\approx \frac{100\%}{\sqrt{p}}} \qquad \text{(C.12)}$$

Obviously, this improved reliability has come at the expense of the spectral resolution, which has been reduced by a factor of p. A compromise is therefore always required, in order to balance an acceptable spectral resolution against an acceptable reliability. A duration of 15–30 min and a value of $p = 20$–30 are typical for observations at sea. The corresponding frequency resolution is then $\delta f \approx 0.01$–0.02 Hz and the error in the spectral densities is about 20%. The reliability may also be quantified with a confidence interval. This is the interval, within which the expected value is located with a certain probability, e.g., the 90% confidence interval (see Figs. C.3 and C.4).

3c The discrete wave record

In practice, the wave records are discretised by sampling the original signal of the wave sensor at a fixed time *interval* Δt. This interval is usually 0.5 s for wave observations at sea. A direct consequence of this discretisation is that the integrals in the above Fourier transforms are replaced with discrete sums, giving an error that is not so obvious. To illustrate this, consider a harmonic wave with frequency f_1 that is sampled at a constant interval Δt

---

[5] The distribution of this ensemble average $\langle \frac{1}{2}a_i^2\rangle$ is a χ^2-distribution with $2p$ degrees of freedom.

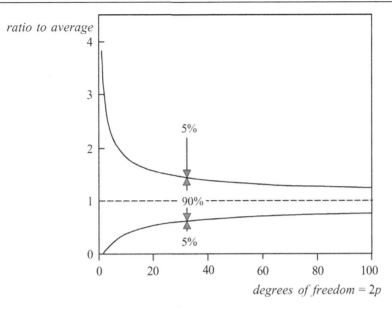

Figure C.4 The 90% confidence interval from the χ^2-distribution.

(the solid line, long wave in Fig. C.5). The only data available in the discrete record of this wave are the values at equidistant times (indicated with dots). However, it is entirely possible to have another harmonic wave (with frequency f_2: the dashed line, short wave in Fig. C.5) with the same values at the same discrete moments in time (the same dots). A Fourier analysis, using these sampled elevations, therefore cannot distinguish the two wave components.

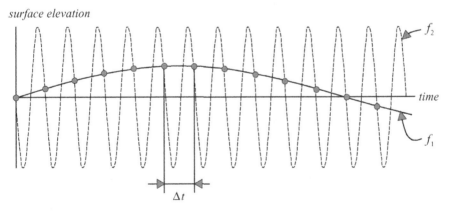

Figure C.5 Two harmonic waves with frequencies f_1 and f_2 that are given at discrete, constant time intervals $\Delta t = 1/(f_1 + f_2)$ are indistinguishable at these discrete times (as indicated by the dots).

NOTE C2   The Nyquist frequency

The discretised wave record can be seen as the 'true' surface elevation multiplied by a delta series, i.e., a series of delta functions at interval Δt (see the left-hand column in the figure below). Since *multiplication* in the time domain corresponds to *convolution* in the spectral domain, the wave spectrum should be convoluted with the spectrum of the delta series. Briefly stated, convolution is that each value of the one function is distributed in the shape of the other. Since the spectrum of the surface elevation is, strictly speaking, an even function (see Eq. 3.5.18) and the spectrum of a delta series is another delta series with interval $\Delta f = 1/\Delta t$, the result is a repetition of the (even) spectrum of the waves, with interval $\Delta f = 1/\Delta t$ (see the right-hand column in the figure below). The effect is that the tails of the repeating spectra overlap, giving the impression that the frequencies $1/(2\Delta t)$, $1/\Delta t$, $3/(2\Delta t)$, $2/\Delta t$, $5/(2\Delta t)$, ... etc. (multiples of the Nyquist frequency $1/(2\Delta t)$) are mirror frequencies.

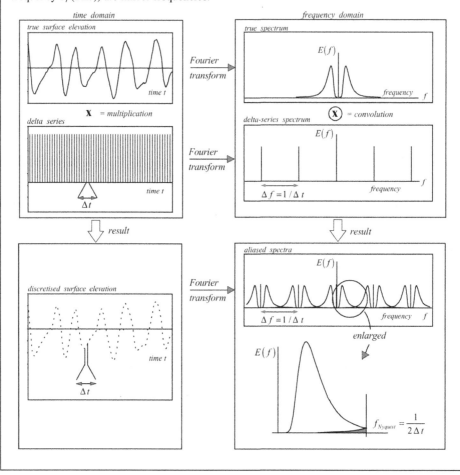

The consequence is that, in the spectral analysis, the energy density of the (high) frequency f_2 is added to the energy density of the (low) frequency f_1. It is as if the energy density of these high frequencies were mirrored around a frequency called the Nyquist frequency (or 'mirror' frequency; see Fig. C.6). The value of this Nyquist frequency is

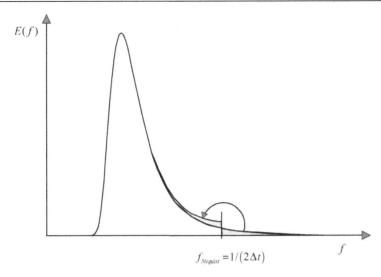

$$f_{Nyquist} = 1/(2\Delta t)$$

Figure C.6 Aliasing in the spectrum of a wave record with discrete time intervals Δt is equivalent to mirroring the spectrum around the Nyquist frequency $f_{Nyquist} = 1/(2\,\Delta t)$.

(see Note C2)

$$\boxed{f_{Nyquist} = f_N = \frac{1}{2\,\Delta t}}$$ (C.13)

The high-frequency energy thus appears at other frequencies than those at which it should, in other words, under an 'alias'. The phenomenon is therefore called 'aliasing'.

The aliasing phenomenon will always cause a relatively large error (about 100%) near the Nyquist frequency, but, with the rapidly decreasing energy densities in the tail of ocean wave spectra, it usually does not seriously affect the main part of the spectrum *if* the Nyquist frequency is chosen to be much higher than a characteristic frequency of the spectrum. The Nyquist frequency should therefore be chosen wisely, for instance more than four or five times the mean frequency (for measurements at sea, usually $f_{Nyquist} = 1$ Hz, corresponding to $\Delta t = 0.5$ s).

It can be shown with a formal analysis that the aliasing effect is due to a periodic repetition in the frequency domain of the true spectrum (see Note C2). It can also be illustrated with a phenomenon that is well known from old movies in which the wheels of a car have spokes. In such movies the wheels often seem to turn in the wrong direction. In Fig. C.7, four consecutive frames in such a movie are shown with the wheel slightly turned from one frame to the next (the time interval between the frames is 1/24 s). One would expect spoke A in the first frame to be recognised as the same turning spoke A in frames 2, 3 and 4 (the *real* forward-turning wheel in Fig. C.7). However, the human brain identifies spoke B in frame 2 with spoke A in frame 1 (because they are nearest to one another). The same happens between the other frames. The effect is that the wheel is interpreted as turning slowly backwards instead of quickly forwards (the *perceived* backward-turning wheel in Fig. C.7).

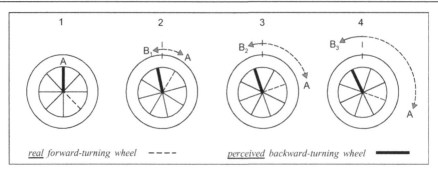

Figure C.7 The illusion of a backward-turning wheel in four frames of an old movie (aliasing). The spoke indicated with B, jumping from one spoke to the next, counter-clockwise (indicated with B_1, B_2 and B_3), is erroneously interpreted as the continually clockwise-rotating spoke A.

3d *Instrument and processing noise*

Measurements of the sea-surface elevation are always based on some physical characteristic of the water (surface) that is transformed by some instrument into numbers. These numbers do not exactly give the sea-surface elevation; they are always contaminated to some extent by the measurement technique, the instrument and the processing of the original signal. The extent of this contamination is often not precisely known. At best, a comparison with superior instruments or processing is available. Owing to such contamination, the observed time series differs from the actual surface elevation, sometimes even considerably. The variance density spectrum, which is estimated from such a contaminated time series, is therefore always, one hopes only slightly, different from the true variance density spectrum. This difference is referred to as 'observation noise' or 'instrument noise'. This subject is important for measuring ocean waves, but it is mentioned here only to make the reader aware of the problem.

Appendix D

Tides and currents

1 Introduction

Variations in time of the water depth or the presence of an ambient current (e.g., due to tides or a storm surge) may change the amplitude, frequency and direction of waves. This is generally due to energy bunching, transfer of energy between waves and currents, frequency shifting (including Doppler shifting) and current-induced refraction. Energy bunching (the shoaling-like effect of the current on the waves) is readily accounted for in the energy balance equation of the waves by using the proper velocity of wave energy propagating across a current (see Section 7.3.5). The energy transfer between waves and currents is readily accounted for by replacing the energy balance by the action balance (see Section 8.4.1). In the following, we consider only current-induced refraction and the frequency shifting.

2 Refraction

In modelling waves in the presence of an ambient current, we must make a distinction amongst various directions: the direction of the current (along the streamline; see Fig. D.1), the direction of the wave orthogonal (normal to the wave crest) and the direction of energy propagation (along the wave ray). In the *absence* of an ambient current, the energy travels in the same direction as the wave (wave ray = wave orthogonal) but, in the *presence* of an ambient current, this is not the case (wave ray $\neq$ wave orthogonal); the energy is then transported in a direction given by the vector sum of the relative group velocity $\vec{c}_g$ (along the orthogonal) and the ambient current velocity $\vec{U}$ (along the streamline). Wave energy is therefore generally not transported in the wave direction if an ambient current is present (some energy travels along the crest).

The propagation of a wave across an ambient current can be seen as the sum of the propagation of the wave in the absence of the current and the bodily transport of the wave by the current (during which the wave does not propagate relative to the current). Here, we need to consider only this bodily transport, because the propagation in the absence of the current (including refraction) has been treated in Section 7.3.2 (only the result will be used here).

If the ambient current is not uniform, the bodily transport of the wave (i.e., the wave propagates only *with* the water and not *through* the water) induces a change in wave direction. This is called current-induced refraction. The current moves the water particles, and hence the iso-phase lines (e.g., a crest), in the direction of the current, i.e., in the direction of the streamline (see Fig. D.2). If the motion is parallel to the crest, the wave direction does not change. If the motion has a component normal to the crest, the direction generally does change. More specifically, the *variation* along the crest of the current *normal* to the crest changes the wave direction, i.e., current-induced refraction is due to the along-crest variation of $U_n = U \cos \alpha$ (where U is the current speed and α is the angle between the wave direction and the current direction, see Fig. D.1). This variation of the current along

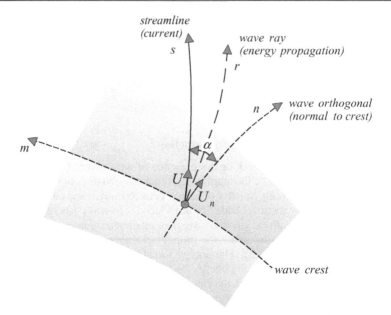

Figure D.1 An ambient current generally deflects the propagation direction of the wave energy away from the wave direction and a distinction must be made between the *wave* direction (normal to the wave crest; the wave orthogonal) and the direction of *energy* propagation (which is affected by the ambient current; the wave ray).

the crest makes the wave change direction in precisely the same manner as a variation in water depth along the crest does in the case of depth-induced refraction. To find the corresponding current-induced turning rate, consider a crest along which the current speed varies (see Fig. D.2).

For the derivation, we will use a local, left-turning system of orthogonal m, n co-ordinates (counter-clockwise rotations are positive), with m along an iso-phase line (call it a crest) and n along a line oriented normal to the crest (the wave orthogonal). The crest moves in the direction of the current but we consider a point A that moves with the crest along an orthogonal (the point A thus shifts along the crest). This point moves during a time interval Δt over a distance $\Delta n_A = U_n \Delta t$. During the same time interval, a similar point B on the crest moves over a distance $\Delta n_B = (U_n + \Delta U_n)\Delta t$. If the distance along the crest between these two points is Δm, then the corresponding directional turning of the crest $\Delta \theta$ is $\Delta \theta = -(\Delta n_B - \Delta n_A)/\Delta m = -\Delta U_n \Delta t/\Delta m$. During the same time interval, the energy travels with the crest along a streamline with the same turning of wave direction $\Delta \theta$ (assuming a locally straight crest), so the rate of change of the wave direction while travelling with the wave energy (i.e., with the current) is $\Delta \theta/\Delta t = -\Delta U_n/\Delta m$, or, for infinitesimal differences,

$$\frac{d\theta}{dt} = c_{\theta,ref} = -\frac{\partial U_n}{\partial m} \qquad (D.1)$$

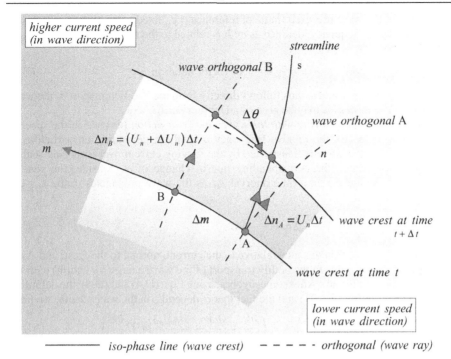

Figure D.2 The turning of a wave crest towards the region with lower current speed (in the wave direction).

Superimposing the propagation of the wave *through* the water (depth-induced refraction, see Eq. 7.3.13) and the above bodily transport *with* the water (current-induced refraction, see Eq. D.1) gives the total rate of directional change:

$$c_{\theta,ref,depth+current} = -\frac{c_g}{c}\frac{\partial c}{\partial m} - \frac{\partial U_n}{\partial m} \tag{D.2}$$

A more physically based derivation is given by Jonsson (1990). A mathematically more formal derivation is referred to at the end of Section 3 of this appendix.

3 Frequency-shifting

We consider the effect of the ambient current on the frequency of the waves in three frames of reference: one that is moving with the current, one that is fixed to the bottom and one that moves with the wave energy or action; and we assume that the parameters of the harmonic wave vary only slowly (in space and time).

In a frame of reference *moving with the current*, all results of the linear wave theory can be applied without modification (within the approximations of the linear theory, of course). The frequency of the wave in this moving frame of reference is then called the *relative* frequency, denoted as σ. The dispersion relationship is accordingly written as (see Eq. 7.3.29)

$$\sigma^2 = gk\tanh(kd) \tag{D.3}$$

The frequency of the wave in a *fixed* frame of reference (i.e., fixed to the stationary bottom) is called the *absolute* frequency, denoted as ω. It is related to the relative frequency by (see Eq. 7.3.30)

$$\omega = \sigma + kU_n \qquad \text{or} \qquad \omega = \sigma + \vec{k} \cdot \vec{U} \text{ in vector notation} \tag{D.4}$$

The difference between σ and ω thus follows directly from the bodily transport of the wave by the current. The term kU_n may be seen as a Doppler shift.

In a frame of reference *moving with the wave energy or action* (as used in the spectral energy or action balance), the relative frequency σ evolves, depending on variations in depth and current as the wave is transported by the varying current over a varying seabed topography. To determine the corresponding rate of change of σ, consider the general expression for a change $\Delta\sigma$ in a time interval Δt as the wave propagates in the n, s-coordinates system:

$$\Delta\sigma = \frac{d\sigma}{dt}\Delta t = \frac{\partial\sigma}{\partial t}\Delta t + \frac{\partial\sigma}{\partial n}\Delta n + \frac{\partial\sigma}{\partial s}\Delta s \tag{D.5}$$

where Δn is the propagation distance relative to the current, normal to the crest, and Δs is the propagation distance due to the bodily transport of the wave energy (or action) with the current (along the streamline). After some algebra, using Eqs. (D.4) and (D.5), the definition of the group velocity Eq. (5.4.31) and the fact that σ depends on the water depth, we find

$$\frac{d\sigma}{dt} = c_\sigma = \frac{\partial\sigma}{\partial d}\left(\frac{\partial d}{\partial t} + U\frac{\partial d}{\partial s}\right) + c_g\left(\frac{\partial k}{\partial t} + \frac{\partial\omega}{\partial n} - \frac{\partial(kU_n)}{\partial n} + U\frac{\partial k}{\partial s}\right) \tag{D.6}$$

where $c_g = \partial\sigma/\partial k$. To simplify this expression, we need to invoke the concept of conservation of wave crests (or wave-number density): $\partial k/\partial t + \partial\omega/\partial n = 0$ (see Eq. b in Note D). Substituting this into Eq. (D.6) gives

$$c_\sigma = \frac{\partial\sigma}{\partial d}\left(\frac{\partial d}{\partial t} + U\frac{\partial d}{\partial s}\right) + c_g\left(-\frac{\partial(kU_n)}{\partial n} + U\frac{\partial k}{\partial s}\right) \tag{D.7}$$

The term $U\partial k/\partial s$ in Eq. (D.7) can be written differently, as $U\partial k/\partial s = (\vec{k}/k)U\partial\vec{k}/\partial s$ in vector notation or, in terms of components, as

$$U\frac{\partial k}{\partial s} = \frac{k_x}{k}\left(U_x\frac{\partial k_x}{\partial x} + U_y\frac{\partial k_x}{\partial y}\right) + \frac{k_y}{k}\left(U_x\frac{\partial k_y}{\partial x} + U_y\frac{\partial k_y}{\partial y}\right) \tag{D.8}$$

Substituting $\partial k_y/\partial x = \partial k_x/\partial y$ (which follows from the irrotationality of the wave-number vector field; see Note D) into Eq. (D.8) gives

$$U\frac{\partial k}{\partial s} = \frac{k_x}{k}\left(U_x\frac{\partial k_x}{\partial x} + U_y\frac{\partial k_y}{\partial x}\right) + \frac{k_y}{k}\left(U_x\frac{\partial k_x}{\partial y} + U_y\frac{\partial k_y}{\partial y}\right) \tag{D.9}$$

which may also be written as $U\partial k/\partial s = \vec{U}\cdot\partial\vec{k}/\partial n$ in vector notation. Substituting this result into the second term on the right-hand side of Eq. (D.7) gives

$$\begin{aligned} -\frac{\partial(kU_n)}{\partial n} + U\frac{\partial k}{\partial s} &= -\frac{\partial(\vec{k}\cdot\vec{U})}{\partial n} + \vec{U}\cdot\frac{\partial\vec{k}}{\partial n} \\ &= -\vec{k}\cdot\frac{\partial\vec{U}}{\partial n} - \vec{U}\cdot\frac{\partial\vec{k}}{\partial n} + \vec{U}\cdot\frac{\partial\vec{k}}{\partial n} \\ &= -\vec{k}\cdot\frac{\partial\vec{U}}{\partial n} \end{aligned} \tag{D.10}$$

So Eq. (D.7) may also be written as

$$c_\sigma = \frac{\partial \sigma}{\partial d}\left(\frac{\partial d}{\partial t} + U\frac{\partial d}{\partial s}\right) - c_g \vec{k} \cdot \frac{\partial \vec{U}}{\partial n}$$ (D.11)

which is used in spectral wave models (Rivero *et al.*, 1997; Booij *et al.*, 1999). The first term in the brackets represents the effect of the time variation of the depth. The second term in the brackets represents the effect of the current bodily moving the wave over a horizontally varying depth. The second term on the right-hand side represents the effect of the wave moving with a horizontally varying current. The corresponding variations in absolute frequency ω and wave number k follow directly from the variation in the relative frequency σ with Eqs. (D.3) and (D.4), without any additional computations (the time variation of the current is accounted for).

NOTE D   Conservation and rotation of wave number

Conservation of wave number (also called conservation of wave-number density or conservation of crests)

The fact that a harmonic wave does not create crests leads to a convenient relationship between the time derivative of wave number $\partial k/\partial t$ and the spatial derivative of frequency $\partial \omega/\partial x$ (if we consider a one-dimensional situation with the wave propagating in the x-direction). There are many ways to derive this relationship. Here follow a few.

Balance of number of waves
To arrive at the relationship, consider a one-dimensional situation in which a (quasi-)harmonic wave with a slowly varying wave length and period travels in the positive x-direction through a box with length Δx (see the figure below).

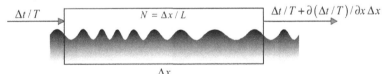

The number of waves in the box is $N = \Delta x/L$ (where L is the average wave length in the box). During a time interval Δt this number changes by $[\partial N/\partial t]\Delta t = [\partial(\Delta x/L)/\partial t]\,\Delta t$. Since a harmonic wave does not create wave crests, the number of waves in the box can change only by a net import of waves. This net import is equal to the number of waves $\Delta t/T$ which *enter* through the left-hand side of the box during the time interval Δt (where T is the average wave period), minus the number of waves which *leave* the box through the right-hand side of the box during this interval, $\Delta t/T + [\partial(\Delta t/T)/\partial x]\,\Delta x$. The net import of waves is then $-[\partial(\Delta t/T)/\partial x]\,\Delta x$. Since the change of number of waves in the box during the time interval Δt is equal to the net import of waves, it follows that

$$\frac{\partial(\Delta x/L)}{\partial t}\Delta t = -\frac{\partial(\Delta t/T)}{\partial x}\Delta x$$ (a)

Dividing by Δt and Δx, multiplying by 2π and recognising that $k = 2\pi/L$ and $\omega = 2\pi/T$ gives

$$\frac{\partial k}{\partial t} + \frac{\partial \omega}{\partial x} = 0$$ (b)

which represents the conservation of wave crests.

Wave-number balance

The derivation can also be formulated in terms of a balance equation for the wave number k, propagating with the phase speed c (k may be interpreted as *wave-number density*, since it is the number of waves per unit horizontal distance, multiplied by 2π). Substituting $\mu = k$ and $u_x = c$ into the one-dimensional version of the general balance equation (Eq. 5.3.8 of Chapter 5) gives

$$\frac{\partial k}{\partial t} + \frac{\partial (ck)}{\partial x} = 0 \qquad\qquad\qquad (c)$$

or, since $c = \omega/k$,

$$\frac{\partial k}{\partial t} + \frac{\partial \omega}{\partial x} = 0 \qquad\qquad\qquad (d)$$

which is identical to Eq. (b).

Mathematical formalism

The phase of a harmonic wave travelling in the x-direction can be written (see the main text, Eq. 5.4.1) as $\psi = \omega t - kx$. The frequency and the wave number are then $\omega = \partial \psi / \partial t$ and $k = -\partial \psi / \partial x$, respectively. Taking the space derivative of the former and the time derivative of the latter gives

$$\frac{\partial^2 \psi}{\partial x\, \partial t} = \frac{\partial \omega}{\partial x} = -\frac{\partial k}{\partial t} \qquad\qquad\qquad (e)$$

so

$$\frac{\partial k}{\partial t} + \frac{\partial \omega}{\partial x} = 0 \qquad\qquad\qquad (f)$$

which is also identical to Eq. (b).

Ambient current

In the presence of an ambient current, the absolute propagation speed of the wave is $c + U_n$ (where U_n is the component of the current speed in the wave direction), so, from Eq. (c), we find

$$\frac{\partial k}{\partial t} + \frac{\partial \left[(c + U_n)k\right]}{\partial x} = 0 \qquad\qquad\qquad (g)$$

or, since $\omega = \sigma + kU_n$ and $c = \sigma/k$, so $(c + U_n)k = \omega$, we find

$$\frac{\partial k}{\partial t} + \frac{\partial \omega}{\partial x} = 0 \qquad\qquad\qquad (h)$$

which shows that this relationship also holds for situations with ambient currents.

These results are readily extended to a two-dimensional situation of a harmonic wave propagating in an arbitrary direction. The phase can then be written as $\psi = \omega t - k_x x - k_y y$ and the result is

$$\frac{\partial k_x}{\partial t} + \frac{\partial \omega}{\partial x} = 0$$

and $\qquad\qquad\qquad (i)$

$$\frac{\partial k_y}{\partial t} + \frac{\partial \omega}{\partial y} = 0$$

or

$$\frac{\partial k}{\partial t} + \frac{\partial \omega}{\partial n} = 0 \qquad\qquad\qquad (j)$$

if n is the horizontal co-ordinate, as in the main text of this appendix.

Rotation of wave number

Another convenient property of the (two-dimensional) wave-number field $\vec{k} = (k_x, k_y)$ is that it is irrotational. For two-dimensional situations, this relates the wave-number component in the x-direction k_x to the wave-number component in the y-direction k_y. The simplest derivation is equivalent to the above mathematical formalism. On writing the phase ψ of a propagating harmonic wave in a two-dimensional situation as $\psi = \omega t - k_x x - k_y y$, the wave number in the x-direction is $k_x = -\partial \psi / \partial x$ and the wave number in the y-direction is $k_y = -\partial \psi / \partial y$. It follows then that $-\partial^2 \psi / \partial x \partial y = \partial k_y / \partial x = \partial k_x / \partial y$, so

$$\frac{\partial k_y}{\partial x} - \frac{\partial k_x}{\partial y} = 0 \qquad \text{(k)}$$

which expresses the fact that the wave-number vector field $\vec{k} = (k_x, k_y)$ is irrotational (see Appendix B).

A similar derivation is given by Christoffersen (1982). An alternative derivation can be based on considering the wave rays as characteristics (in a mathematical sense)[1] of the dispersion relationship $\omega = [gk \tanh(kd)]^{1/2} + kU_n$. This derivation is rather formal: writing a harmonic wave component as a wave with phase $\psi = \omega t - k_x x - k_y y$ allows us to define the absolute radian frequency ω and the wave-number vector $\vec{k} = (k_x, k_y)$ as derivatives of the phase function, i.e., $\omega = \partial \psi / \partial t$, $k_x = -\partial \psi / \partial x$ and $k_y = -\partial \psi / \partial y$. Substituting these derivatives into the dispersion relationship makes this relationship a nonlinear, first-order, partial differential equation. The change of frequency per unit time along a characteristic (i.e., a wave ray) of this equation is then given by the same expression as above (Eq. D.11; for a general treatment of such equations, see for instance Webster, 1955, his Eq. 45, Section 24). The directional rate of turning of the wave along the characteristic (the current-induced refraction) too is a universal property of the partial differential equation and its expression is identical to Eq. (D.2).

---

[1] A characteristic is a line (in this case called a wave ray) along which a partial differential equation reduces to an ordinary differential equation.

Appendix E

Shallow-water equations

1 Introduction

In very shallow water, where vertical accelerations in the water can be ignored (the waves are called *long waves*), the wave profile and its propagation can be computed with vertically integrated mass- and momentum-balance equations. These equations can be derived formally from the three-dimensional equations of Section 5.3.2 or they can be derived from basic considerations in which the vertical variations are ignored *a priori* (only vertical averages are considered). The latter approach is shown here.

2 The vertically integrated balance equation (general)

Consider a body of fluid with a free surface and some arbitrary conservative property of that fluid, represented by its density μ (it could be salinity, mass, heat etc.) in a vertical column with horizontal width Δx in the x-direction, unit width in the y-direction and height $D = \eta + d$, where d is the still-water depth and η is the surface elevation above the still-water level (see Fig. E.1).

Deriving the (one-dimensional) balance equation of the property μ in the column is analogous to deriving the three-dimensional balance equations in Section 5.3.2. It is essentially the bookkeeping of the quantity of this property in the column over a time interval Δt:

storage of μ in the column during time interval Δt

$$= \text{net import of } \mu + \text{local production of } \mu \text{ during time interval } \Delta t \qquad (E.1)$$

The first term on the *left*-hand side of this balance equation is equal to the quantity of μ in the column at the *end* of the interval minus the quantity of μ at the *start* of the interval (per unit column width):

$$\text{storage of } \mu \text{ in column} = \left(\mu D \, \Delta x + \frac{\partial(\mu D)}{\partial t} \Delta x \Delta t \right) - \mu D \Delta x$$

$$= \frac{\partial(\mu D)}{\partial t} \Delta x \Delta t \qquad (E.2)$$

The first term on the right-hand side of the balance equation Eq. (E.1) is the net import of the property μ in the x-direction (during the interval Δt). It is equal to the import in the x-direction through the left-hand side of the column minus the export in the x-direction through the right-hand side of the column (per unit width; see Fig. E.1):

$$\text{net import of } \mu \text{ in the } x\text{-direction} = \bar{u}_x \mu D \, \Delta t - \left(\bar{u}_x \mu D + \frac{\partial(\bar{u}_x \mu D)}{\partial x} \Delta x \right) \Delta t$$

$$= -\frac{\partial(\bar{u}_x \mu D)}{\partial x} \Delta x \Delta t \qquad (E.3)$$

where $\bar{u}_x$ is the water velocity, averaged over the total depth D. The second term on the right-hand side of the balance equation Eq. (E.1) is the local production of property μ in

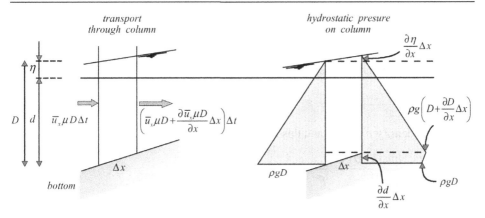

Figure E.1 A property μ being transported by the water in the x-direction through a column, and the hydrostatic pressure on that column.

the column during the time interval Δt (per unit width):

$$local\ production\ of\ \mu = S\,\Delta x\,\Delta t \tag{E.4}$$

where S is the production of μ per unit time, per unit horizontal surface area. Substituting Eqs. (E.2)–(E.4) into Eq. (E.1) gives the balance equation for the property μ for the column over the time interval Δt:

$$\frac{\partial(\mu D)}{\partial t}\Delta x\,\Delta t = -\frac{\partial(\bar{u}_x\mu D)}{\partial x}\Delta x\,\Delta t + S\,\Delta x\,\Delta t \tag{E.5}$$

Dividing by $\Delta x\,\Delta t$ and moving the transport term to the left-hand side gives the (one-dimensional) balance equation for the property μ:

$$\frac{\partial(\mu D)}{\partial t} + \frac{\partial(\bar{u}_x\mu D)}{\partial x} = S \tag{E.6}$$

The first term on the left-hand side is the local rate of change of the quantity of property μ. The term following this is the advective term. It represents the effect of transporting the property μ. Lastly, the term on the right-hand side is called the source term since it represents the generation of the property μ (per unit horizontal surface area, per unit time; if negative, it represents dissipation).

3 The vertically integrated mass-balance equation

If we want to obtain the *mass balance equation*, we take μ as the density of water, $\mu = \rho$, and substitute this into Eq. (E.6):

$$\frac{\partial(\rho D)}{\partial t} + \frac{\partial(\bar{u}_x\rho D)}{\partial x} = S \tag{E.7}$$

If we assume that the water density is constant ($\partial\rho/\partial t = 0$ and $\partial\rho/\partial x = 0$), the bottom is fixed ($\partial d/\partial t = 0$, so that $\partial D/\partial t = \partial(d + \eta)/\partial t = \partial\eta/\partial t$), and there is no production of water ($S = 0$), this equation reduces to the continuity equation:

$$\frac{\partial\eta}{\partial t} + \frac{\partial(\bar{u}_x D)}{\partial x} = 0 \qquad \text{one-dimensional continuity equation} \tag{E.8}$$

4 The vertically integrated momentum-balance equation

If we want to obtain the *momentum balance equation*, we take μ as the momentum density of the water, $\mu = \rho \bar{u}_x$, substitute this into Eq. (E.6), and interpret S as a force in the x-direction F_x, with the result that (see Section 5.3.2 for the relation between force and momentum)

$$\frac{\partial(\rho \bar{u}_x D)}{\partial t} + \frac{\partial \bar{u}_x(\rho \bar{u}_x D)}{\partial x} = F_x \tag{E.9}$$

If the water density ρ is constant, this reduces to

$$\frac{\partial(\bar{u}_x D)}{\partial t} + \frac{\partial \bar{u}_x(\bar{u}_x D)}{\partial x} = \frac{F_x}{\rho} \tag{E.10}$$

and, if we then apply the product rule for differentiation, we find

$$\bar{u}_x \frac{\partial D}{\partial t} + D \frac{\partial(\bar{u}_x)}{\partial t} + \bar{u}_x \frac{\partial(\bar{u}_x D)}{\partial x} + \bar{u}_x D \frac{\partial \bar{u}_x}{\partial x} = \frac{F_x}{\rho} \tag{E.11}$$

Since, for a fixed bottom, $\partial D/\partial t = \partial \eta/\partial t$ (see above), Eq. (E.11) may be written as

$$\bar{u}_x \frac{\partial \eta}{\partial t} + D \frac{\partial \bar{u}_x}{\partial t} + \bar{u}_x \frac{\partial(\bar{u}_x D)}{\partial x} + \bar{u}_x D \frac{\partial \bar{u}_x}{\partial x} = \frac{F_x}{\rho} \tag{E.12}$$

Subtracting the equation of continuity, multiplied by $\bar{u}_x$,

$$\bar{u}_x \frac{\partial \eta}{\partial t} + \bar{u}_x \frac{\partial(\bar{u}_x D)}{\partial x} = 0 \tag{E.13}$$

gives the equation of motion:

$$D \frac{\partial \bar{u}_x}{\partial t} + \bar{u}_x D \frac{\partial \bar{u}_x}{\partial x} = \frac{F_x}{\rho} \tag{E.14}$$

or

$$\frac{\partial \bar{u}_x}{\partial t} + \bar{u}_x \frac{\partial \bar{u}_x}{\partial x} = \frac{F_x}{\rho D} \qquad \text{one-dimensional shallow-water} \tag{E.15}$$
$$\text{equation of motion}$$

The continuity equation Eq. (E.8) and the equation of motion Eq. (E.15) are called the *one-dimensional shallow-water equations*. These equations are readily extended to two-dimensional x, y-space:

$$\frac{\partial \eta}{\partial t} + \frac{\partial(\bar{u}_x D)}{\partial x} + \frac{\partial(\bar{u}_y D)}{\partial y} = 0 \qquad \text{two-dimensional shallow-water}$$
$$\text{continuity equation} \tag{E.16}$$

$$\frac{\partial \bar{u}_x}{\partial t} + \bar{u}_x \frac{\partial \bar{u}_x}{\partial x} + \bar{u}_y \frac{\partial \bar{u}_x}{\partial y} = \frac{F_x}{\rho D} \qquad \text{the two-dimensional shallow-water}$$
$$\text{equation of motion in the } x\text{-direction} \tag{E.17}$$

$$\frac{\partial \bar{u}_y}{\partial t} + \bar{u}_x \frac{\partial \bar{u}_y}{\partial x} + \bar{u}_y \frac{\partial \bar{u}_y}{\partial y} = \frac{F_y}{\rho D} \qquad \text{the two-dimensional shallow-water}$$
$$\text{equation of motion in the } y\text{-direction} \tag{E.18}$$

which are the *two-dimensional shallow-water equations* on which many numerical hydrody-namic models are based. These models are usually driven by hydrostatic pressure gradients but also by other forces, such as wind stress, bottom friction, the Coriolis force (accelera-tion), atmospheric-pressure gradients and gradients of the wave-induced radiation stresses.

5 Wave-induced set-up and set-down

The above equations can be used to estimate the wave-induced set-up and set-down (but not the wave-induced currents) with a relatively simple extension of a conventional *wave* model, without the help of a hydrodynamic model. To that end, the force in Eq. (E.18) should represent the hydrostatic force and the radiation-stress gradients.

Consider first the one-dimensional situation of Fig. E.1. The *net* horizontal force acting on the column is equal to the hydrostatic force (per unit width) on the left-hand side of the column, minus the hydrostatic force on the right-hand side and the horizontal component of the hydrostatic force along the bottom (for a bottom sloping upwards in the x-direction, $\partial d/\partial x$ is negative):

$$F_x = \frac{1}{2}\rho g D^2 - \left[\frac{1}{2}\rho g\left(D + \frac{\partial D}{\partial x}\Delta x\right)^2 - \rho g\left(D + \frac{1}{2}\frac{\partial D}{\partial x}\Delta x\right)\frac{\partial d}{\partial x}\Delta x\right] \tag{E.19}$$

which, after some algebra and ignoring all second-order terms (i.e., terms with $(\Delta x)^2$) and dividing by Δx, gives

$$\frac{F_x}{\rho D} = -g\frac{\partial \eta}{\partial x} \tag{E.20}$$

Substituting this into Eq. (E.15) and adding the wave-induced force (the gradient of the radiation stress) gives (compare with Eq. 7.4.20)

$$\frac{\partial \bar{u}_x}{\partial t} + \bar{u}_x\frac{\partial \bar{u}_x}{\partial x} + g\frac{\partial \bar{\eta}}{\partial x} = -\frac{1}{\rho D}\frac{\partial S_{xx}}{\partial x} \tag{E.21}$$

where $\bar{\eta}$ is now (in the presence of waves) the surface elevation averaged over a duration that is long compared with the wave period but short compared with the time scale of variations in S_{xx}. The corresponding *two-dimensional* equations (including the continuity equation) are

$$\frac{\partial \bar{\eta}}{\partial t} + \frac{\partial (\bar{u}_x D)}{\partial x} + \frac{\partial (\bar{u}_y D)}{\partial y} = 0 \qquad \text{continuity equation} \tag{E.22}$$

$$\frac{\partial \bar{u}_x}{\partial t} + \bar{u}_x\frac{\partial \bar{u}_x}{\partial x} + \bar{u}_y\frac{\partial \bar{u}_x}{\partial y} + g\frac{\partial \bar{\eta}}{\partial x} = -\frac{1}{\rho D}\left(\frac{\partial S_{xx}}{\partial x} + \frac{\partial S_{xy}}{\partial y}\right)$$

equation of motion
in the x-direction $\tag{E.23}$

$$\frac{\partial \bar{u}_y}{\partial t} + \bar{u}_x\frac{\partial \bar{u}_y}{\partial x} + \bar{u}_y\frac{\partial \bar{u}_y}{\partial y} + g\frac{\partial \bar{\eta}}{\partial y} = -\frac{1}{\rho D}\left(\frac{\partial S_{yy}}{\partial y} + \frac{\partial S_{yx}}{\partial x}\right)$$

equation of motion
in the y-direction $\tag{E.24}$

where S_{xx}, S_{yy}, S_{xy} and S_{yx} are the wave-induced radiation stresses (see also Eqs. 7.4.16. and 7.4.17). If we substitute the equations of motion for *stationary* conditions (Eqs. E.23 and E.24 with time removed) into the divergence of the forces $(\text{div}(\vec{F}) = \partial F_x/\partial x +$ $\partial F_y/\partial y)$, and neglect the divergence of all advective acceleration terms (i.e., $\text{div (acceleration)} = \partial \text{acceleration}/\partial x + \partial \text{acceleration}/\partial y \approx 0$), we find (see Eq. 9.4.5)

$$\frac{\partial}{\partial x}\left(\rho g D \frac{\partial \overline{\eta}}{\partial x}\right) + \frac{\partial}{\partial y}\left(\rho g D \frac{\partial \overline{\eta}}{\partial y}\right) = -\frac{\partial}{\partial x}\left(\frac{\partial S_{xx}}{\partial x} + \frac{\partial S_{xy}}{\partial y}\right)$$
$$-\frac{\partial}{\partial y}\left(\frac{\partial S_{yy}}{\partial y} + \frac{\partial S_{yx}}{\partial x}\right) \tag{E.25}$$

This shows that, if we ignore the divergence of the acceleration (i.e., consider only slowly varying current fields), the wave-induced set-up (on the left-hand side of Eq. E.25) is determined by the divergence of the driving force field (i.e., the right-hand side of this equation, which is the divergence of the radiation stress gradients; see Dingemans *et al.*, 1987, and Section 9.4). This divergence of the driving force field is the rotation-free part of the radiation stress gradients.[1] This approximation leads to a numerically entirely different type of model from that normally used for the shallow-water equations (Eqs. E.16–E.18). For instance, the wave-induced set-up can now be computed without considering the wave-induced currents, which in a full hydrodynamic model is not possible. Equation (E.25) is a Poisson equation for which standard computational techniques are available. This allows estimating the wave-induced set-up with a relatively small extension of a numerical wave model with a Poisson-solver (but it is an approximation for slow variations; it is not valid near sharp features in the coastline or obstacles such as headlands or breakwaters). For *one-dimensional* cases (in which all $\partial./\partial y$ are zero), Eq. (E.25) reduces to

$$\frac{d\overline{\eta}}{dx} = -\frac{1}{\rho g D}\frac{dS_{xx}}{dx} \tag{E.26}$$

(which can also be arrived at with a simpler derivation; see Eq. 7.4.20) which can be solved with a simple numerical scheme (e.g., the trapezoidal rule). In SWAN, the above radiation stresses are replaced with their spectral versions (see Section 9.4),

---

[1] A force field $\vec{F}(x, y)$ can always be seen as the sum of one part $\vec{F}_1(x, y)$ with divergence only (i.e., without rotation: $rotation = curl\ (\vec{F}_1) = \partial F_{1,x}/\partial y - \partial F_{1,y}/\partial x = 0$), and another part $\vec{F}_2(x, y)$ with rotation only (i.e., without divergence: $\text{div}(\vec{F}_2) = \partial F_{2,x}/\partial x + \partial F_{2,y}/\partial y = 0$). The part with divergence only is therefore the rotation-free part and vice versa.

References

Aage, C., T. D. Allen, D. J. T., Carter, G. Lindgren and M. Olagnon, 1998, *Oceans from Space*, Plouzané, Édition Ifremer, 162 pp.

Abbott, M. B., M. H. Peterson and O. Skovgaard, 1978, On the numerical modelling of short waves in shallow water, *J. Hydraul. Res.*, IAHR, **16**, 3, 173–204

Abbott, M. B. and D. R. Basco, 1989, *Computational Fluid Dynamics*, New York, John Wiley & Sons, Inc., 425 pp.

Abramowitz, M. and I. A. Stegun (eds.), 1965, *Handbook of Mathematical Functions*, New York, Dover Publications, Inc., 1043 pp.

Abreu, M., A. Larraza and E. Thornton, 1992, Nonlinear transformations of directional wave spectra in shallow water, *J. Geophys. Res.*, **97**, C10, 15 579–15 589

Agnon, Y., A. Sheremet, J. Gonsalves and M. Stiassnie, 1993, Nonlinear evolution of a unidirectional shoaling wave field, *Coastal Engineering*, **20**, 29–58

Agnon, Y. and A. Sheremet, 1997, Stochastic nonlinear shoaling of directional spectra, *J. Fluid Mech.*, **345**, 79–99

Ahn, K., 2000, Statistical distribution of wave heights in finite water depth, *Proc. 27th Int. Conf. Coastal Engineering* (Sydney), Reston, VA, ASCE, pp. 533–544

Airy, G. B., 1845, Tides and waves, *Encyclopaedia Metropolitana*, London, Scientific Department, pp. 241–396

Allender, J., T. Audunson, S. F. Barstow, S. Bjerken, H. E. Krogstad, P. Steinbakke, L. Vartdal, L. E. Borgman and C. Graham, 1989, The WADIC project; a comprehensive field evaluation of directional wave instrumentation, *Ocean Engineering*, **16**, 5/6, 505–536

Allsop, N. W. H. and S. S. L. Hettiarachchi, 1988, Reflections from coastal structures, *Proc. 21st Int. Conf. Coastal Engineering* (Malaga), New York, ASCE, pp. 782–794

Alpers, W. R, D. B. Ross and C. L. Rufenach, 1981, On the detectability of ocean surface waves by real and synthetic aperture radar, *J. Geophys. Res.*, **86**, C7, 6481–6498

Alves, J. H. G. M. and M. L. Banner, 2003, Performance of a saturation-based dissipation-rate source term in modeling the fetch-limited evolution of wind waves, *J. Phys. Oceanogr.*, **33**, 6, 1274–1298

Alves, J. H. G. M., M. L. Banner and I. R. Young, 2003, Revisiting the Pierson–Moskowitz asymptotic limits for fully developed wind waves, *J. Phys. Oceanogr.*, **33**, 6, 1301–1323

Apel, J. R., 1994, An improved model of the ocean surface wave vector spectrum and its effects on radar backscatter, *J. Geophys. Res.*, **99**, C8, 16 269–16 291

Arcilla, A. S. and C. M. Lemos, 1990, *Surf-Zone Hydrodynamics*, Barcelona, Centro Internacional de Métodos Numéricos en Ingenieria, 310 pp.

Ardhuin, F., T. H. C. Herbers and W. C. O'Reilly, 2001, A hybrid Eulerian–Lagrangian model for spectral wave evolution with application to bottom friction on the continental shelf, *J. Phys. Oceanogr.*, **31**, 6, 1498–1516

Ardhuin, F. and T. H. C. Herbers, 2002, Bragg scattering of random surface gravity waves by irregular seabed topography, *J. Fluid Mech.*, **451**, 1–33

Arhan, M. and R. Ezraty, 1978, Statistical relations between successive wave heights, *Oceanol. Acta*, **1**, 151–158

Aris, R., 1962, *Vectors, Tensors, and the Basic Equations of Fluid Mechanics*, Englewood Cliffs, NJ, Prentice-Hall, Inc., 286 pp. (reprinted New York, Dover Publications, Inc., 1989)

Arthur, R. S., W. H. Munk and J. D. Isaacs, 1952, The direct construction of wave rays, *Trans. Am. Geophys. Union.*, **33**, 855–865

Atkins, J. E., 1977, Special reports on freak waves, *The Marine Observer*, January 1977, 32–35

Babanin, A. V., I. R. Young and M. L. Banner, 2001, Breaking probabilities for dominant surface waves on water of finite depth, *J. Geophys. Res.*, **106**, C6, 11 659–11 676

Baldock, T. E., P. Holmes, S. Bunker and P. van Weert, 1998, Cross-shore hydrodynamics within an unsaturated surf zone, *Coastal Engineering*, **34**, 173–196

Banner, M. L. and W. K. Melville, 1976, On the separation of air flow over water waves, *J. Fluid Mech.*, **77**, 4, 825–842

Banner, M. L., I. S. F. Jones and J. C. Trinder, 1989, Wavenumber spectra of short gravity waves, *J. Fluid Mech.*, **198**, 321–344

Banner, M. L., 1990a, Equilibrium spectra of wind-waves, *J. Phys. Oceanogr.*, **20**, 7, 966–984

—1990b, The influence of wave breaking on the surface pressure distribution in wind–wave interactions, *J. Fluid Mech.*, **211**, 463–495

Banner, M. L. and R. H. J. Grimshaw (eds.), 1991, *Breaking Waves*, Berlin, International Union of Theoretical and Applied Mechanics and Springer-Verlag, 387 pp.

Banner, M. L. and I. R. Young, 1994, Modelling spectral dissipation in the evolution of wind waves. Part I: assessment of existing model performance, *J. Phys. Oceanogr.*, **24**, 7, 1550–1571

Banner, M. L., A. V. Babanin and I. R. Young, 2000, Breaking probability for dominant waves on the sea surface, *J. Phys. Oceanogr.*, **30**, 12, 3145–3160

Barber, N. F. and F. Ursell, 1948, The generation and propagation of ocean waves and swell: part I, wave periods and velocities, *Phil. Trans. Roy. Soc. London*, A, **240**, 527–560

Barber, N. F., 1969, *Water Waves*, London, Wykeham Publications, 142 pp.

Barnett, T. P., 1968, On the generation, dissipation and prediction of ocean wind waves, *J. Geophys. Res.*, **73**, 2, 513–529

Barrick, D. E., 1968, Rough surface scattering based on the specular point theory, *IEEE Trans. Antennas Propagat.*, **AP-14**, 4, 449–454

Barstow, S. F., 1996, World Wave Atlas, *AVISO Altimeter Newsletter*, **4**, 24–25

Battjes, J. A., 1968, Refraction of water waves, *J. Waterways and Harbors Div.*, New York, ASCE, WW4, 437–451

—1972a, Long-term wave height distributions at seven stations around the British Isles, *Deutsch. Hydrogr. Z.*, **25**, 4, 179–189

—1972b, Radiation stresses in short-crested waves, *J. Mar. Res.*, **30**, 1, 56–64

—1972c, Set-up due to irregular waves, *Proc. 13th Conf. Coastal Engineering* (Vancouver), New York, ASCE, pp. 1993–2004

—1974a, Computations of set-up, longshore currents, run-up and overtopping due to wind-generated waves, Ph.D. thesis, published as *Communications on Hydraulics*, Delft University of Technology, Dept. of Civil Engineering, Report No. 74–2, 244 pp.

—1974b, Surf similarity, *Proc. 14th Conf. Coastal Engineering* (Copenhagen), New York, ASCE, pp. 466–480

Battjes, J. A. and J. P. F. M. Janssen, 1978, Energy loss and set-up due to breaking of random waves, *Proc. 16th Conf. Coastal Engineering* (Hamburg), New York, ASCE, pp. 569–587

Battjes, J. A., 1984, A review of methods to establish the wave climate for breakwater design, *Coastal Engineering*, **8**, 141–160

Battjes, J. A. and G.Ph. van Vledder, 1984, Verification of Kimura's theory for wave group statistics, *Proc. 19th Int. Conf. Coastal Engineering* (Houston), New York, ASCE, pp. 642–648

Battjes, J. A. and M. J. F. Stive, 1985, Calibration and verification of a dissipation model for random breaking waves, *J. Geophys. Res.*, **90**, C5, 9159–9167

Battjes, J. A., T. J. Zitman and L. H. Holthuijsen, 1987, A reanalysis of the spectra observed in JONSWAP, *J. Phys. Oceanogr.*, **17**, 8, 1288–1295

Battjes. J. A., 1988, Surf-zone dynamics, *Ann. Rev. Fluid Mech.*, **20**, 257–293

Battjes, J. A. and S. Beji, 1992, Breaking waves propagating over a shoal, *Proc. 23rd Int. Conf. Coastal Engineering* (Venice), New York, ASCE, pp. 42–50

Battjes, J. A., Y. Eldeberky and Y. Won, 1993, Spectral Boussinesq modelling of breaking waves, *Proc. 2nd Int. Symp. on Ocean Wave Measurement and Analysis WAVES 93* (New Orleans), New York, ASCE, pp. 813–820

Battjes, J. A., 1994, Shallow water wave modelling, *Proc. Int. Symp.: Waves – Physical and Numerical Modeling*, eds. M. Isaacson and M. Quick, Vancouver, University of British Columbia, **I**, pp. 1–23

Battjes, J. A. and H. W. Groenendijk, 2000, Wave height distributions on shallow foreshores, *Coastal Engineering*, **40**, 3, 161–182

Bauer, E. and C. Staabs, 1998, Statistical properties of global significant wave heights and their use for validation, *J. Geophys. Res.*, **103**, C1, 1153–1166

Becq, F., M. Benoit and Ph. Forget, 1998, Numerical simulations of directionally spread shoaling surface gravity waves, *Proc. 26th Int. Conf. Coastal Engineering* (Copenhagen), Reston, VA, ASCE, pp. 523–536

Becq-Girard, F., P. Forget and M. Benoit, 1999, Nonlinear propagation of unidirectional wave fields over varying topography, *Coastal Engineering*, **38**, 2, 91–113

Beji, S. and J. A. Battjes, 1993, Experimental investigation of wave propagation over a bar, *Coastal Engineering*, **19**, 1–2, 151–162

Beji, S. and K. Nadaoka, 1997, A time-dependent nonlinear mild-slope equation for water waves, *Proc. Roy. Soc. London* A, **453**, 319–332

Belcher, S. E., J. A. Harris and R. L. Street, 1994, Linear dynamics of wind waves in coupled turbulent air–water flow. Part 1. Theory, *J. Fluid Mech.*, **271**, 119–151

Bendat, J. S. and A. G. Piersol, 1971, *Random Data: Analysis and Measurement Procedures*, New York, Wiley-Interscience, 407 pp.

Bender, L. C., 1996, Modification of the physics and numerics in a third-generation ocean wave model, *J. Atm. Ocean. Technol.*, **13**, 3, 726–750

Benoit, M., F. Marcos and F. Becq, 1996, Development of a third generation shallow-water wave model with unstructured spatial meshing, *Proc. 25th Int. Conf. Coastal Engineering* (Orlando), New York, ASCE, pp. 465–478

Benoit, M., P. Frigaard and H. A. Schaffer, 1997, Analyzing multidirectional wave spectra: a tentative classification of available methods, *Proc. Seminar on Multidirectional Waves and their Interaction with Structures*, San Francisco, CA, Baltimore, The Johns Hopkins University Press, IAHR, pp. 131–158

Berkhoff, J. C. W., 1972, Computations of combined refraction–diffraction, *Proc. 13th Conf. Coastal Engineering* (Vancouver), New York, ASCE, pp. 471–490

Bertotti, L and L. Cavaleri, 1985, Coastal set-up and wave breaking, *Oceanol. Acta*, **8**, 2, 237–242

Bertotti, L and L. Cavaleri, 1994, Accuracy of wind and wave evaluation in coastal regions, *Proc. 24ᵗʰ Int. Conf. Coastal Engineering* (Kobe), New York, ASCE, pp. 57–67

Bidlot, J.-R., B. Hansen and P. A. E. M. Janssen, 1996, Wave modelling and operational forecasting at ECMWF, *Proc. 1ˢᵗ Int. Conf. EuroGOOS*, Amsterdam, Elsevier, pp. 206–213

Bishop, C. T. and M. A. Donelan, 1987, Measuring waves with pressure transducers, *Coastal Engineering*, **11**, 309–328

Blackman, R. B. and J. W. Tukey, 1958, *The Measurement of Power Spectra*, New York, Dover Publications Inc., 190 pp.

Booij, N., 1981, Gravity waves on water with non-uniform depth and current, Ph.D. Thesis, published as *Communications on Hydraulics*, Delft University of Technology, Department of Civil Engineering, Report No. 81–1, 130 pp.

Booij. N. and L. H. Holthuijsen, 1987, Propagation of ocean waves in discrete spectral wave models, *J. Comput. Phys.*, **68**, 2, 307–326

Booij, N., L. H. Holthuijsen, N. Doorn and A. T. M. M. Kieftenburg, 1997, Diffraction in a spectral wave model, *Proc. 3rd Int. Symp. on Ocean Wave Measurement and Analysis WAVES 97* (Virginia Beach), Reston, VA, ASCE, pp. 243–255

Booij, N., R. C. Ris and L. H. Holthuijsen, 1999, A third-generation wave model for coastal regions, Part I, Model description and validation, *J. Geophys. Res.*, **104**, C4, 7649–7666

Booij, N., L. H. Holthuijsen and IJ. Haagsma, 2001, The effect of swell on the generation and dissipation of waves, *Proc. 4ᵗʰ Int. Symp. on Ocean Wave Measurements and Analysis WAVES 2001* (San Francisco), Reston, VA, ASCE, pp. 501–506

Borge, J. C. N., K. Reichert and J. Dittmer, 1999, Use of nautical radar as a wave monitoring instrument, *Coastal Engineering*, **37**, 331–342

Borge, J. C. N., S. Lehner, A. Niedermeier and J. Shulz-Stellenfleth, 2004, Detection of ocean wave groupiness from spaceborne synthetic aperture radar, *J. Geophys. Res.*, **109**, C07005, doi:10.1029/2004JC00298, 18 p.

Borgman, L. E., 1973, Probabilities for highest wave in hurricane, *J. Waterways, Harbors and Coastal Engineering Div.* ASCE, **99**, WW2, 185–207

Bortkovskii, R. S., 1983, *Air–Sea Exchange of Heat and Moisture during Storms*, Dordrecht, D. Reidel Publishing Company, 247 pp.

Boussinesq, J., 1872, Théorie des ondes et des remous qui se propagent le long d'un canal rectangulaire horizontal, en communiquant au liquide contenu dans ce canal des vitesses sensiblement pareilles de la surface au fond, *J. Math. Pures Appl., Série 2*, **17**, 55–108

Bouws, E, 1978, *Wind and Wave Climate in the Netherlands Sector of the North Sea between 53° and 54° North Latitude*, De Bilt, Koninklijk Nederlands Meteorologisch Instituut, Scientific Report W.R. 78–9

Bouws, E. and J. A. Battjes, 1982, A Monte-Carlo approach to the computation of refraction of water waves, *J. Geophys. Res.*, **87**, C8, 5718–5722

Bouws, E. and G. J. Komen, 1983, On the balance between growth and dissipation in an extreme, depth-limited wind-sea in the southern North Sea, *J. Phys. Oceanogr.*, **13**, 9, 1653–1658

Bouws, E., H. Günther, W. Rosenthal and C. L. Vincent, 1985, Similarity of the wind wave spectrum in finite depth water. 1. Spectral form, *J. Geophys. Res.*, **90**, C1, 975–986

Bowen, A. J., D. L. Inman and V. P. Simmons, 1968, Wave "set-down" and set-up, *J. Geophys. Res.*, **73**, 8, 2569–2577

Bretherton, F. P. and C. J. R. Garrett, 1969, Wave trains in inhomogeneous moving media, *Proc. Roy. Soc. London*, A, **302**, 529–554

Bretschneider, C. L., 1952, The generation and decay of wind waves in deep water, *Trans. Am. Geophys. Union*, **33**, 3, 381–389

—1958, Revisions in wave forecasting: deep and shallow water, *Proc. 6ᵗʰ Conf. Coastal Engineering* (Gainsville, Palm Beach and Miami Beach, Florida), Richmond, CA, Council on Wave Research, University of California, 30–67

—1959, Hurricane design-wave practices, *Trans. ASCE*, **124**, 39–62

Breugem, W. A. and L. H. Holthuijsen, 2006, Generalised wave growth from Lake George, *J. Waterway, Port, Coastal, and Ocean Engineering*, Reston, VA, ASCE (in press).

Brevik, I. and B. Aas, 1980, Flume experiment on waves and currents. I. Rippled bed, *Coastal Engineering*, **3**, 149–177

Briggs, M. J. and P. L.-F. Liu, 1993, Experimental study of monochromatic wave–ebb current interaction, *Proc. 2ⁿᵈ Int. Symp. on Ocean Wave Measurement and Analysis WAVES 93* (New Orleans), New York, ASCE, pp. 474–488

Briggs, M. J., E. F. Thompson and C. L. Vincent, 1995, Wave diffraction around breakwater, *J. Waterway, Port, Coastal, and Ocean Engineering*, New York, ASCE, **121**, 1, 23–35

Brink-Kjær, O., 1984, Depth–current refraction of wave spectra, *Proc. Symp. Description and Modelling of Directional Seas*, paper No. C-7, Lyngby, Technical University of Denmark, 12pp.

Broeze, J., E. F. G. van Daalen and P. J. Zandbergen, 1993, A three-dimensional panel method for nonlinear free surface waves on vector computers, *Comput. Mech.*, **13**, 1/2, 12–28

Buccino, M. and M. Calabrese, 2002, Wave heights distribution in the surf zone: analysis of experimental data, *Proc. 28ᵗʰ Int. Conf. Coastal Engineering* (Cardiff), Reston, VA, ASCE, pp. 209–221

Buckley, W. H., 1983, A study of extreme waves and their effects on ship structures, U.S. Coast Guard Report No. SR-1281, Ship Structure Committee Report No. SSC-320, U.S. National Technical Information Service, VA 22161, 82 pp.

Buckley, W. H., R. D. Pierce, J. B. Peters and M. J. Davis, 1984, Use of the half-cycle analysis method to compare measured wave height and simulated Gaussian data having the same variance spectrum, *Ocean Engineering*, **11**, 423–445

Burgers, G. and V. K. Makin, 1993, Boundary-layer model results for wind-sea growth, *J. Phys. Oceanogr.*, **23**, 2, 372–385

Cai, M., D. R. Basco and J. Baumer, 1992, Bar/trough effects on wave height probability distributions and energy losses in surf zones, *Proc. 23ʳᵈ Int. Conf. Coastal Engineering* (Venice), New York, ASCE, pp. 103–115

Caires, S. and A. Sterl, 2003, On the estimation of return values of significant wave height data from the reanalysis of the European Centre for Medium-range Weather Forecasts, in *Safety and Reliability*, eds. T. Bedford and P. H. A. J. M. van Gelder, Lisse Swets and Zeitlinger, pp. 353–361

—2005, 100-year return value estimates for ocean wind speeds and significant wave height from the ERA-40 data, *J. Climate*, **18**, 1032–1048

Cardone, V. J., W. J. Pierson and E. G. Ward, 1976, Hindcasting the directional spectra of hurricane-generated waves, *J. Petroleum Technol.*, **28**, 385–394

Cartwright, D. E. and M. S. Longuet-Higgins, 1956, The statistical distribution of the maxima of a random function, *Proc. Roy. Soc. London*, A, **237**, 212–232

Cartwright, D. E., 1958, On estimating the mean energy of sea waves from the highest wave in a record, *Proc. Roy. Soc. London*, A, **247**, 22–48

Castillo, E., 1988, *Extreme Value Theory in Engineering*, Boston, MA, Academic Press, Inc., 389 pp.

Cavaleri, L., 1979, Resistance wave staff, accuracy of the measurements, *L'energia elettrica*, **6**, 299–306

Cavaleri, L. and P. Malanotte-Rizzoli, 1981, Wind wave prediction in shallow water: theory and application, *J. Geophys. Res.*, **86**, C11, 10 961–10 973

Cavaleri, L., 1984, The CNR meteo-oceanographic spar buoy, *Deep-Sea Res.*, **31**, 4, 427–437

Cavaleri, L., L. Bertotti and P. Lionello, 1989, Shallow water application of the third-generation WAM wave model, *J. Geophys. Res.*, **94**, C6, 8111–8124

Cavaleri, L. and P. Lionello, 1990, Linear and nonlinear approaches to bottom friction in wave motion: a critical intercomparison, *Estuarine, Coastal and Shelf Sci.*, **30**, 355–367

Cavaleri, L., 1994, Wind variability, in *Dynamics and Modelling of Ocean Waves*, eds. G. J. Komen, L. Cavaleri, M. Donelan, K. Hasselmann, S. Hasselmann and P. A. E. M. Janssen, Cambridge University Press, pp. 320–331

—2000, The oceanographic tower *Acqua Alta* – activity and prediction of sea states at Venice, *Coastal Engineering*, **39**, 29–70

Cavanié, M., M. Arhan and R. Ezraty, 1976, A statistical relationship between individual heights and periods of storm waves, *Proc. Behaviour of Offshore Structures*, Trondheim, The Norwegian Institute of Technology, pp. 354–363

CEM, 2002, Coastal Engineering Manual, U.S. Army Corps of Engineers, http://users. coastal.ufl.edu/~sheppard/eoc6430/Coastal_Engineering_Manual.htm

Chalikov, D. V., 1986, Numerical simulation of the boundary layer above waves, *Boundary-Layer Meteorol.*, **34**, 63–98

Chappelear, J. E., 1962, Shallow water waves, *J. Geophys. Res.*, **67**, 12, 4693–4704

Charnock, H., 1955, Wind stress on a water surface, *Q. J. Roy. Meteorol. Soc.*, **81**, 639

Chawla, A. and J. T. Kirby, 2002, Monochromatic and random waves breaking at blocking points, *J. Geophys. Res.*, **107**, C7, doi:10.1029/2001JC001042, 19 p.

Chen, Y. and H. Wang, 1983, Numerical model for nonstationary shallow water wave spectral transformations, *J. Geophys. Res.*, **88**, C14, 9851–9863

Chen, Y., R. T. Guza and S. Elgar, 1997, Modelling spectra of breaking surface waves in shallow water, *J. Geophys. Res.*, **102**, C11, 25 035–25 046

Christoffersen, J. B., 1982, *Current Depth Refraction of Dissipative Water Waves*, Lyngby, Institute of Hydrodynamics and Hydraulic Engineering, 177 pp.

Cokelet, E. D., 1977, Steep gravity waves in water of arbitrary uniform depth, *Phil. Trans. Roy. Soc. London*, A, **286**, 1335, 183–230

Coles, S., 2001, *An Introduction to Statistical Modeling of Extreme Values*, London, Springer-Verlag, 208 pp.

Collins, J. I., 1970, Probabilities of breaking wave characteristics, *Proc. 13th Conf. Coastal Engineering* (Washington), New York, ASCE, pp. 399–414

Collins, J. I., 1972, Prediction of shallow-water spectra, *J. Geophys. Res.*, **77**, 15, 2693–2707

COST, 2005, *Measuring and Analysing the Directional Spectra of Ocean Waves*, eds. D. Hauser, K. Kahma, H. E. Krogstad, S. Lehner, J. A. J. Monbaliu and L. R. Wyatt, Luxembourg, Office for Official Publications of the European Communities, 465 pp.

Cote, L. J., J. O. Davis, W. Marks, R. J. McGough, E. Mehr, W. J. Pierson, J. F. Ropek, G. Stephenson and R. C. Vetter, 1960, The directional spectrum of wind generated sea as determined from data obtained by the Stereo Wave Observation Project, in *Meteoro-logical Papers*, **2**, 6, New York, New York University, College of Engineering, 88 pp.

Cramer, H., 1946, *Mathematical Methods of Statistics*, Princeton, NJ, Princeton University Press, 575 pp.

Crapper, G. D., 1984, *Introduction to Water Waves*, Chichester, Ellis Horwood Ltd., 224 pp.

Dacunha, N. M. C., N. Hogben and K. S. Andrews, 1984, Ocean wave statistics: a new look, *Proc. Oceanology International Conference*, Brighton, Society of Underwater Technology, OI 2.15/1–13

Dally, W. R., R. G. Dean and R. A. Dalrymple, 1984, A model for breaker decay on beaches, *Proc. 19th Int. Conf. Coastal Engineering* (Houston), New York, ASCE, pp. 82–98

Dally, W. R., R. G. Dean and R. A. Dalrymple, 1985, Wave height variation across beaches of arbitrary profile, *J. Geophys. Res.*, **90**, C6, 11 917–11 927

Dalrymple, R. A. and J. T. Kirby, 1988, Models for very wide-angle water waves and wave diffraction, *J. Fluid Mech.*, **192**, 33–50

Dalrymple, R. A., K. D. Suh, J. T. Kirby and J. W. Chae, 1989, Models for very wide-angle water waves and wave diffraction. Part 2. Irregular bathymetry, *J. Fluid Mech.*, **201**, 299–322

Davis, R. E. and L. A. Regier, 1977, Methods for estimating directional wave spectra from multi-element arrays, *J. Mar. Res.*, **35**, 3, 453–477

Deacon, G. E. R., 1949, Recent studies of waves and swell, in *Ocean Surface Waves, Annals of the New York Academy of Sciences*, **51**, 3, 475–482

Dean, R. G., 1965, Stream function representation of nonlinear ocean waves, *J. Geophys. Res.*, **70**, 18, 4561–4572

—1974, *Evaluation and Development of Water Wave Theories for Engineering Application*, Technical Report No. 4, I + II, Ft. Belvoir, VA, U.S. Army Corps of Engineers, Coastal Engineering Research Center, 133 + 534 pp.

Dean, R. G. and R. A. Dalrymple, 1998, *Water Wave Mechanics for Engineers and Scientists*, Singapore, World Scientific, **2**, 353 pp.

de Vries, J. J., J. Waldron and V. Cunningham, 2003, Field tests of the new Datawell DWR-G GPS wave buoy, *Sea Technol.*, **44**, 12, 50–55

Ding, L. and D. M. Farmer, 1994, Observations of breaking surface wave statistics, *J. Phys. Oceanogr.*, **24**, 6, 1368–1387

Dingemans, M. W., A. C. Radder and H. J. de Vriend, 1987, Computation of the driving forces of wave-induced currents, *Coastal Engineering*, **11**, 539–563

Dingemans, M. W., 1997a, *Water Wave Propagation over Uneven Bottoms, Part 1 – Linear Wave Propagation*, Singapore, World Scientific, **13**, pp. 1–471

—1997b, *Water Wave Propagation over Uneven Bottoms, Part 2 – Nonlinear Wave Propagation*, Singapore, World Scientific, **13**, pp. 473–967

Doering, J. C. and A. J. Bowen, 1986, Shoaling surface gravity waves: a bispectral analysis, *Proc. 20th Int. Conf. Coastal Engineering* (Taipei), New York, ASCE, pp. 150–162

Doering, J. C. and A. J. Bowen, 1995, Parameterization of orbital velocity asymmetries of shoaling and breaking waves using bispectral analysis, *Coastal Engineering*, **26**, 1–2, 15–33

Donelan, M. A. and Y. Yuan, 1984, Wave dissipation by surface processes, in *Dynamics and Modelling of Ocean Waves*, eds. G. J. Komen, L. Cavaleri, M. Donelan, K. Hasselmann, S. Hasselmann and P. A. E. M. Janssen, Cambridge, Cambridge University Press, pp. 143–155

Donelan, M. A., J. Hamilton and W. H. Hui, 1985, Directional spectra of wind-generated waves, *Phil. Trans. Roy. Soc. London*, A, **315**, 509–562

Donelan, M. A. and W. H. Hui, 1990, Mechanics of ocean surface waves, in *Surface Waves and Fluxes*, *I*, eds. G. L. Geernaert and W. J. Plant, Dordrecht, Kluwer Academic Publishers, pp. 209–246

Dorrestein, R., 1960, Simplified method of determining refraction coefficients for sea waves, *J. Geophys. Res.*, **65**, 2, 637–642

Douglass, S. L. and J. R. Weggel, 1988, Laboratory experiments on the influence of wind on nearshore wave breaking, *Proc. 21st Int. Conf. Coastal Engineering* (Malaga), New York, ASCE, pp. 632–643

Douglass, S. L., 1990, Influence of wind on breaking waves, *J. Waterway, Port, Coastal, and Ocean Engineering*, New York, ASCE, **116**, 6, 651–663

Draper, L., 1965, "Freak" ocean waves, *Mar. Obs.*, **35**, 193–195

Earle, M. D., 1975, Extreme wave conditions during hurricane Camille, *J. Geophys. Res.*, **80**, 3, 377–379

Earle, M. D. and A. Malahoff (eds.), 1979, *Ocean Wave Climate*, *Proc. Ocean Wave Climate Symp.*, Herndon, Va., *1977*, New York, Plenum Press, 368 pp.

Ebersole, B. A., 1985, Refraction–diffraction model for linear water waves, *J. Waterway, Port, Coastal, and Ocean Engineering*, New York, ASCE, **111**, 6, 939–953

Eckart, C., 1952, The propagation of gravity waves from deep to shallow water, *Proc. NBS Semicentennial Symp. on Gravity Waves*, 1951, Washington, National Bureau of Standards, Circular 521, pp. 165–173

Edgeworth, F. Y., 1908, The law of error, *Trans. Camb. Phil. Soc.*, **20**, 36–65

Eldeberky, Y. and J. A. Battjes, 1995, Parameterization of triad interactions in wave energy models, *Proc. Coastal Dynamics '95* (Gdańsk), New York, ASCE, pp. 140–148

Eldeberky, Y. and J. A. Battjes, 1996, Spectral modelling of wave breaking: application to Boussinesq equations, *J. Geophys. Res.*, **101**, C1, 1253–1264

Eldeberky, Y., 1996, Nonlinear transformation of wave spectra in the nearshore zone, Ph.D. thesis, published as *Communications on Hydraulic and Geotechnical Engineering*, Delft University of Technology, Faculty of Civil Engineering, Report No. 96–4, 203 pp.

Eldeberky, Y. and P. A. Madsen, 1998, Deterministic and stochastic evolution equations for fully dispersive and weakly nonlinear waves, *Coastal Engineering*, **38**, 1, 1–25

Elfouhaily, T., B. Chapron and K. Katsaros, 1997, A unified directional spectrum for long and short wind-driven waves, *J. Geophys. Res.*, **102**, C7, 15 781–15 796

Elgar, S., R. T. Guza and R. J. Seymour, 1984, Groups of waves in shallow water, *J. Geophys. Res.*, **89**, C3, 3623–3634

Elgar, S. and R. T. Guza, 1985a, Shoaling gravity waves: comparisons between field observations, linear theory, and a nonlinear model, *J. Fluid Mech.*, **158**, 47–70

—1985b, Observations of bispectra of shoaling surface gravity waves, *J. Fluid Mech.*, **161**, 425–448

—1986, Nonlinear model predictions of bispectra of shoaling surface gravity waves, *J. Fluid Mech.*, **167**, 1–18

Elgar, S., M. H. Freilich and R. T. Guza, 1990, Model–data comparisons of moments of nonbreaking shoaling surface gravity waves, *J. Geophys. Res.*, **95**, C9, 16 055–16 063

Elgar, S., R. T. Guza and M. Freilich, 1993, Observations of nonlinear interactions in directionally spread shoaling surface gravity waves, *J. Geophys. Res.*, **98**, 20 299–20 305

Elgar, S., T. H. C. Herbers, V. Chandran and R. T. Guza, 1995, Higher-order spectral analysis of nonlinear ocean surface gravity waves, *J. Geophys. Res.*, **100**, C3, 4977–4983

Elgar, S., R. T. Guza, B. Raubenheimer, T. H. C. Herbers and E. L. Gallagher, 1997, Spectral evolution of shoaling and breaking waves on a barred beach, *J. Geophys. Res.*, **102**, C7, 15 797–15 805

Elliott, J. A., 1972a, Microscale pressure fluctuations measured within the lower atmospheric boundary layer, *J. Fluid Mech.*, **53**, 2, 351–384

—1972b, Microscale pressure fluctuations near waves being generated by wind, *J. Fluid Mech.*, **54**, 3, 427–448

Elwany, M. H. S., W. C. O'Reilly, R. T. Guza and R. E. Flick, 1995, Effects of Southern California kelp beds on waves, *J. Waterway, Port, Coastal, and Ocean Engineering*, New York, ASCE, **121**, 2, 143–150

Ewans, K. C., 1998, Observations of the directional spectrum of fetch-limited waves, *J. Phys. Oceanogr.*, **28**, 3, 495–512

Ewing, J. A., 1971, A numerical wave prediction method for the North Atlantic Ocean, *Deutsch. Hydrogr. Z.*, **24**, 6, 241–261

—1973, Mean length of runs of high waves, *J. Geophys. Res.*, **78**, 1933–1936

Ewing, J. A., T. J. Weare and B. A. Worthington, 1979, A hindcast study of extreme wave conditions in the North Sea, *J. Geophys. Res.*, **84**, C9, 5739–5747

Ewing, J. A. and A. K. Laing, 1987, Directional spectra of seas near full development, *J. Phys. Oceanogr.*, **17**, 10, 1696–1706

Ewing, J. A. and R. C. Hague, 1993, A second-generation wave model for coastal wave prediction, *Proc. 2nd Int. Symp. Ocean Wave Measurement and Analysis WAVES 93* (New Orleans), New York, ASCE, pp. 576–589

Farge, M., 1992, Wavelet transforms and their application to turbulence, *Annu. Rev. Fluid Mech.*, **24**, 395–457

Fenton, J. D., 1985, A fifth-order Stokes theory for steady waves, *J. Waterway, Port, Coastal, and Ocean Engineering*, New York, ASCE, **111**, 2, 216–234

—1988, The numerical solution of steady water wave problems, *Computers Geosci.*, Pergamon Press, **14**, 3, 357–368

—1990, Nonlinear wave theories, in *The Sea*, eds. B. Le Méhauté and D. Hanes, New York, John Wiley & Sons Inc., **9A**, pp. 3–25

Fenton, J. D. and W. D. McKee, 1990, On calculating the lengths of water waves, *Coastal Engineering*, **14**, 499–513

Fenton, J. D., 1999, Numerical methods for nonlinear waves, *Advances in Coastal and Ocean Engineering*, ed. P. L.-F. Liu, Singapore, World Scientific, **5**, pp. 241–324

Ferreira, J. A. and C. G. Soares, 1998, An application of the peaks over threshold method to predict extremes of significant wave height, *J. Offshore Mechanics and Arctic Engineering*, ASME, **120**, 165–176

Ferziger, J. H. and M. Perić, 2002, *Computational Methods for Fluid Dynamics*, 3rd edn., Berlin, Springer-Verlag, 423 pp.

Fisher, F. H. and F. N. Spiess, 1963, Flip-Floating Instrument Platform, *J. Acoust. Soc. Am.*, **35**, 10, 1633–1644

Forristall, G. Z., 1978, On the statistical distribution of wave heights in a storm, *J. Geophys. Res.*, **83**, C5, 2353–2358

—1981, Measurements of a saturated range in ocean wave spectra, *J. Geophys. Res.*, **86**, C9, 8075–8084

—1984, The distribution of measured and simulated wave heights as a function of spectral shape, *J. Geophys. Res.*, **89**, C6, 10 547–10 552

Forristall, G. Z. and K. C. Ewans, 1998, Worldwide measurements of directional wave spreading, *J. Atm. Ocean Technol.*, **15**, 2, 440–469

Forristall, G. Z., 2000, Wave crest distributions: observations and second-order theory, *J. Phys. Oceanogr.*, **30**, 1931–1943

Foufoula-Georgiou, E. and P. Kumar (eds.), 1994, *Wavelets in Geophysics*, San Diego, CA, Academic Press, 372 pp.

Freilich, M. H. and R. T. Guza, 1984, Nonlinear effects on shoaling surface gravity waves, *Phil. Trans. Roy. Soc. London*, A, **311**, 1–41

Freilich, M. H., R. T. Guza and S. L. Elgar, 1990, Observations of nonlinear effects in directional spectra of shoaling gravity waves, *J. Geophys. Res.*, **95**, C6, 9645–9656

Galvin, C. J., 1968, Breaker type classification on three laboratory beaches, *J. Geophys. Res.*, **73**, 12, 3651–3659

—1972, Wave breaking in shallow water, in *Waves on Beaches and Resulting Sediment Transport*, ed. R. E. Meyer, New York, Academic Press, pp. 413–456

Garrett, C. J. R., 1967, The adiabatic invariant for wave propagation in a nonuniform moving medium, *Proc. Roy. Soc. London*, A, **299**, 26–27

Geernaert, G. L. and W. J. Plant, 1990, *Surface Waves and Fluxes, Vol. I – Current Theory*, Dordrecht, Kluwer Academic Publishers, 337 pp.

Gelci, C. J. and E. Devillaz, 1970, Le calcul numérique de l'état de la mer, *La houille blanche*, **25**, 2, 117–136

Gelci, R., J. Cazale and J. Vassal, 1956, Utilisation des diagrammes de propagation à la provision énergétique de la houle, *Bulletin d'Information du Comité Central d'Océanographie et d'Études des Côtes*, **8**, 4, 169–197

Gelci, R., E. Devillaz and P. Chavy, 1964, Évolution de l'état de la mer, calcul numérique des advections, *Notes de l'Établissement d'Études et de Recherches Météorologiques*, **166**, 14 pp.

Georges, T. M. and J. A. Harlan, 1994, New horizons for over-the-horizon radar?, *IEEE Antennas and Propagation Magazine*, **36**, 4, 14–24

Gerson, M, 1975, The techniques and uses of probability plotting, *The Statistician, J. Roy. Statist. Soc.*, Series D, **24**, 4, 235–257

Goda, Y., H. Takeda and Y. Moriya, 1967, *Laboratory Investigation on Wave Transmission over Breakwaters*, Report of the Port and Harbour Research Institute, No. 13, pp. 1–38

Goda, Y., 1975, Irregular wave deformation in the surf zone, *Coastal Engineering in Japan*, **18**, 13–26

—1978, The observed joint probability distribution of periods and heights of sea waves, *Proc. 16th Conf. Coastal Engineering* (Hamburg), New York, ASCE, pp. 227–246

Goda, Y., T. Takayama and Y. Suzuki, 1978, Diffraction diagrams for directional random waves, *Proc. 16th Conf. Coastal Engineering* (Hamburg), New York ASCE, pp. 628–650

Goda, Y., 1986, Effect of wave tilting on zero-crossing wave heights and periods, *Coastal Engineering in Japan*, **29**, 79–90

—1988a, Statistical variability of sea state parameters as a function of a wave spectrum, *Coastal Engineering in Japan*, **31**, 1, 39–52

—1988b, On the methodology of selecting design wave height, *Proc. 21st Int. Conf. Coastal Engineering* (Malaga), New York, ASCE, pp. 899–913

Goda, Y. and K. Kobune, 1990, Distribution function fitting for storm wave data, *Proc. 22nd Int. Conf. Coastal Engineering* (Delft), New York, ASCE, pp. 18–31

Goda, Y., 1992, Uncertainty of design parameters from viewpoint of extreme statistics, *J. Offshore Mechanics and Arctic Engineering*, ASME, **114**, 76–82

Goda, Y., M. P. Hawkes, E. Mansard, M. J. Martin, M. Mathiesen, E. Peltier, E. Thompson and G. van Vledder, 1993, Intercomparison of extremal wave analysis methods using numerically simulated data, *Proc. 2nd Int. Symp. Ocean Wave Measurement and Analysis WAVES 93* (New Orleans), New York, ASCE, pp. 963–977

Goda, Y., 1997, Directional wave spectrum and its engineering applications, in *Advances in Coastal and Ocean Engineering*, ed. P. L.-F. Liu, Singapore, World Scientific, **3**, pp. 67–102

Goda, Y. and K. Morinobu, 1998, Breaking wave heights on horizontal bed affected by approach slope, *Coastal Engineering J.*, **40**, 4, 307–326

Goda, Y., 2000, *Random Seas and Design of Maritime Structures*, Singapore, World Scientific, 443 pp.

Godden, J., 1977, Are episodic waves responsible for ship disappearances?, *Shipping*, 9(37)–11(39)

Golding, B. W., 1983, A wave prediction system for real-time sea state forecasting, *Q. J. Roy. Meteorol. Soc.*, **109**, 393–416

Golub, G. H. and C. F. van Loan, 1986, *Matrix Computations*, London, North Oxford Academic, 476 pp.

González, F. I., 1984, A case-study of wave-current–bathymetry interactions at the Columbia River entrance, *J. Phys. Oceanogr.*, **14**, 6, 1065–1078

Gorshkov, S. G., 1986, *World Ocean Atlas*, Vol. **1**: Pacific Ocean; Vol. **2**: Atlantic and Indian Oceans; Vol. **3**: Arctic Ocean, Oxford, Pergamon Press

Gourlay, M. R., 1992, Wave set-up, wave run-up and beach water table: interaction between surf zone hydraulics and groundwater hydraulics, *Coastal Engineering*, **17**, 93–144

Graber, H. C. and O. S. Madsen, 1988, A finite-depth wind-wave model. Part I: model description, *J. Phys. Oceanogr.*, **18**, 11, 1465–1483

Graham, C., 1982, The parameterisation and prediction of wave height and wind speed persistence statistics for oil industry operational planning purposes, *Coastal Engineering*, **6**, 303–329

Grant, W. D. and O. S. Madsen, 1982, Movable bed roughness in unsteady oscillatory flow, *J. Geophys. Res.*, **87**, C1, 469–481

Gringorten, I. I., 1963, A plotting rule for extreme probability paper, *J. Geophys. Res.*, **68**, 3, 813–814

Gumbel, E. J., 1958, *Statistics of Extremes*, New York, Columbia University Press, 375 pp.

Gumbley, J., 1977, Holes in the sea swallow up ships, *Freighting World*, 16, March

Günther, H., W. Rosenthal, T. J. Weare, B. A. Worthington, K. Hasselmann and J. A. Ewing, 1979, A hybrid parametrical wave prediction model, *J. Geophys. Res.*, **84**, C9, 5727–5738

Günther, H., S. Hasselmann and P. A. E. M. Janssen, 1992, *The WAM Model Cycle 4 (Revised Version)*, Technical Report No. 4, Hamburg, Deutsches Klimatisches Rechenzentrum

Günther, H. and W. Rosenthal, 2002, Singular waves, propagation and prognosis, *Proc. 7th Int. Workshop on Wave Hindcasting and Forecasting* (Banff, Alberta), ed. V. R. Swail, published on CD only, Toronto, Ontario, Canada, Environment Canada, 8 pp.

Gutshabash, Ye. Sh. and I. V. Lavrenov, 1986, Swell transformation in the Cape Agulhas Current, *Izv. Acad. Sci. USSR, Atmospheric and Oceanic Phys.*, **22**, 6, 494–497

Haine, R. A., 1980, Second generation shipborne wave recorder, *Transducer Technol.*, **2**, 25–28

Hansen, C., K. B. Katsaros, S. A. Kitaigorodskii and S. E. Larsen, 1990, The dissipation range of wind-wave spectra observed on a lake, *J. Phys. Oceanogr.*, **20**, 9, 1264–1277

Hardy, T. A. and I. R. Young, 1996, Field study of wave attenuation on an offshore coral reef, *J. Geophys. Res.*, **101**, C6, 14 311–14 326

Haring, R. E., A. R. Osborne and L. P. Spencer, 1976, Extreme wave parameters based on continental shelf storm wave records, *Proc. 15th Conf. Coastal Engineering* (Honolulu), New York, ASCE, pp. 151–170

Harlow, F. H. and J. E. Welch, 1965, Numerical calculation of time-dependent viscous incompressible flow of fluid with free surface, *Phys. Fluids*, **8**, 12, 2182–2189

Hashimoto, N., M. Mitsui, Y. Goda, T. Nakai and T. Takahashi, 1996, Improvement of submerged Doppler-type directional wave meter and its application to field observation, *Proc. 25th Int. Conf. Coastal Engineering* (Orlando), New York, ASCE, pp. 629–642

Hashimoto, N., 1997, Analysis of the directional wave spectrum from field data, *Advances in Coastal and Ocean Engineering*, ed. P. L.-F. Liu, Singapore, World Scientific, **3**, pp. 103–143

Hasselmann, D. E., M. Dunckel and J. A. Ewing, 1980, Directional wave spectra observed during JONSWAP 1973, *J. Phys. Oceanogr.*, **10**, 8, 1264–1280

Hasselmann, K., 1960, Grundgleichungen der Seegangsvoraussage, *Schiffstechnik*, **1**, 191–195

—1962, On the nonlinear energy transfer in a gravity-wave spectrum. Part 1. General theory, *J. Fluid Mech.*, **12**, 481–500

—1963a, On the nonlinear energy transfer in a gravity-wave spectrum. Part 2. Conservation theorems; wave–particle analogy; irreversibility, *J. Fluid Mech.*, **15**, 273–281

—1963b, On the nonlinear energy transfer in a gravity-wave spectrum. Part 3. Evaluation of the energy flux and swell-sea interaction for a Neumann spectrum, *J. Fluid Mech.*, **15**, 385–398

Hasselmann, K., W. Munk and G. MacDonald, 1963, Bispectra of ocean waves, in *Time Series Analysis*, ed. M. Rosenblatt, New York, John Wiley and Sons, pp. 125–139

Hasselmann, K. and J. I. Collins, 1968, Spectral dissipation of finite-depth gravity waves due to turbulent bottom friction, *J. Mar. Res.*, **26**, 1, 1–12

Hasselmann, K., 1968, Weak-interaction theory of ocean waves, *Basic Developments in Fluid Dynamics*, ed. M. Holt, New York, Academic Press, **2**, pp. 117–182

Hasselmann, K., T. P. Barnett, E. Bouws, H. Carlson, D. E. Cartwright, K. Enke, J. A. Ewing, H. Gienapp, D. E. Hasselmann, P. Kruseman, A. Meerburg, P. Müller, D. J. Olbers, K. Richter, W. Sell and H. Walden, 1973, Measurements of wind-wave growth and swell decay during the Joint North Sea Wave Project (JONSWAP), *Deutsch. Hydrogr. Z.*, Suppl., **A8**, 12, 95 pp.

Hasselmann, K., 1974, On the spectral dissipation of ocean waves due to white capping, *Boundary-Layer Meteorol.*, **6**, 1–2, 107–127

Hasselmann, K., D. B. Ross, P. Müller and W. Sell, 1976, A parametric wave prediction model, *J. Phys. Oceanogr.*, **6**, 200–228

Hasselmann, S. and K. Hasselmann, 1981, A symmetrical method of computing the nonlinear transfer in a gravity-wave spectrum, *Hamburger Geophys. Einzelschr.*, A, **52**, 138 pp.

Hasselmann, S., K. Hasselmann, J. H. Allender and T. P. Barnett, 1985a, Computations and parameterizations of the nonlinear energy transfer in a gravity wave spectrum. Part II: parameterizations of the nonlinear transfer for application in wave models, *J. Phys. Oceanogr.*, **15**, 11, 1378–1391

Hasselmann, K., R. K. Raney, W. J. Plant, W. Alpers, R. A. Shuchman, D. R. Lyzenga, C. L. Rufenach and M. J. Tucker, 1985b, Theory of synthetic aperture radar imaging: a MARSEN view, *J. Geophys. Res.*, **90**, C3, 4659–4686

Hasselmann, K. and S. Hasselmann, 1991, On the nonlinear mapping of an ocean wave spectrum into a synthetic aperture radar image spectrum and its inversion, *J. Geophys. Res.*, **96**, C6, 10 713–10 729

Hasselmann, S., C. Brüning, K. Hasselmann and P. Heimbach, 1996, An improved algorithm for the retrieval of ocean wave spectra from synthetic aperture radar image spectra, *J. Geophys. Res.*, **101**, C7, 16 615–16 629

Haver, S. and O. J. Andersen, 2000, Freak waves: rare realisations of a typical population or typical realisations of a rare population?, *Proc. 10th Offshore and Polar Engineering Conf.*, ISOPE, **III**, pp. 123–130

Heathershaw, A. D., M. W. L. Blackley and P. J. Hardcastle, 1980, Wave direction estimates in coastal waters using radar, *Coastal Engineering*, **3**, 249–267

Herbers, T. H. C. and R. T. Guza, 1991, Wind-wave nonlinearity observed at the sea floor, Part I: forced-wave energy, *J. Phys. Oceanogr.*, **21**, 12, 1740–1761

Herbers, T. H. C. and M. C. Burton, 1997, Nonlinear shoaling of directionally spread waves on a beach, *J. Geophys. Res.*, **102**, C9, 21 101–21 114

Herbers, T. H. C., S. Elgar and R. T. Guza, 1999, Directional spreading of waves in the nearshore, *J. Geophys. Res.*, **104**, C4, 7683–7693

Herbers, T. H. C., E. J. Hendrickson and W. C. O'Reilly, 2000a, Propagation of swell across a wide continental shelf, *J. Geophys. Res.*, **105**, C8, 19 729–19 737

Herbers, T. H. C., N. R. Russnogle and S. Elgar, 2000b, Spectral energy balance of breaking waves within the surf zone, *J. Phys. Oceanogr.*, **30**, 11, 2723–2737

Herbers, T. H. C., M. Orzech, S. Elgar and R. T. Guza, 2003, Shoaling transformation of frequency-directional spectra, *J. Geophys. Res.*, **108**, C1, 3013, doi:10.1029/2001JC001304, 17 p.

Herbich, J. B. (ed.), 1990, *Handbook of Coastal and Ocean Engineering, 1, Wave Phenomena and Coastal Structures*, Houston, TX, Gulf Publishing Company, 1155 pp.

Herman, R., 1992, Solitary waves, *Am. Scientist*, **80**, 350–361

Herterich, K. and K. Hasselmann, 1980, A similarity relationship for the nonlinear transfer in a finite-depth gravity-wave spectrum, *J. Fluid Mech.*, **97**, 1, 215–224

Hessner, K., K. Reichert, J. Dittmer, J. C. Nieto Borge and H. Günther, 2001, Evaluation of WAMOS II wave data, *Proc. 4th Int. Symp. Ocean Wave Measurement and Analysis WAVES 2001* (San Francisco), Reston, VA, ASCE, pp. 221–230

Hirt, C. W. and B. D. Nichols, 1981, Volume of fluid (VOF) method for the dynamics of free boundaries, *J. Comput. Phys.*, **39**, 201–225

Hogben, N. and F. E. Lumb, 1967, *Ocean Wave Statistics*, London, Her Majesty's Stationery Office

Hogben, N., N. M. Dacunha and G. F. Olliver, 1985, *Global Wave Statistics*, London, British Maritime Technology Ltd and Unwin Brothers

Hogben, N., 1988, Experience from compilation of Global Wave Statistics, *Ocean Engineering*, **15**, 1–31

—1990a, Overview of global wave statistics, *Encyclopedia of Fluid Mechanics*, Houston, TX, Gulf Publishing Company, **10**, pp. 391–427

—1990b, Long term wave statistics, in *The Sea*, eds. B. Le Méhauté and D. Hanes, New York, John Wiley & Sons Inc., **9A**, pp. 293–333

Holthuijsen, L. H., 1983a, Stereophotography of ocean waves, *Appl. Ocean Res.*, **5**, 4, 204–209

—1983b, Observations of the directional distribution of ocean wave energy in fetch-limited conditions, *J. Phys. Oceanogr.*, **13**, 2, 192–207

Holthuijsen, L. H. and T. H. C. Herbers, 1986, Statistics of breaking waves observed as whitecaps in the open sea, *J. Phys. Oceanogr.*, **16**, 2, 290–297

Holthuijsen, L. H. and S. de Boer, 1988, Wave forecasting for moving and stationary targets, *Proc. Computer Modelling in Ocean Engineering* (Venice), eds. B. A. Schrefler and O. C. Zienkiewicz, Rotterdam, Balkema, pp. 231–234

Holthuijsen, L. H., N. Booij and T. H. C. Herbers, 1989, A prediction model for stationary, short-crested waves in shallow water with ambient currents, *Coastal Engineering*, **13**, 23–54

Holthuijsen, L. H. and H. L. Tolman, 1991, Effects of the Gulf Stream on ocean waves, *J. Geophys. Res.*, **96**, C7, 12 755–12 771

Holthuijsen, L. H. and N. Booij, 1994, Bottom induced scintillation of long- and short-crested waves, *Proc. Int. Symp.: Waves – Physical and Numerical Modelling*, eds. M. Isaacson and M. Quick, Vancouver, University of British Columbia, **II**, pp. 604–613

Holthuijsen L. H., A. Herman and N. Booij, 2003, Phase-decoupled refraction–diffraction for spectral wave models, *Coastal Engineering*, **49**, 4, 291–305

Houmb, O. G. and H. Rye, 1973, Analyses of wave data from the Norwegian continental shelf, *Proc. 2nd Int. on Conf. Port and Ocean Engineering under Arctic Conditions*, Reykjavík, University of Iceland, Department of Engineering and Science, pp. 780–788

Hsiao, S. V. and O. H. Shemdin, 1978, Bottom dissipation in finite-depth water waves, *Proc. 16th Conf. Coastal Engineering* (Hamburg), New York, ASCE, pp. 434–448

—1983, Measurements of wind velocity and pressure with a wave follower during MARSEN, *J. Geophys. Res.*, **88**, C14, 9841–9849

Hsu, T.-W., S.-H. Ou and J.-M. Liau, 2005, Hindcasting nearshore wind waves using a FEM code for SWAN, *Coastal Engineering*, **52**, 177–195

Huang, N. E., S. R. Long, C.-C. Tung, Y. Yuen and L. F. Bliven, 1981, A unified two-parameter wave spectral model for a general sea state, *J. Fluid Mech.*, **112**, 203–224

Huang, N. E., C.-C. Tung and S. R. Long, 1990a, Wave spectra, in *The Sea*, eds. B. Le Méhauté and D. Hanes, New York, John Wiley & Sons Inc., in **9A**, pp. 197–237

—1990b, The probability structure of the ocean surface, in *The Sea*, eds. B. Le Méhauté and D. Hanes, New York, John Wiley & Sons Inc., **9A**, pp. 335–366

Hughes, S. A. and L. E. Borgman, 1987, Beta-Rayleigh distribution for shallow water waves, *Proc. Coastal Hydrodynamics* (Newark), New York, ASCE, pp. 17–31

Hunt, J. N., 1979, Direct solution of wave dispersion equation, *J. Waterways, Port, Coastal and Ocean Div.*, New York, ASCE, **105**, WW4, 457–459

Hurdle, D. P. and R. J. H. Stive, 1989, Revision of SPM 1984 wave hindcast model to avoid inconsistencies in engineering applications, *Coastal Engineering*, **12**, 339–351

Hwang, P. A., E. J. Walsh, W. B. Krabill, R. N. Swift, S. S. Manizade, J. F. Scott and M. D. Earle, 1998, Airborne remote sensing applications to coastal wave research, *J. Geophys. Res.*, **103**, C9, 18 791–18 800

Hwang, P. A., D. W. Wang, E. J. Walsh, W. B. Krabill and R. N. Swift, 2000a, Airborne measurements of the wave number spectra of ocean surface waves. Part I: Spectral slope and dimensionless spectral coefficient, *J. Phys. Oceanogr.*, **30**, 11, 2753–2767

—2000b, Airborne measurement of the wavenumber spectra of ocean surface waves. Part II: Directional distribution, *J. Phys. Oceanogr.*, **30**, 11 2768–2787

Hwang, P. A. and D. W. Wang, 2001, Directional distributions and mean square slopes in the equilibrium and saturation ranges of the wave spectrum, *J. Phys. Oceanogr.*, **31**, 5, 1346–1360

Iribarren, R. and C. Nogales, 1949, Protection des ports, XVIIth Int. *Naval Congress*, Lisbon, Section II, Communication, **4**, pp. 31–80

Isobe, M., 1985, Calculation and application of first-order cnoidal wave theory, *Coastal Engineering*, **9**, 309–325

Jackson, F. C., W. T. Walton and C. Y. Peng, 1985, A comparison of *in situ* and airborne radar observations of ocean wave directionality, *J. Geophys. Res.*, **90**, C1, 1005–1018

James, I. D., 1986, A note on the theoretical comparison of wave staffs and Waverider buoys in steep gravity waves, *Ocean Engineering*, **13**, 2, 209–214

Janssen, P. A. E. M. and G. J. Komen, 1985, Effect of atmospheric stability on the growth of surface gravity waves, *Boundary-Layer Meteorol.*, **32**, 85–96

Janssen, P. A. E. M., 1991a, Quasi-linear theory of wind-wave generation applied to wave forecasting, *J. Phys. Oceanogr.*, **21**, 11, 1631–1642

—1991b, Consequences of the effect of surface gravity waves on the mean air flow, *Breaking Waves*, IUTAM Symposium, Sydney, eds. M. L. Banner and R. H. J. Grimshaw, Berlin, Springer-Verlag, pp. 193–206

Janssen, P. A. E. M. and P. Viterbo, 1996, Ocean waves and the atmospheric climate, *J. Climate*, **9**, 6, 1269–1287

Janssen, P. A. E. M., 2003, Nonlinear four-wave interactions and freak waves, *J. Phys. Oceanogr.*, **33**, 4, 863–884

Janssen, P., 2004, *The Interaction of Ocean Waves and Wind*, Cambridge, Cambridge University Press, 300 pp.

Janssen, T. T., T. H. C. Herbers and J. A. Battjes, 2004, A discrete spectral evolution model for nonlinear waves over 2D topography, *Proc. 29th Int. Conf. Coastal Engineering* (Lisbon), Singapore, World Scientific, pp. 119–131

Jeans, G., C. Primrose, N. Descusse, B. Strong and P. van Weert, 2003a, A comparison between directional wave measurements from the RDI Workhorse with waves and the Datawell directional Waverider, *Proc. IEEE/OES 7th Working Conf. on Current Measurement Technology*, pp. 148–151

Jeans, G., I. Bellamy, J. J. de Vries and P. van Weert, 2003b, Sea trial of the new Datawell GPS directional Waverider, *Proc. IEEE/OES 7th Working Conf. on Current Measurement Technology*, pp. 145–147

Jeffreys, H., 1925, On the formation of waves by wind, *Proc. Roy. Soc. London*, A, **107**, 189–206

—1926, On the formation of waves by wind (second paper), *Proc. Roy. Soc. London*, A, **110**, 241–247

Jenkins, G. M. and D. G. Watts, 1968, *Spectral Analysis*, San Francisco, Holden-Day, CA, 525 pp.

Johnson, J. W., 1947, The refraction of surface waves by currents, *Trans. Am. Geophys. Union*, **28**, 6, 867–874

Johnson, R. R., E. P. D. Mansard and J. Ploeg, 1978, Effects of wave grouping on break-water stability, *Proc. 16th Conf. Coastal Engineering* (Hamburg), New York, ASCE, pp. 2228–2243

Jones, I. S. F. and Y. Toba (eds.), 2001, *Wind Stress over The Ocean*, Cambridge, Cambridge University Press, 307 pp.

Jonsson, I. G., 1966, Wave boundary layers and friction factors, *Proc. 10th Conf. Coastal Engineering* (Tokyo), New York, ASCE, pp. 127–148

Jonsson, I. G. and N. A. Carlsen, 1976, Experimental and theoretical investigations in an oscillatory turbulent boundary layer, *J. Hydraul. Res.*, IAHR, **14**, 1, 45–60

Jonsson, I. G., 1980, A new approach to oscillatory rough turbulent boundary layers, *Ocean Engineering*, **7**, 1, 109–152

Jonsson, I. G. and J. D. Wang, 1980, Current-depth refraction of water waves, *Ocean Engineering*, **7**, 1, 153–171

Jonsson, I. G., 1990, Wave–current interaction, in *The Sea*, eds. B. Le Méhauté and D. Hanes, New York, John Wiley & Sons Inc., **9A**, pp. 65–120

Kahma, K. K., 1981, A study of the growth of the wave spectrum with fetch, *J. Phys. Oceanogr.*, **11**, 11, 1503–1515

Kahma, K. K. and C. J. Calkoen, 1992, Reconciling discrepancies in the observed growth of wind-generated waves, *J. Phys. Oceanogr.*, **22**, 12, 1389–1405

Kahma, K. K. and H. Pettersson, 1994, Wave growth in a narrow fetch geometry, *The Global Atmosphere and Ocean System*, **2**, 253–263

Kaihatu, J. M. and J. T. Kirby, 1995, Nonlinear transformation of waves in finite depth water, *Phys. Fluids*, **7**, 8, 1903–1914

Kaminsky, G. M. and N. C. Kraus, 1993, Evaluation of depth-limited wave breaking criteria, *Proc. 2nd Int. Symp. Ocean Wave Measurement and Analysis WAVES 93* (New Orleans), New York, ASCE, pp. 180–193

Kamphuis, J. W., 2000, *Introduction to Coastal Engineering and Management*, Singapore, World Scientific, **16**, 437 pp.

Karlsson, T., 1969, Refraction of continuous ocean wave spectra, *J. Waterways and Harbors Div.*, New York, ASCE, **95**, WW4, 437–448

Katsaros, K. B. and S. S. Atatürk, 1991, Dependence of wave-breaking statistics on wind stress and wind wave development, *Breaking Waves IUTAM Symp.* (Sydney), eds. M. L. Banner and R. H. J. Grimshaw, Berlin, Springer-Verlag, pp. 119–132

Kawai, S., K. Okada and Y. Toba, 1977, Field data support of three-seconds power law and $gu_*\sigma^{-4}$-spectral form for growing wind waves, *J. Oceanogr. Soc. Japan*, **33**, 3, 137–150

Kawai, S., 1982, Structure of air flow separation over wind wave crests, *Boundary-Layer Meteorol.*, **23**, 503–521

Kenyon, K. E., 1971, Wave refraction in ocean currents, *Deep-Sea Res.*, **18**, 1023–1034

Khandekar, M. L., 1989, *Operational Analysis and Prediction of Ocean Wind Waves*, New York, Springer-Verlag, 214 pp.

Kim, Y. C. and E. J. Powers, 1979, Digital bispectral analysis and its application to nonlinear wave interactions, *IEEE Trans. Plasma Sci.*, **PS-7**, 2, 120–131

Kim, Y. C., J. M. Beal, E. J. Powers and R. W. Miksad, 1980, Bispectrum and nonlinear wave coupling, *Phys. Fluids*, **23**, 2, 258–263

Kimura, A., 1980, Statistical properties of random wave groups, *Proc. 17th Conf. Coastal Engineering* (Sydney), New York, ASCE, pp. 2955–2973

—1981, Joint distribution of the wave heights and periods of random sea waves, *Coastal Engineering in Japan*, **24**, 77–92

Kinsman, B., 1965, *Wind Waves, Their Generation and Propagation on The Ocean Surface*, Englewood Cliffs, NJ, Prentice-Hall, 676 pp.

Kirby, J. T., 1984, A note on linear surface wave–current interaction over slowly varying topography, *J. Geophys. Res.*, **89**, C1, 745–747

—1986, Higher-order approximations in the parabolic equation method for water waves, *J. Geophys. Res.*, **91**, C1, 933–952

—1990, Modelling shoaling directional wave spectra in shallow water, *Proc. 22nd Int. Conf. Coastal Engineering* (Delft), New York, ASCE, pp. 109–122

Kirby, J. T. and J. M. Kaihatu, 1996, Structure of frequency domain models for random waves breaking, *Proc. 25th Int. Conf. Coastal Engineering* (Orlando), New York, ASCE, pp. 1144–1155

Kitaigorodskii, S. A., V. P. Krasitskii and M. M. Zaslavskii, 1975, On Phillips' theory of equilibrium range in the spectra of wind-generated gravity waves, *J. Phys. Oceanogr.*, **5**, 7, 410–420

Kitaigorodskii, S. A., 1983, On the theory of the equilibrium range in the spectrum of wind-generated gravity waves, *J. Phys. Oceanogr.*, **13**, 5, 816–827

Kitano, T., H. Mase and W. Kioka, 2001, Theory of significant wave period based on spectral integrals, *Proc. 4th Int. Symp. Ocean Wave Measurement and Analysis WAVES 2001* (San Francisco), Reston, VA, ASCE, pp. 414–423

Klopman, G. and M. J. F. Stive, 1989, Extreme waves and wave loading in shallow water, *Proc. Workshop "Wave and Current Kinematics and Loading"* (Paris), London, The Oil Industry International Exploration & Production Forum, Report No. 3.12/156, pp. 161–170

Kobayashi, T., A. Kawai, M. Koduka and T. Yasuda, 2001, Application of nautical radar to the field observation of waves and currents, *Proc. 4th Int. Symp. Ocean Wave Measurement and Analysis WAVES 2001* (San Francisco), Reston, VA, ASCE, pp. 76–85

Kofoed-Hansen, H. and J. H. Rasmussen, 1998, Modelling of nonlinear shoaling based on stochastic evolution equations, *Coastal Engineering*, **33**, 203–232

Komatsu, K. and A. Masuda, 1996, A new scheme of nonlinear energy transfer among wind waves: RIAM method – algorithm and performance, *J. Oceanogr.*, **52**, 4, 509–537

Komen, G. J., S. Hasselmann and K. Hasselmann, 1984, On the existence of a fully developed wind-sea spectrum, *J. Phys. Oceanogr.*, **14**, 8, 1271–1285

Komen, G. J., L. Cavaleri, M. Donelan, K. Hasselmann, S. Hasselmann and P. A. E. M. Janssen, 1994, *Dynamics and Modelling of Ocean Waves*, Cambridge, Cambridge University Press, 532 pp.

Korteweg, D. J. and G. de Vries, 1895, On the change of form of long waves advancing in a rectangular canal, and a new type of long stationary waves, *Phil. Mag.*, Series 5, **39**, 422–443

Krogstad, H. E., S. F. Barstow, O. Haug, P. Ø. Markussen, G. Ueland and I. Rodriguez, 1997, SMART-800: a GPS based directional wave buoy, *Proc. 3rd Int. Symp. Ocean Wave Measurement and Analysis WAVES 97* (Virginia Beach), Reston, VA, ASCE, pp. 1182–1195

Krogstad, H. E., S. F. Barstow, S. E. Aasen and I. Rodriguez, 1999, Some recent developments in wave buoy measurement technology, *Coastal Engineering*, **37**, 309–329

Kuik, A. J., G. Ph. van Vledder and L. H. Holthuijsen, 1988, A method for the routine analysis of pitch-and-roll buoy wave data, *J. Phys. Oceanogr.*, **18**, 7, 1020–1034

Kuriyama, Y., 1994, Numerical model for longshore current distribution on a bar-trough beach, *Proc. 24th Int. Conf. Coastal Engineering* (Kobe), New York, ASCE, pp. 2237–2251

Kuwashima, S. and N. Hogben, 1986, The estimation of wave height and wind speed persistence statistics from cumulative probability distributions, *Coastal Engineering*, **9**, 563–590

Lai, R. J., S. R. Long and N. E. Huang, 1989, Laboratory studies of wave–current interaction: kinematics of the strong interactions, *J. Geophys. Res.*, **94**, C11, 16 201–16 214

Laitone, E. V., 1960, The second approximation to cnoidal and solitary waves, *J. Fluid Mech.*, **9**, 430–444

Lacombe, H., 1951, The diffraction of a swell. A practical approximate solution and its justification, *Proc. NBS Semicentennial Symp. on Gravity Waves*, Washington, National Bureau of Standards, Circular 521, pp. 129–140

—1965, *Cours d'océanographie physique*, Paris, Gauthiers-Villars, 392 pp.

Lamb, H., 1932, *Hydrodynamics*, 6th edn, New York, Dover Publications, Inc., 738 pp.

Larson, T. R. and J. W. Wright, 1975, Wind-generated gravity–capillary waves: laboratory measurements of temporal growth rates using micro-wave backscatter, *J. Fluid Mech.*, **70**, 3, 417–436

Larson, M., 1995, Model for decay of random waves in surf zone, *J. Waterway, Port, Coastal, and Ocean Engineering*, New York, ASCE, **121**, 1, 1–12

Lavrenov, I. V., 2003, *Wind-waves in Oceans. Dynamics and Numerical Simulation*, Berlin, Springer-Verlag, 376 pp.

LeBlond, P. H. and L. A. Mysak, 1978, *Waves in The Ocean*, Amsterdam, Elsevier Scientific Publishing Company, 602 pp.

Leadbetter, M. R., G. Lindgren and H. Rootzén, 1983, *Extremes and Related Properties of Random Sequences and Processes*, New York, Springer-Verlag, 336 pp.

Lehner, S., J. Schulz-Stellenfleth, A. Niedermeier, J. Horstmann and W. Rosenthal, 2001, Extreme waves observed by synthetic aperture radar, *Proc. 4th Int. Symp. Ocean Wave Measurement and Analysis WAVES 2001* (San Francisco), Reston, VA, ASCE, pp. 125–134

LeMéhauté, B., 1962, On non-saturated breakers and the wave run-up, *Proc. 8th Conf. Coastal Engineering* (Mexico City), Council on Wave Research, the Engineering Foundation, Berkeley, University of California, pp. 77–92

LeMéhauté, B. and R. C. Y. Koh, 1967, On the breaking of waves arriving at an angle to the shore, *J. Hydraul. Res.*, IAHR, **5**, 1, 67–88

LeMéhauté, B., 1976, *An Introduction to Hydrodynamics and Water Waves*, New York, Springer-Verlag, 315 pp.

LeMéhauté, B. and J. D. Wang, 1982, Wave spectrum changes on sloped beach, *J. Waterway, Port, Coastal and Ocean Div.*, New York, ASCE, **108**, 1, 33–47

Lewis, A.W., and R. N. Allos, 1990, JONSWAP's parameters: sorting out the inconsistencies, *Ocean Engineering*, **17**, 4, 409–415

Li, C. W. and M. Mao, 1992, Spectral modelling of typhoon-generated waves in shallow waters, *J. Hydraul. Res.*, IAHR, **30**, 5, 611–622

Lighthill, J., 1978, *Waves in Fluids*, Cambridge, Cambridge University Press, 504 pp.

Lin, P. and P. L.-F. Liu, 1999, Free surface tracking methods and their applications to wave hydrodynamics, *Advances in Coastal and Ocean Engineering*, ed. P. L.-F. Liu, Singapore, World Scientific, **5**, pp. 213–240

Liu, P. C., 1994, Wavelet spectrum analysis and ocean wind waves, in *Wavelets in Geophysics*, eds. E. Foufoula-Georgiou and P. Kumar, San Diego, CA, Academic Press, pp. 151–166

Liu, P. C. and N. Mori. 2000, Characterizing freak waves with wavelet transform analysis, in *Rogue Waves 2000*, eds. M. Olagnon and G. A. Athanassoulis, Plouzané, IFREMER, pp. 151–155

Liu, P. C. and A. V. Babanin, 2004, Using wavelet spectrum analysis to resolve breaking events in the wind wave time series, *Annales Geophysicae*, **22**, 3335–3345

Liu, P. L.-F., S. B. Yoon and J. T. Kirby, 1985, Nonlinear refraction–diffraction of waves in shallow water, *J. Fluid Mech.*, **153**, 185–201

Liu, P. L.-F., 1990, Wave transformation, in *The Sea*, eds. B. LeMéhaute and D. Hanes, New York, John Wiley & Sons Inc., **9A**, pp. 27–63

—2001, Numerical modelling of breaking waves in nearshore environment, *Proc. 4th Int. Symp. Ocean Wave Measurement and Analysis WAVES 2001* (San Francisco), Reston, VA, ASCE, pp. 1–12

Longuet-Higgins, M. S., 1952, On the statistical distributions of the heights of sea waves, *J. Mar. Res.*, **11**, 3, 245–265

—1957, The statistical analysis of a random, moving surface, *Phil. Trans. Roy. Soc. London*, A, **250**, 321–387

Longuet-Higgins, M. S. and R. W. Stewart, 1960, Changes in the form of short gravity waves on long waves and tidal currents, *J. Fluid Mech.*, **8**, 4, 565–583

—1961, The changes in amplitude of short gravity waves on steady non-uniform currents, *J. Fluid Mech.*, **10**, 3, 529–549

—1962, Radiation stress and mass transport in gravity waves, with applications to "surf beats", *J. Fluid Mech.*, **13**, 481–504

—1963, A note on wave set-up, *J. Mar. Res.*, **21**, 4–10

Longuet-Higgins, M. S., 1963, The effect of nonlinearities on statistical distribution in the theory of sea waves, *J. Fluid Mech.*, **17**, 459–480

Longuet-Higgins, M. S., D. E. Cartwright and N. D. Smith, 1963, Observations of the directional spectrum of sea waves using the motions of a floating buoy, in *Ocean Wave Spectra*, New York, Prentice Hall, pp. 111–136

Longuet-Higgins, M. S. and R. W. Stewart, 1964, Radiation stresses in water waves; physical discussion, with applications, *Deep-Sea Res.*, **11**, 4, 529–562

Longuet-Higgins, M. S., 1969, On wave breaking and the equilibrium spectrum of wind-generated waves, *Proc. Roy. Soc. London* A, **310**, 151–159

—1973, The mechanics of the surfzone, *Proc. 13ᵗʰ Int. Congress of Theoretical and Applied Mechanics*, eds. E. Becker and G. K. Mikhailov, Moscow, Springer-Verlag, pp. 213–228

—1975, On the joint distribution of the periods and amplitudes of sea waves, *J. Geophys. Res.*, **80**, 18, 2688–2693

Longuet-Higgins, M. S. and E. D. Cokelet, 1976, The deformation of steep surface waves on water, I. A numerical method of computation, *Proc. Roy. Soc. London*, A, **350**, 1–26

Longuet-Higgins, M. S., 1976, On the nonlinear transfer of energy in the peak of a gravity-wave spectrum: a simplified model, *Proc. Roy. Soc. London*, A, **347**, 311–328

—1980, On the distribution of the heights of sea waves: some effects of nonlinearity and finite band-width, *J. Geophys. Res.*, **85**, C3, 1519–1523

—1983, On the joint distribution of wave periods and amplitudes in a random wave field, *Proc. Roy. Soc. London*, A, **389**, 241–258

—1984, Statistical properties of wave groups in a random sea, *Phil. Trans. Roy. Soc. London*, A, **312**, 219–250

—1987, A stochastic model of sea-surface roughness, I. Wave crests, *Proc. Roy. Soc. London*, A, **410**, 19–34

Luo, W. and J. Monbaliu, 1994, Effects of the bottom friction formulation on the energy balance for gravity waves in shallow water, *J. Geophys. Res.*, **99**, C9, 18501–18511

Lygre, A. and H. E. Krogstad, 1986, Maximum entropy estimation of the directional distribution in ocean wave spectra, *J. Phys. Oceanogr.*, **16**, 12, 2052–2060

Maa, J. P.-Y., T.-W. Hsu and D.-Y. Lee, 2002, The RIDE model: an enhanced computer program for wave transformation, *Ocean Engineering*, **29**, 11, 1441–1458

MacLaren Plansearch Ltd., 1991, *Wind and Wave Climate Atlas of Canada*, Halifax, MacLaren Plansearch Ltd.

Madsen, O. S., Y.-K. Poon and H. C. Graber, 1988, Spectral wave attenuation by bottom friction: theory, *Proc. 21ˢᵗ Int. Conf. Coastal Engineering* (Malaga), New York, ASCE, pp. 492–504

Madsen, P. A. and O. R. Sørensen, 1992, A new form of the Boussinesq equations with improved linear dispersion characteristics. Part 2: a slowly-varying bathymetry, *Coastal Engineering*, **18**, 3–4, 183–205

—1993, Bound waves and triad interactions in shallow water, *Ocean Engineering*, **20**, 4, 359–388

Madsen, P. A. and H. A. Schäffer, 1999, A review of Boussinesq-type equations for surface gravity waves, in *Advances in Coastal and Ocean Engineering*, ed. P. L.-F. Liu, Singapore, World Scientific, **5**, pp. 1–94

Mallat, S., 1998, *A Wavelet Tour of Signal Processing*, San Diego, CA, San Diego Academic Press, 577 pp.

Mardia, K. V., 1972, Statistics of directional data, in *Probability and Mathematical Statistics*, eds. Z. W. Birnbaum and E. Lukacs, London, Academic Press, 357 pp.

Mase, H. and Y. Iwagaki, 1982, Wave height distributions and wave grouping in surf zone, *Proc. 18ᵗʰ Conf. Coastal Engineering* (Cape Town), New York, ASCE, pp. 58–76

Mase, H. and J. T. Kirby, 1992, Hybrid frequency-domain KdV equation for random wave transformation, *Proc. 23ʳᵈ Int. Conf. Coastal Engineering* (Venice), New York, ASCE, pp. 474–487

Mase, H., T. Takayama and T. Kitano, 1997, Transformation of double peak spectral waves, *Proc. 3ʳᵈ Int. Symp. Ocean Wave Measurement and Analysis WAVES 97* (Virginia Beach), Reston, VA, ASCE, pp. 232–242

Massel, S. R., 1996, *Ocean Surface Waves: Their Physics and Prediction*, Advanced Series on Ocean Engineering, Singapore, World Scientific, **11**, 491 pp.

Mastenbroek, C., G. Burgers and P. A. E. M. Janssen, 1993, The dynamical coupling of a wave model and a storm surge model through the atmospheric boundary layer, *J. Phys. Oceanogr.*, **23**, 8, 1856–1866

Masuda, A. and Y. Y. Kuo, 1981, Bispectra for the surface displacement of random gravity waves in deep water, *Deep-Sea Res.*, **28A**, 3, 2223–237

Mathiesen, M., P. Hawkes, M. J. Martin, E. Thompson, Y. Goda, E. Mansard, E. Peltier and G. van Vledder, 1994, Recommended practice for extreme wave analysis, *J. Hydraulic Res.*, IAHR, **32**, 6, 803–814

McLeish, W. and D. B. Ross, 1983, Imaging radar observations of direction properties of ocean waves, *J. Geophys. Res.*, **88**, C7, 4407–4419

Mei, C. C., 1985, Resonant reflection of surface water waves by periodic sandbars, *J. Fluid Mech.*, **152**, 315–335

—1989, *The Applied Dynamics of Ocean Surface Waves*, Singapore, World Scientific, 740 pp.

Mei, C. C., M. Stiassnie and D. K.-P. Yue, 2006, *Theory and Applications of Ocean Surface Waves, Part 1: Linear Aspects, Part 2: Nonlinear Aspects*, Advanced Series on Ocean Engineering, 2, Singapore, World Scientific, **23**, 1071 pp.

Mendez, F. J., I. J. Losada and R. Medina, 2004, Depth-limited distribution of the highest wave in a sea state, *Proc. 29ᵗʰ Int. Conf. Coastal Engineering* (Lisbon), Singapore, pp. 1022–1031

Miche, R., 1944, Mouvements ondulatoires des mers en profondeur constante ou décroissante, *Annales des ponts et chaussées*, **114**, 369–406

Miles, J. W., 1957, On the generation of surface waves by shear flows, *J. Fluid Mech.*, **3**, 185–204

—1960, On the generation of surface waves by turbulent shear flow, *J. Fluid Mech.*, **7**, 469–478

Miller, H. C. and C. L. Vincent, 1990, FRF spectrum: TMA with Kitaigorodskii's f^{-4} scaling, *J. Waterway, Port, Coastal, and Ocean Engineering*, New York, ASCE, **116**, 1, 57–78

Mitsuyasu, H. 1968, On the growth of spectrum of wind-generated waves (I), *Reports of Research Institute for Applied Mechanics, Kyushu University*, **16**, 55, 459–482

—1969, On the growth of spectrum of wind-generated waves (II), *Reports of Research Institute for Applied Mechanics, Kyushu University*, **17**, 59, 235–248

Mitsuyasu, H., F. Tasai, T. Suhara, S. Mizuno, M. Ohkusu, T. Honda and K. Rikiishi, 1975, Observations of the directional spectrum of ocean waves using a cloverleaf buoy, *J. Phys. Oceanogr.*, **5**, 10, 750–760

Mitsuyasu, H., 1977, Measurement of the high-frequency spectrum of ocean surface waves, *J. Phys. Oceanogr.*, **7**, 6, 882–891

Mitsuyasu, H., F. Tasai, T. Suhara, S. Mizuno, M. Ohkusu, T. Honda and K. Rikiishi, 1980, Observation of the power spectrum of ocean waves using a cloverleaf buoy, *J. Phys. Oceanogr.*, **10**, 2, 286–296

Monahan, E. C. and G. MacNoicaill (eds.), 1986, *Oceanic Whitecaps and Their Role in Air–Sea Exchange Processes*, Dordrecht, D. Reidel Publishing Company, 294 pp.

Monbaliu, J., R. Padilla-Hernández, J. C. Hargreaves, J. C. Carretero Albiach, W. Luo, M. Sclavo and H. Günther, 2000, The spectral wave model, WAM, adapted for applications with high spatial resolution, *Coastal Engineering*, **41**, 41–62

Muir, L. R. and A. H. El-Shaarawi, 1986, On the calculation of extreme wave heights: a review, *Ocean Engineering*, **13**, 1, 93–118

Munk, W. H. and M. A. Traylor, 1947, Refraction of ocean waves: a process linking underwater topography to beach erosion, *J. Geol.*, **LV**, 1, 1–26

Munk, W. H., 1949a, The solitary wave theory and its applications to surf problems, in *Ocean Surface Waves, Annals New York Acad. Sci*, **51**, 3, 376–424

—1949b, Surf beats, *Trans. Am. Geophys. Union*, **30**, 6, 849–854

—1950, Origin and generation of waves, *Proc. 1ˢᵗ Conf. Coastal Engineering* (Long Beach), New York, ASCE, pp. 1–4

Munk, W. H. and R. S. Arthur, 1952, Wave intensity along a refracted ray, *Proc. NBS Semicentennial Symp. on Gravity Waves*, Washington, National Bureau of Standards, Circular 521, pp. 95–108

Munk, W. H., G. R. Miller, F. E. Snodgrass and N. F. Barber, 1963, Directional recording of swell from distant storms, *Phil. Trans. Roy. Soc. London*, A, **255**, 505–584

Nadaoka, K., S. Beji and Y. Nakagawa, 1997, A fully dispersive weakly nonlinear model for water waves, *Proc. Roy. Soc. London*, A, **453**, 303–318

Nagai, T., H. Ogawa, Y. Terada, T. Kato and M. Kudaka, 2004, GPS buoy application to offshore wave, tsunami and tide observation, *Proc. 29ᵗʰ Int. Conf. Coastal Engineering* (Lisbon), Singapore, World Scientific, pp. 1093–1105

Nakamura, S. and K. Katoh, 1992, Generation of infra-gravity waves in breaking process of wave groups, *Proc. 23ʳᵈ Int. Conf. Coastal Engineering* (Venice), New York, ASCE, pp. 990–1003

Nelson, R. C., 1987, Design wave heights on very mild slopes – an experimental study, *Civil Engineering Trans.*, Institution of Engineers Australia, **29**, 157–161

—1994, Depth limited wave heights in very flat regions, *Coastal Engineering*, **23**, 43–59

—1997, Height limits in top down and bottom up wave environments, *Coastal Engineering*, **32**, 247–254

Nepf, H. M., C. H. Wu and E. S. Chan, 1998, A comparison of two- and three-dimensional wave breaking, *J. Phys. Oceanogr.*, **28**, 1496–1510

Neu, H. J. A., 1984, Interannual variations and longer-term changes in the sea state of the North Atlantic from 1970 to 1982, *J. Geophys. Res.*, **89**, C4, 6397–6402

Neumann, G. and W. J. Pierson Jr, 1966, *Principles of Physical Oceanography*, Englewood Cliffs, NJ, Prentice-Hall, Inc., 545 pp.

NMRI, 2005, *Statistical Database of Winds and Waves around Japan*, Tokyo, Seakeeping Group, Marine Safety Department, National Maritime Research Institute of Japan, http://www.nmri.go.jp/wwjapan/namikaze_main_e.html

Nordenstrøm, N., 1969, *Methods for Predicting Long Term Distribution of Wave Loads and Probability of Failure for Ships. Appendix II, Relations between Visually Estimated and Theoretical Wave Heights and Periods*, Oslo, Det Norske Veritas, Research Department, Report No. 69-22-S

Norheim, C. A., T. H. C. Herbers and S. Elgar, 1998, Nonlinear evolution of surface wave spectra on a beach, *J. Phys. Oceanogr.*, **28**, 7, 1534–1551

Nwogu, O., 1993, Alternative form of Boussinesq equations for nearshore wave propagation, *J. Waterway, Port, Coastal, and Ocean Engineering*, New York, ASCE, **119**, 6, 618–638

—1994, Nonlinear evolution of directional wave spectra in shallow water, *Proc. 24ᵗʰ Int. Conf. Coastal Engineering* (Kobe), New York, ASCE, pp. 467–481

Ochi, M. K. and C-H. Tsai, 1983, Prediction of occurrence of breaking waves in deep water, *J. Phys. Oceanogr.*, **13**, 11, 2008–2019

Ochi, M. K. and W.-C. Wang, 1984, Non-Gaussian characteristics of coastal waves, *Proc. 19ᵗʰ Int. Conf. Coastal Engineering* (Houston), New York, ASCE, pp. 516–531

Ochi, M. K., 1992, New approach for estimating the severest sea state from statistical data, *Proc. 23ʳᵈ Int. Conf. Coastal Engineering*, (Venice), New York, ASCE, pp. 512–525

—1998, *Ocean Waves, The Stochastic Approach*, Cambridge, Cambridge University Press, 319 pp.

—2003, *Hurricane-generated Seas*, Elsevier Ocean Engineering Book Series, eds. R. Bhattacharyya and M. E. McCormick, Amsterdam, Elsevier, **8**, 140 pp.

Oh, S.-H., N. Mizutani, N. Hashimoto and K.-D. Suh, 2004, Laboratory observation of breaking criteria of wind-generated deep water waves, *Proc. 29ᵗʰ Int. Conf. Coastal Engineering* (Lisbon), Singapore, World Scientific, pp. 428–440

Onorato, M., A. R. Osborne, M. Serio and S. Bertone, 2001, Freak waves in random oceanic sea states, *Phys. Rev. Lett.*, **86**, 25, 5831–5834

Onorato, M., A. R. Osborne and M. Serio, 2002, Extreme wave events in directional, random oceanic sea states, *Phys. Fluids*, **14**, 14, L25–L28

Onorato, M., A. R. Osborne, M. Serio, L. Cavaleri, C. Brandini and C. T. Stansberg, 2004, Observations of strongly non-Gaussian statistics for random sea surface gravity waves in wave flume experiments, *Phys. Rev.*, E, **70** (2), 6, 1–4

O'Reilly, W. C. and R. T. Guza, 1991, Comparison of spectral refraction and refraction–diffraction wave models. *J. Waterway, Port, Coastal, and Ocean Engineering*, New York, ASCE, **117**, 3, 199–215

Osborne, A. R., M. Onorato and M. Serio, 2000, The nonlinear dynamics of rogue waves and holes in deep-water gravity wave trains, *Phys. Lett.*, A **275**, 386–393

Padilla-Hernández, R. and J. Monbaliu, 2001, Energy balance of wind waves as a function of the bottom friction formulation, *Coastal Engineering*, **43**, 131–148

Pandey, M. D., P. H. A. J. M. van Gelder and J. K. Vrijling, 2004, Dutch case studies of the estimation of extreme quantiles and associated uncertainty by bootstrap simulations, *Environmetrics*, **15**, 687–699

Pawka, S. S., 1983, Island shadows in wave directional spectra, *J. Geophys. Res.*, **88**, C4, 2579–2591

Peirson, W. L. and S. E. Belcher, 2004, Growth response of waves to the wind stress, *Proc. 29ᵗʰ Int. Conf. Coastal Engineering* (Lisbon), Singapore, World Scientific, pp. 667–676

Penney, W. G. and A. T. Price, 1952, The diffraction theory of sea waves and the shelter afforded by breakwaters, *Phil. Trans. Roy. Soc. London*, A, **244**, 236–253

Peregrine, D. H., 1967, Long waves on a beach, *J. Fluid Mech.*, **27**, 4, 815–827

—1976, Interaction of water waves and currents, *Adv. Appl. Mech.*, **16**, 9–117

—1990, Computations of breaking waves, in *Water Wave Kinematics*, eds. A. Tørum and O. T. Gudmestad, Amsterdam, Kluwer Academic Publishers, pp. 475–490

—1999, Large-scale vorticity generation by breakers in shallow and deep water, *Eur. J. Mech. B/Fluids*, **18**, 3, 403–408

Perrie, W. and V. Zakharov, 1999, The equilibrium range cascades of wind-generated waves, *Eur. J. Mech. B/Fluids*, **18**, 3, 365–371

Peters, D. J., C. J. Shaw, C. K. Grant, J. C. Heideman and D. Szabo, 1993, Modelling the North Sea through the North Sea Storm Study (NESS), *Proc. 25ᵗʰ Annual Offshore Technology Conf.*, Report No. OTC 7130, pp. 479–493

Petruaskas, C. and P. M. Aagaard, 1970, Extrapolation of historical storm data for estimating design wave heights, *Proc. 2ⁿᵈ Annual Offshore Technology Conf.*, Report No. OTC1190, pp. 409–427

Phillips, O. M., 1957, On the generation of waves by turbulent wind, *J. Fluid Mech.*, **2**, 417–445

—1958, The equilibrium range in the spectrum of wind-generated waves, *J. Fluid Mech.*, **4**, 426–434

—1960, On the dynamics of unsteady gravity waves of finite amplitude. Part 1. The elementary interactions, *J. Fluid Mech.*, **9**, 193–217

—1977, *The Dynamics of The Upper Ocean*, Cambridge, Cambridge University Press, 336 pp.

—1981, Wave interactions – the evolution of an idea, *J. Fluid Mech.*, **106**, 215–227

—1985, Spectral and statistical properties of the equilibrium range in wind-generated gravity waves, *J. Fluid Mech.*, **156**, 505–531

Pierson, W. J. Jr., J. J. Tuttell and J. A. Wooley, 1952, The theory of the refraction of a short crested Gaussian sea surface with application to the northern New Jersey coast, *Proc. 3ʳᵈ Conf. Coastal Engineering* (Cambridge, MA), New York, ASCE, pp. 86–108

Pierson, W. J., G. Neumann and R. W. James, 1955, *Practical Methods for Observing and Forecasting Ocean Waves by Means of Wave Spectra and Statistics*, Washington, U.S. Navy Hydrographic Office, Publication No. 603 (reprinted 1960), 284 pp.

Pierson, W. J. and L. Moskowitz, 1964, A proposed spectral form for fully developed wind seas based on the similarity theory of S. A. Kitaigorodskii, *J. Geophys. Res.*, **69**, 24, 5181–5190

Pierson, W. J., L. J. Tick and L. Baer, 1966, Computer based procedures for preparing global wave forecasts and wind field analysis capable of using wave data obtained by a spacecraft, *Proc. 6ᵗʰ Naval Hydrodynamics Symp.*, Washington, Office of Naval Research, article 20–1, 42 pp.

Piest, J., 1965, Seegangsbestimmung und Seegangsrefraktion in einem Meer mit nicht ebenen Boden – eine theoretische Untersuchung, *Deutsch. Hydrogr. Z.*, Sonderdruck, **18**, 6, 253–260

Plant, W. J., 1982, A relationship between wind stress and wave slope, *J. Geophys. Res.*, **C87**, NC3, 1961–1967

Pontes, M. T., G. A. Athanassoulis, S. Barstow, L. Cavaleri, B. Holmes, D. Mollison and H. Oliviera Pires, 1996, *WERATLAS – Atlas of Wave Energy Resource in Europe*, Lisbon, INETI

Price, W. G. and R. E. D. Bishop, 1974, *Probabilistic Theory of Ship Dynamics*, London, Chapman and Hall, 311 pp.

Putnam, J. A. and J. W. Johnson, 1949, The dissipation of wave energy by bottom friction, *Trans. Am. Geophys. Union*, **30**, 1, 67–74

Radder, A. C., 1979, On the parabolic equation method for water-wave propagation, *J. Fluid Mech.*, **95**, 1, 159–176

Rahman, M., 1995, *Water Waves: Relating Modern Theory to Advanced Engineering Applications*, Oxford, Clarendon Press, 343 pp.

Raichlen, F., 1993, Waves propagating on an adverse jet, *Proc. 2ⁿᵈ Int. Symp. Ocean Wave Measurement and Analysis WAVES 93* (New Orleans), New York, ASCE, pp. 657–670

Rasmussen, J. H., 1998, Deterministic and stochastic modelling of surface gravity waves in finite depth, Unpublished Ph.D. thesis, Lyngby, Technical University of Denmark, Department of Hydrodynamics and Water Resources (ISVA), 245 pp.

Raubenheimer, B., R. T. Guza and S. Elgar, 1996, Wave transformation across the inner surf zone, *J. Geophys. Res.*, **101**, C11, 25 589–25 597

Rayleigh, Lord, 1880, On the resultant of a large number of vibrations of the same pitch and of arbitrary phase, *Phil. Mag.*, Series 5, **10**, 60, 73–78

Repko, A., P. H. A. J. M. van Gelder, H. G. Voortman and J. K. Vrijling, 2000, Bivariate statistical analysis of wave climates, *Proc. 27th Int. Conf. Coastal Engineering* (Sydney), Reston, VA, ASCE, pp. 583–596

Resch, F. J., J. S. Darrozes and G. M. Afèti, 1986, Marine liquid aerosol production from bursting air bubbles, *J. Geophys. Res.*, **91**, C1, 1019–1029

Resio, D. T., 1987, Shallow-water waves. I: theory, *J. Waterway, Port, Coastal, and Ocean Engineering*, New York, ASCE, **113**, 3, 264–281

—1988, Shallow-water waves. II: data comparison, *J. Waterway, Port, Coastal, and Ocean Engineering*, New York, ASCE, **114**, 1, 50–65

Resio, D. T. and W. Perrie, 1991, A numerical study of nonlinear energy fluxes due to wave–wave interactions. Part 1: methodology and basic results, *J. Fluid Mech.*, **223**, 609–629

Resio, D. T., V. R. Swail, R. E. Jensen and V. J. Cardone, 1999, Wind speed scaling in fully developed seas, *J. Phys. Oceanogr.*, **29**, 8, 1801–1811

Resio, D. T., J. H. Phil, B. A. Tracy and C. L. Vincent, 2001, Nonlinear fluxes and the finite depth equilibrium range in wave spectra, *J. Geophys. Res.*, **106**, C4, 6985–7000

Resio, D. T., C. E. Long and C. L. Vincent, 2004, Equilibrium-range constant in wind-generated wave spectra, *J. Geophys. Res.*, **109**, C01018, doi:10.1029/2003JC001788 (14 pp.)

Rice, S. O., 1944, Mathematical analysis of random noise, *Bell System Technol. J.*, **23**, 282–332

—1945, Mathematical analysis of random noise, *Bell System Technol. J.*, **24**, 46–156

—1954, Mathematical analysis of random noise, in *Selected Papers on Noise and Stochastic Processes*, ed. N. Wax, New York, Dover Publications Inc., pp. 133–294

Ris, R. C. and L. H. Holthuijsen, 1996, Spectral modelling of current induced wave-blocking, *Proc. 25th Int. Conf. Coastal Engineering* (Orlando), New York, ASCE, pp. 1247–1254

Rivero, F. J., A. S. Arcilla and E. Carci, 1997, An analysis of diffraction in spectral wave models, *Proc. 3rd Int. Symp. Ocean Wave Measurement and Analysis WAVES 97* (Virginia Beach), Reston, VA, ASCE, pp. 431–445

Roelvink, J. A., 1993, Dissipation in random wave groups incident on a beach, *Coastal Engineering*, **19**, 127–150

Rogers, W. E., J. M. Kaihatu, H. A. H. Petit, N. Booij and L. H. Holthuijsen, 2002, Diffusion reduction in an arbitrary scale third generation wind wave model, *Ocean Engineering*, **29**, 11, 1357–1390

Rogers, B. D. and R. A. Dalrymple, 2004, SPH modelling of breaking waves, *Proc. 29th Int. Conf. Coastal Engineering* (Lisbon), Singapore, World Scientific, pp. 415–427

Ross, D. B., V. J. Cardone and J. W. Conaway, 1970, Laser and microwave observations of sea-surface conditions for fetch-limited 17- to 25-m/s winds, *IEEE Trans. Geosci. Electronics*, **GE-8**, 4, 326–336

Ross, D. B. and V. Cardone, 1974, Observations of oceanic whitecaps and their relation to remote measurements of surface wind speed, *J. Geophys. Res.*, **79**, 3, 444–452

Russell, T. L., 1963, A step-type recording wave gage, *Proc. Conf. Ocean Wave Spectra*, Englewood Cliffs, NJ, Prentice Hall Inc., pp. 251–257

Rye, H., 1974, Wave group formation among storm waves, *Proc. 14th Conf. Coastal Engineering* (Copenhagen), New York, ASCE, pp. 164–183

Sainflou, G., 1928, Essai sur des digues maritimes verticales, *Annales des ponts et chaussées*, **98**, 4, 5–48

Sakai, T. and J. A. Battjes, 1980, Wave shoaling calculated from Cokelet's theory, *Coastal Engineering*, **4**, 65–84

Sakai, T., M. Koseki and Y. Iwagaki, 1983, Irregular wave refraction due to current, *J. Hydraul. Eng.*, New York, ASCE, **109**, 9, 1203–1215

Sakai, S. and H. Saeki, 1984, Effects of opposing current on wave transformation on a sloping bed, *Proc. 19th Int. Conf. Coastal Engineering* (Houston), New York, ASCE, pp. 1132–1148

Salih, B. A., R. Burrows and R. G. Tickell, 1988, Storm statistics in the North Sea, *Proc. 21st Int. Conf. Coastal Engineering* (Malaga), New York, ASCE, pp. 956–970

Sallenger, A. H. Jr., P. C. Howard, C. H. Fletcher III and P. A. Howd, 1983, A system for measuring bottom profile, waves and currents in the high-energy nearshore environment, *Mar. Geol.*, **51**, 63–76

Sand, S. E., N. E. Ottesen-Hansen, P. Klinting, O. T. Gudmestad and M. J. Sterndorff, 1990, Freak wave kinematics, in *Water Wave Kinematics*, eds. A. Tørum and O. T. Gudmestad, Dordrecht, Kluwer Academic Publishers, pp. 535–549

Sanders, J. W., 1976, A growth-stage scaling model for the wind-driven sea, *Deutsch. Hydrogr Z.*, **29**, 4, 136–161

Sasaki, W., I. Iwasaki, T. Matsuura, S. Iizuka and I. Watabe, 2005. Changes in wave climate off Hiratsuka, Japan, as affected by storm activity over the western North Pacific, *J. Geophys. Res.*, **110**, C09008, doi: 10.1029/2004 JC002730

Sawaragi, T, 1995, *Coastal Engineering – Waves, Beaches, Wave–Structure Interactions*, Amsterdam, Elsevier, 479 pp.

Schule, J. J., L. S. Simpson and P. S. DeLeonibus, 1971, A study of fetch-limited wave spectra with an airborne laser, *J. Geophys. Res.*, **76**, 18, 4160–4171

Schulz-Stellenfleth, J. and S. Lehner, 2004, Measurement of 2-D sea surface elevation fields using complex synthetic aperture radar data, *IEEE Trans. Geosci. Remote Sensing*, **GRS-42**, 6, 1149–1160

Schumann, E. H., 1976, Changes in energy of surface gravity waves in the Agulhas Current, *Deep-Sea Res.*, **23**, 6, 509–518

Seelig, W. N. and J. P. Ahrens, 1981, *Estimation of Wave Reflection and Energy Dissipation Coefficients for Beaches, Revetments, and Breakwaters*, Technical Paper No. 81-1, Ft. Belvoir, U.S. Army Corps of Engineers, Coastal Engineering Research Center

Sénéchal, N., P. Bonneton and H. Dupuis, 2001, Generation of secondary waves due to wave propagation over a bar: a field investigation, *Proc. 4th Int. Symp. Ocean Wave Measurement and Analysis WAVES 2001* (San Francisco), Reston, VA, ASCE, pp. 764–772

Seymour, R. J., 1977, Estimating wave generation on restricted fetches, *J. Waterway, Port, Coastal and Ocean Div.*, ASCE, **103**, WW2, 251–264

Shemdin, O. H. and E. Y. Hsu, 1967, Direct measurements of aerodynamic pressure above a simple progressive gravity wave, *J. Fluid Mech.*, **30**, 2, 403–416

Shemdin, O., K. Hasselmann, S. V. Hsiao and K. Herterich, 1977, Nonlinear and linear bottom interaction effects in shallow water, *Proc. NATO Conf. on Turbulent Fluxes through the Sea Surface, Wave Dynamics, and Prediction*, New York, Plenum Press, pp. 347–372

Sheremet, A. and G. W. Stone, 2003, Observations of nearshore wave dissipation over muddy sea beds, *J. Geophys. Res.*, **108**, C11, 3357

Shum, K. T. and W. K. Melville, 1984, Estimates of the joint statistics of amplitudes and periods of ocean waves using an integral transform technique, *J. Geophys. Res.*, **89**, C4, 6467–6476

Singleton, R. C., 1969, An algorithm for computing the mixed-radix fast Fourier transform, *IEEE Trans. Audio Electro-acoustics*, **AU-17**, 2, 93–103

Skjelbreia, L. and J. A. Hendrickson, 1960, Fifth order gravity wave theory, *Proc. 7th Conf. Coastal Engineering*, Berkeley, CA, Council on Wave Research, Engineering Foundation, University of California, pp. 184–196

Skourup, J., N.-E. O. Hansen and K. K. Andreasen, 1997, Non-Gaussian extreme waves in the central North Sea, *J. Offshore Mechanics and Arctic Engineering*, ASME, **119**, 3, 146–150

Skovgaard, O. and M. H. Petersen, 1977, Refraction of cnoidal waves, *Coastal Engineering*, **1**, 43–61

Snodgrass, F. E., G. W. Groves, K. F. Hasselmann, G. R. Miller, W. H. Munk and W. H. Powers, 1966, Propagation of ocean swell across the Pacific, *Phil. Trans. Roy. Soc. London*, A, **259**, 431–497

Smith, J. M. and C. L. Vincent, 1992, Shoaling and decay of two wave trains on beach, *J. Waterway, Port, Coastal, and Ocean Engineering*, New York, ASCE, **118**, 5, 517–533

Smith, J. M., H. E. Bermudez and B. A. Ebersole, 2000, Modelling waves at Willapa, Washington, *Proc. 27th Int. Conf. Coastal Engineering* (Sydney), Reston, VA, ASCE, pp. 826–839

Smith, J. M., 2001, Breaking in a spectral wave model, *Proc. 4th Int. Symp. Ocean Wave Measurement and Analysis WAVES 2001* (San Francisco), Reston, VA, ASCE, pp. 1022–1031

Smith, J. M. and C. L. Vincent, 2002, Application of spectral equilibrium ranges in the surf zone, *Proc. 28th Int. Conf. Coastal Engineering* (Cardiff), Singapore, World Scientific, pp. 269–279

—2003, Equilibrium ranges in surf zone wave spectra, *J. Geophys. Res.*, **108**, C11, 3366, doi: 1029/2003 JC001930

Smith, J. M., 2004, Shallow-water spectral shapes, *Proc. 29th Int. Conf. Coastal Engineering* (Lisbon), Singapore, World Scientific, pp. 206–217

Snyder, R. L., F. W. Dobson, J. A. Elliott and R. B. Long, 1981, Array measurement of atmospheric pressure fluctuations above surface gravity waves, *J. Fluid Mech.*, **102**, 1–59

Soares, C. G., 2003, Probabilistic models of waves in the coastal zone, in *Advances in Coastal Modeling*, ed. C. Lakhan, Amsterdam, Elsevier Science, pp. 159–187

Sobey, R. J., 1986, Wind-wave prediction, *Ann. Rev. Fluid Mech.*, **18**, 149–172

Sobey, R. J. and I. R. Young, 1986, Hurricane wind waves – a discrete spectral model, *J. Waterway, Port, Coastal, and Ocean Engineering*, New York, ASCE, **112**, 3, 370–389

Sommerfeld, A., 1896, Mathematische Theorie der Diffraktion, *Mathematische Annalen*, **47**, 317–374

Sorensen, R. M., 1993, *Basic Wave Mechanics: For Coastal and Ocean Engineers*, New York, John Wiley & Sons, 284 pp.

Sørensen, O. R., H. Kofoed-Hansen, M. Rugbjerg and L. S. Sørensen, 2004, A third-generation spectral wave model using an unstructured finite volume technique, *Proc. 29th Int. Conf. Coastal Engineering* (Lisbon), Singapore, World Scientific, pp. 894–906

Southgate, H. N., 1984, Techniques of ray averaging, *Int. J. Num. Meth. Fluids*, **4**, 725–747

Southgate, H. N. and R. B. Nairn, 1993, Deterministic profile modelling of nearshore processes, Part 1. Waves and currents, *Coastal Engineering*, **19**, 27–56

SPM, 1973, 1984, *Shore Protection Manual*, U.S. Army Coastal Engineering Research Center, **I**

Srokosz, M. A. and P. G. Challenor, 1987, Joint distributions of wave height and period; a critical comparison, *Ocean Engineering*, **14**, 4, 295–311

Srokosz, M. A., 1988, A note on the joint distribution of wave height and period during the growth phase of a storm, *Ocean Engineering*, **15**, 4, 379–387

—1990, Wave statistics, in *Surface Waves and Fluxes*, eds. G. L. Geernaert and W. J. Plant, Dordrecht, Kluwer Academic Publishers, **1**, pp. 285–332

Stansell, P., 2005, Distributions of extreme wave, crest and trough heights measured in the North Sea, *Ocean Engineering*, **32**, 8–9, 1015–1036

Stefanakos, Ch. N., G. A. Athanassoulis, L. Cavaleri, L. Bertotti and J. M. Lefèvre, 2004a, Wind and wave climatology of the Mediterranean Sea, part 1: wind statistics, *Proc. 14ᵗʰ Int. Offshore and Polar Engineering Conf.* (Toulon), Cupertino, CA, ISOPE, pp. 177–186

Stefanakos, Ch. N., G. A. Athanassoulis, L. Cavaleri, L. Bertotti and J. M. Lefèvre, 2004b, Wind and wave climatology of the Mediterranean Sea, part 2: wave statistics, *Proc. 14ᵗʰ Int. Offshore and Polar Engineering Conf.* (Toulon), Cupertino, CA, ISOPE, pp. 187–196

Stelling, G. S. and J. J. Leendertse, 1992, Approximation of convective processes by cyclic AOI methods, *Proc. 2ⁿᵈ Int. Conf. Estuarine and Coastal Modeling* (Tampa), New York, ASCE, pp. 771–782

Stelling, G. S. and M. Zijlema, 2003, An accurate and efficient finite-difference algorithm for non-hydrostatic free-surface flow with application to wave propagation, *Int. J. Num. Meth. Fluids*, **43**, 1, 1–23

Stoker, J. J., 1957, *Water Waves, The Mathematical Theory with Applications*, New York, Interscience Publishers, Inc.; 567 pp.

Stokes, G. G., 1847, On the theory of oscillatory waves, *Trans. Camb. Phil. Soc.*, **8**, 441–455 (reprinted in *Mathematical and Physical Papers*, London, **1**, pp. 314–326)

Suh, K. D., Y. Y. Kim and D. Y. Lee, 1994, Equilibrium range spectrum of waves propagating on currents, *J. Waterway, Port, Coastal, and Ocean Engineering*, New York, ASCE, **120**, 5, 434–450

Suzuki, Y., I. Isozaki and T. Tanahashi, 1994, On the development of a global ocean wave model JWA3G, *Proc. Int. Conf. on Port and Harbour Construction, Hydro-Port '94*, Yokosuka, Port and Harbour Research Institute, pp. 227–237

Svendsen, I. A., 1984, Wave heights and set-up in a surf zone, *Coastal Engineering*, **8**, 303–329

Svendsen, I. A., 2006, *Introduction to nearshore hydrodynamics*, Advanced Series on Ocean Engineering, Singapore, World Scientific, **24**, 722 p.

Svendsen, I. A., W. Qin and B. A. Ebersole, 2003, Modelling waves and currents at the LSTF and other laboratory facilities, *Coastal Engineering*, **50**, 19–45

Sverdrup, H. V. and W. H. Munk, 1946, Empirical and theoretical relations between wind, sea and swell, *Trans. Am. Geophys. Union*, **27**, 823–827

—1947, *Wind, Sea and Swell: Theory of Relations for Forecasting,* Washington, U.S. Navy Hydrographic Office, Publication. No. 601

SWAMP (Sea Wave Modelling Project) Group (24 authors), 1985, *Ocean Wave Modelling*, New York, Plenum Press, 256 pp.

Takayama, T., N. Hashimoto, T. Nagai, T. Takahashi, H. Sasaki and Y. Ito, 1994, Development of a submerged Doppler-type directional wave meter, *Proc. 24ᵗʰ Int. Conf. Coastal Engineering* (Kobe), New York, ASCE, pp. 624–634

Tayfun, M. A., 1981, Breaking-limited wave heights, *J. Waterway, Port, Coastal and Ocean Div.*, New York, ASCE, **107**, WW2, 59–69

—1990, Distribution of large wave heights, *J. Waterway, Port, Coastal, and Ocean Engineering*, New York ASCE, **116**, 6, 686–707

—2004, Statistics of wave crests in storms, *J. Waterway, Port, Coastal, and Ocean Engineering*, New York, ASCE, **130**, 4, 155–161

Taylor, P. A. and R. J. Lee, 1984, Simple guidelines for estimating wind speed variations due to small-scale topographic features, *Climatol. Bull.*, **18**, 3–32

Thijsse, J. Th., 1948, Dimensions of wind-generated waves, *General Assembly of Association d'Océanographie Physique (IGGU), Oslo, Procès-Verbaux*, **4**, 80–81

Thijsse, J. Th. and J. B. Schijf, 1949, no title, *XVIIth Int. Naval Congress*, Lisbon, Section II, Communication 4, pp. 151–171

Thijsse, J. Th., 1952, Growth of wind-generated waves and energy transfer, *Proc. NBS Semicentennial Symp. on Gravity Waves*, Washington, National Bureau of Standards, Circular 521, pp. 281–287

Thornton, E. B., 1977, Rederivation of the saturation range in the frequency spectrum of wind-generated gravity waves, *J. Phys. Oceanogr.*, **7**, 1, 137–140

Thornton, E. B. and G. Schaeffer, 1978, Probability density functions of breaking waves, *Proc. 18th Conf. Coastal Engineering* (Hamburg), New York, ASCE, pp. 507–519

Thornton, E. B. and R. T. Guza, 1983, Transformation of wave height distribution, *J. Geophys. Res.*, **88**, C10, 5925–5938

Toba, Y., 1972, Local balance in the air–sea boundary processes, I. On the growth process of wind waves, *J. Oceanogr. Soc. Japan*, **28**, 3, 109–120

—1973, Local balance in the air–sea boundary processes, III. On the spectrum of wind waves, *J. Oceanogr. Soc. Japan*, **29**, 2, 209–220

—1997, The 3/2-power law for ocean wind waves and its applications, in *Advances in Coastal and Ocean Engineering*, ed: P. L.-F. Liu, Singapore, World Scientific, **3**, pp. 31–65

Tolman, H. L., 1990, Wind wave propagation in tidal seas, Ph.D. thesis, published as *Communications on Hydraulic and Geotechnical Engineering*, Delft University of Technology, Faculty of Civil Engineering, Report. No. 90–1, 180 pp.

—1991, A third-generation model for wind waves on slowly varying, unsteady and inhomogeneous depths and currents, *J. Phys. Oceanogr.*, **21**, 6, 782–797

—1992a, Effects of numerics on the physics in a third-generation wind-wave model, *J. Phys. Oceanogr.*, **22**, 10, 1095–1111

—1992b, An evaluation of expressions for the wave energy dissipation due to bottom friction in the presence of currents, *Coastal Engineering*, **16**, 165–179

—1994, Wind waves and movable-bed bottom friction, *J. Phys. Oceanogr.*, **24**, 5, 994–1009

—1995, Subgrid modeling of moveable-bed bottom friction in wind wave models, *Coastal Engineering*, **26**, 57–75

Tolman, H. J. and D. Chalikov, 1996, Source terms in a third-generation wind wave model, *J. Phys. Oceanogr.*, **26**, 11, 2497–2518

Tomiyasu, K., 1978, Tutorial review of synthetic-aperture radar (SAR) with applications to imaging of the ocean surface, *Proc. IEEE*, **66**, 563–583

Tosi, R., L. Cavaleri, G. Grancini and L. Iovenitti, 1984, *Statistica delle onde estreme Mare Tirreno* , Padua, Consiglio Nazionale delle Ricerche, 86 pp.

Tracy, B. A. and D. T. Resio, 1982, *Theory and Calculation of The Nonlinear Energy Transfer between Sea Waves in Deep Water*, WES Report 11, Vicksburg, MD, U.S. Army Engineers Waterways Experiment Station

Tucker, M. J., 1950, Surf beats: sea waves of 1 to 5 min. period, *Proc. Roy. Soc. London*, A, **202**, 565–573

—1956, A shipborne wave recorder, *Trans. Inst. Naval Architects*, **98**, 236–250

—1982, The heave response of a spar buoy, *Ocean Engineering*, **9**, 3, 259–270

—1994, Nearshore wave height during storms, *Coastal Engineering*, **24**, 111–136

Tucker, M. J. and E. G. Pitt, 2001, *Waves in Ocean Engineering*, Amsterdam, Elsevier, 521 pp.

Tukey, J. W. and R. W. Hamming, 1948, *Measuring Noise Color 1*, Murray Hill, NJ, Bell Telephone Lab., Memorandum for File MM-49-110-119, 120 pp.

US Navy, 1974 etc., *Marine Climatic Atlas of the World*, Vol. **I**: North Atlantic Ocean; Vol. **II**: North Pacific Ocean; Vol. **III**: Indian Ocean; Vol. **IV**: South Atlantic Ocean; Vol. **V**: South Pacific Ocean; Vol. **VII**: Arctic, Vol. **VIII**: The World; Vol. **IX**: World-wide Means and Standard Deviations, Washington U.S. Government Printing Office

—1983, *Navy Hindcast Spectral Ocean Wave Model Climatic Atlas: North Atlantic Ocean*, Washington, U.S. Government Printing Office, 375 pp.

van der Vlugt, A. J. M., A. J. Kuik and L. H. Holthuijsen, 1981, The WAVEC directional buoy under development, *Proc. Directional Wave Spectra Applications '81*, (Berkeley), New York, ASCE, pp. 50–60

van der Vorst, H. A., 1992, Bi-CGSTAB: a fast and smoothly converging variant of Bi-CG for the solution of nonsymmetric linear systems, *SIAM J. Sci. Statist. Comput.*, **13**, 2, 631–644

van Gelder, P. H. A. J. M. and J. K. Vrijling, 1999, On the distribution function of the maximum wave height in front of reflecting structures, *Proc. Conf. Coastal Structures*, ed. I. J. Losada, **1**, pp. 37–46

van Vledder, G. Ph., 1992, Statistics of wave group parameters, *Proc. 23rd Int. Conf. Coastal Engineering* (Venice), New York, ASCE, pp. 946–959

van Vledder, G. Ph., Y. Goda, P. Hawkes, E. Mansard, M. J. Martin, M. Mathiesen, E. Peltier and E. Thompson, 1993, A case study of extreme wave analysis: a comparative analysis, *Proc. 2nd Int. Symp. on Ocean Wave Measurement and Analysis WAVES 93* (New Orleans), New York, ASCE, pp. 978–992

van Vledder, G. Ph. and M. Bottema, 2002, Improved modelling of nonlinear four-wave interactions in shallow water, *Proc. 28th Int. Conf. Coastal Engineering* (Cardiff), Singapore, World Scientific, pp. 459–471

van Vledder, G. Ph., 2006, The WRT method for the computation of non-linear four-wave interactions in discrete spectral wave models, *Coastal Engineering*, **53**, 2–3, 223–242

Verhagen, L. A., L. H. Holthuijsen and Y. S. Won, 1992, Modelling ocean waves in the Columbia River entrance, *Proc. 23rd Int. Conf. Coastal Engineering* (Venice), New York, ASCE, pp. 2893–2901

Vincent, C. L., 1985, Depth-controlled wave height, *J. Waterway, Port, Coastal, and Ocean Engineering*, New York, ASCE, **111**, 459–475

Vincent, C. L. and S. A. Hughes, 1985, Wind wave growth in shallow water, *J. Waterway, Port, Coastal, and Ocean Engineering*, New York, ASCE, **111**, 1, 765–770

Vincent, C. L. and M. J. Briggs, 1989, Refraction–diffraction of irregular waves over a mound, *J. Waterway, Port, Coastal, and Ocean Engineering*, New York, ASCE, **115**, 2, 269–284

Vincent, C. L., J. M. Smith and J. Davis, 1994, Parameterization of wave breaking in models, *Proc. Int. Symp.: Waves – Physical and Numerical Modelling*, University of British Columbia, Vancouver, Canada, eds. M. Isaacson and M. Quick, II, pp. 753–762

von Gerstner, F. J., 1802, Theorie der Wellen, *Abhandlungen der königlichen böhmischen Gesellschaft der Wissenschaften*, now listed under *Česka spolecnost nauk* (Prague) and reprinted in *Annalen der Physik* 1809, **32**, 412–440

Voorrips, A. C., V. K. Makin and G. J. Komen, 1994, The influence of atmospheric stratification on the growth of water waves, *Boundary-Layer Meteorol.*, **72**, 287–303

Vuik, C., 1993, Solution of the discretized incompressible Navier–Stokes equations with the GMRES method, *Int. J. Num. Meth. Fluids*, **16**, 507–523

Walsh, E. J., D. W. Hancock III, D. E. Hines, R. N. Swift and J. F. Scott, 1985, Directional wave spectra measured with the surface contour radar, *J. Phys. Oceanogr.*, **15**, 5, 566–592

—1989, An observation of the directional wave spectrum evolution from shoreline to fully developed, *J. Phys. Oceanogr.*, **19**, 5,670–690

WAMDI group (13 authors), 1988, The WAM model – a third generation ocean wave prediction model, *J. Phys. Oceanogr.*, **18**, 12, 1775–1810

Wang, D. W. and P. A. Hwang, 2001a, Evolution of the bimodal directional distribution of ocean waves, *J. Phys. Oceanogr*, **31**, 5, 1200–1221

—2001b, A bimodal directional distribution model for directional buoy measurements, *Proc. 4th Int. Symp. Ocean Wave Measurement and Analysis WAVES 2001* (San Francisco), Reston, VA, ASCE, pp. 163–172

Webb, D, J., 1978, Nonlinear transfers between sea waves, *Deep-Sea Res.*, **25**, 3, 279–298

Weber, N., 1991a, Bottom friction for wind sea and swell in extreme depth-limited situations, *J. Phys. Oceanogr.*, **21**, 1, 149–172

Weber, S. L., 1989, Surface gravity waves and turbulent bottom friction, Unpublished Ph.D. thesis, University of Utrecht, 128 pp.

—1991b, Eddy-viscosity and drag-law models for random ocean wave dissipation, *J. Fluid Mech.*, **232**, 73–98

Webster, A. G., 1955, *Partial Differential Equations of Mathematical Physics*, New York, Dover Publications Inc., 440 pp.

Weggel, J. R., 1972, Maximum breaker height, *J. Waterways, Harbors, and Coastal Engineering Div.*, New York, ASCE, **98**, WW4, 529–548

Wei, G., J. T. Kirby, S. T. Grilli and R. Subramanya, 1995, A fully nonlinear Boussinesq model for surface waves. Part 1. Highly nonlinear unsteady waves, *J. Fluid Mech.*, **294**, 71–92

Whitham, G. B., 1974, *Linear and Nonlinear Waves*, New York, John Wiley and Sons, 636 pp.

Wiegel, R. L., 1960, A presentation of cnoidal wave theory for practical application, *J. Fluid Mech.*, **7**, 2, 273–286

—1964, *Oceanographical Engineering*, Englewood Cliffs, NJ, Prentice Hall Inc., 532 pp.

Willmarth, W. W. and C. E. Wooldridge, 1962, Measurements of the fluctuating pressure at the wall beneath a thick turbulent boundary layer, *J. Fluid Mech.*, **14**, 187–210

Wilson, B. W., 1965, Numerical prediction of ocean waves in the North Atlantic for December 1959, *Deutsch. Hydrogr. Z.*, **18**, 3, 114–130

WMO (World Meteorological Organization), 1998, *Guide to Wave Analysis and Forecasting*, ed. A. K. Laing, Geneva, WMO, 159 pp.

Woolf, D. K. and S. A. Thorpe, 1991, Bubbles and the air–sea exchange of gases in near-saturation conditions, *J. Mar. Res.*, **49**, 3, 435–466

Wu, H.-Y., E.-Y. Hsu and R. L. Street, 1977, The energy transfer due to air-input, nonlinear wave–wave interaction and white cap dissipation associated with wind-generated waves, Technical Report 207, Stanford, CA, Stanford University, 158 pp.

—1979, Experimental study of nonlinear wave–wave interaction and white-cap dissipation of wind-generated waves, *Dynamics Atmos. Oceans*, **3**, 55–78

Wu, J., 1982, Wind-stress coefficients over sea surface from breeze to hurricane, *J. Geophys. Res.*, **87**, C12, 9704–9706

Wyatt, L. R., 1995, The effect of fetch on the directional spectrum of Celtic Sea storm waves, *J. Phys. Oceanogr.*, **25**, 6, 1550–1559

Wyatt, L. R. and L. J. Ledgard, 1996, OSCR wave measurement – some preliminary results, *IEEE J. Oceanic Engineering*, **21**, 1, 64–76

Wyatt, L. R., 1997, The ocean wave directional spectrum, *Oceanography*, Special issue on high frequency radars for coastal oceanography, **10**, 2, 85–89

Wyatt, L. R., S. P. Thompson and R. R. Burton, 1999, Evaluation of high frequency radar wave measurement, *Coastal Engineering*, **37**, 259–282

Wyatt, L. R. and D. Prandle (eds.), 1999, Monitoring current and wave variability in coastal seas, *Coastal Engineering*, Special Issue, **37**, 3 + 4, 193–546

Wyatt, L. R., 2000, Limits to the inversion of HF radar backscatter for ocean wave measurement. *J. Atmos. Ocean Technol.*, **17**, 12, 1651–1666

Yamaguchi, M. 1984, Approximate expressions for integral properties of the JONSWAP spectrum, *Proc. Japanese Society of Civil Engineers*, **345/II-1**, 149–152 [in Japanese]

—1986, A numerical model of nearshore currents based on a finite amplitude wave theory, *Proc. 20th Int. Conf. Coastal Engineering* (Taipei), New York, ASCE, pp. 849–863

—1988, A numerical model of nearshore currents due to irregular waves, *Proc. 21st Int. Conf. Coastal Engineering* (Malaga), New York, ASCE, pp. 1113–1126

Yamaguchi, M., L. H. Holthuijsen, Y. Hatada and M. Hino, 1988, A new hybrid parametrical wave prediction model taking the wave directionality into account, *Proc. Japanese Society of Civil Engineers*, **399/II-10**, 193–202 [in Japanese]

Yamaguchi, M. and Y. Hatada, 1990, A numerical model for refraction computation of irregular waves due to time-varying currents and water depth, *Proc. 22nd Int. Conf. Coastal Engineering* (Delft), New York, ASCE, pp. 205–217

Yamaguchi, M., 1992, Interrelation of cnoidal wave theories, *Proc. 23rd Int. Conf. Coastal Engineering* (Venice), New York, ASCE, pp. 737–750

Yan, L., 1987, *An Improved Wind Input Source Term for Third Generation Ocean Wave Modelling*, Scientific report WR-No 87–8, De Bilt, Royal Netherlands Meteorological Institute (KNMI)

Young, I. R. and R. J. Sobey, 1985, Measurements of the wind-wave energy flux in an opposing wind, *J. Fluid Mech.*, **151**, 427–442

Young, I. R., W. Rosenthal and F. Ziemer, 1985, A three-dimensional analysis of marine radar images for the determination of ocean wave directionality and surface currents, *J. Geophys. Res.*, **90**, C1, 1049–1059

Young, I. R., 1988a, Parametric hurricane wave prediction model, *J. Waterway, Port, Coastal, and Ocean Engineering*, New York, ASCE, **114**, 5, 637–652

—1988b, A shallow water spectral wave model, *J. Geophys. Res.*, **93**, C5, 5113–5129

Young, I. R. and R. J. Sobey, 1988, Deep water swell and spectral wave decay in opposing winds, *J. Waterway, Port, Coastal, and Ocean Engineering*, New York, ASCE, **114**, 6, 732–744

Young, I. R. and M. L. Banner, 1992, Numerical experiments on the evolution of fetch limited waves, in *Breaking Waves*, IUTAM Symposium, Sydney, Australia, eds. M. L. Banner and R. H. J. Grimshaw, Berlin, Springer-Verlag, pp. 267–275

Young, I. R. and G. Ph. van Vledder, 1993, A review of the central role of nonlinear interactions in wind-wave evolution, *Phil. Trans. Roy. Soc. London*, A, **342**, 505–524

Young, I. R., 1994, On the measurement of directional wave spectra, *Appl. Ocean Res.*, **16**, 283–294

Young, I. R., L. A. Verhagen and M. L. Banner, 1995, A note on the bimodal directional spreading of fetch-limited wind waves, *J. Geophys. Res.*, **100**, C1, 773–778

Young, I. R. and L. A. Verhagen, 1996a, The growth of fetch limited waves in water of finite depth. Part 1. Total energy and peak frequency, *Coastal Engineering*, **29**, 47–78

—1996b, The growth of fetch limited waves in water of finite depth. Part 2. Spectral evolution, *Coastal Engineering*, **29**, 79–99

Young, I. R., L. A. Verhagen and S. K. Shatri, 1996, The growth of fetch limited waves in water of finite depth. Part 3. Directional spectra, *Coastal Engineering*, **29**, 101–122

Young, I. R. and G. J. Holland, 1996, *Atlas of the Oceans: Wind and Wave Climate*, Oxford, Pergamon Press, Elsevier Science Inc., 241 pp.

Young, I. R., 1997, Observations of the spectra of hurricane wind generated waves, *Ocean Engineering*, **25**, 4–5, 261–276

—1998, An experimental investigation of the role of atmospheric stability in wind wave growth, *Coastal Engineering*, **34**, 23–33

Young. I. R. and Y. Eldeberky, 1998, Observations of triad coupling of finite depth wind waves, *Coastal Engineering*, **33**, 137–154

Young, I. R. and G. J. Holland, 1998, *Atlas of The Oceans: Wind and Wave Climate*, Version 1.0, CD-ROM, Oxford, Pergamon Press, Elsevier Science Inc.

Young, I. R., 1999, *Wind Generated Ocean Waves*, Amsterdam, Elsevier, **2**, 288 pp.

Young, I. R. and A. V. Babanin, 2006, The form of the asymptotic depth-limited wind-wave frequency spectrum, *J. Geophys. Res.*, 111, C6, C06031

Yu, Y.-X., S.-X. Li, Y. S. Onyx and W. H. Wai, 2000, Refraction and diffraction of random waves through breakwater, *Ocean Engineering*, **27**, 5, 489–509

Zakharov, V. E. and A. N. Pushkarev, 1999, Diffusion model of interacting gravity waves on the surface of deep fluid, *Nonlinear Processes Geophys.*, European Geophysical Society, **6**, 1–10

Zakharov, V. E., 1999, Statistical theory of gravity and capillary waves on the surface of a finite-depth fluid, *Eur. J. Mech. B/Fluids*, **18**, 327–344

Zijlema, M. and Stelling, G. S., 2005, Further experiences with computing non-hydrostatic free-surface flows involving water waves, *Int. J. Num. Meth. Fluids*, **48**, 169–197

Zijlema, M. and A. J. van der Westhuysen, 2005, On convergence behaviour and numerical accuracy in stationary SWAN simulations of nearshore wind wave spectra, *Coastal Engineering*, **52**, 237–256

Index